# MECHANICS
# OF DEFORMABLE BODIES

# MECHANICS
# OF DEFORMABLE
# BODIES

## LECTURES ON THEORETICAL PHYSICS, VOL. II

## BY ARNOLD SOMMERFELD

### UNIVERSITY OF MUNICH

TRANSLATED FROM THE SECOND GERMAN EDITION

## BY G. KUERTI

### HARVARD UNIVERSITY

**ACADEMIC PRESS**    New York   San Francisco   London

A Subsidiary of Harcourt Brace Jovanovich, Publishers

Lectures on Theoretical Physics: Mechanics of Deformable Bodies

Arnold Sommerfeld

ISBN: 9780126546521

Copyright © 1950 Academic Press, Inc. All rights reserved.

索末菲理论物理教程：变形介质力学

ISBN: 9787519296797

Copyright © Elsevier Inc. and Beijing World Publishing Corporation. All rights reserved.

---

**Notice**

Knowledge and best practice in this field are constantly changing. As new research and experience broaden our understanding, changes in research methods, professional practices, or medical treatment may become necessary. Practitioners and researchers must always rely on their own experience and knowledge in evaluating and using any information, methods, compounds or experiments described herein. Because of rapid advances in the medical sciences, in particular, independent verification of diagnoses and drug dosages should be made. To the fullest extent of the law, no responsibility is assumed by Elsevier, authors, editors or contributors in relation to the adaptation or for any injury and/or damage to persons or property as a matter of products liability, negligence or otherwise, or from any use or operation of any methods, products, instructions, or ideas contained in the material herein.

# 导　言

论及近代物理学的构建与物理学教育，这里的物理学教育指的是物理学家的培养，有一个格外突出的人物是非提不可的，那就是德国数学家、物理学家索末菲。

索末菲（Arnold Sommerfeld, 1868–1951）出生于东普鲁士的科尼斯堡（今俄罗斯的加里宁格勒），父亲是位热爱自然科学的医生。1875 年，索末菲进入科尼斯堡的老城学校上学，当时学校的高年级学生中有闵可夫斯基（Hermann Minkowski, 1864–1909）和维恩（Wilhelm Wien, 1864–1928）。1886 年，索末菲进入科尼斯堡大学学习数学和物理，他在该校遇到的老师有希尔伯特（David Hilbert, 1862–1943）、林德曼（Ferdinand von Lindemann, 1852–1939）和胡尔维茨（Adolf Hurwitz, 1859–1919）这些数学巨擘。1891 年，索末菲在林德曼指导下以"数学物理中的任意函数"为题获得博士学位。服完兵役以后，索末菲于 1893 年去往哥廷恩，1894 年做了数学大家克莱因（Felix Klein, 1849–1925）的助手，1895 年以"衍射的数学理论"获得讲师资格。其后，索末菲在亚珅等地任教。1906 年，索末菲到慕尼黑大学担任理论物理教授，建立了理论物理研究所，在那里一直工作到 1939 年退休。

索末菲是一个理论物理学家，但他首先是个数学家、实验物理学家或者说技术物理学家。索末菲对摩擦学、无线电技术、电力传输等技术领域都有可圈可点的贡献。当然，他最值得称道的是对近代物理的贡献。在关于原子中电子的四个量子数（$nlm$;

$m_s$）中的（$l\ m$）这两个量子数都是索末菲 1915 年引入的，而精细结构常数 $\alpha = \dfrac{1}{4\pi\varepsilon_0}\dfrac{e^2}{\hbar c} \sim \dfrac{1}{137}$ 则是索末菲 1916 年引入的。索末菲是无可置疑的量子力学奠基人之一，也是最早接受相对论的学者。

索末菲在哥廷恩给克莱因短时间做过助手，深受克莱因的影响。克莱因的那套研究、教学、主持研讨班、办杂志、编书等作为一代学术宗师的行为模式，索末菲全面继承了下来。1895/1896 冬季学期，克莱因就陀螺理论做了一个系列讲座，后来委托索末菲将讲座内容整理成书。1896 年秋，索末菲开始整理陀螺理论。这期间，索末菲不仅关注相关学问的理论进展，甚至还多次走访海军基地，了解陀螺仪在鱼雷制导上的实际应用。让克莱因和索末菲都没想到的是，因为索末菲作为一个技术物理教授和数学物理教授自己也是诸事缠身，这项工作整整持续了 13 年，直到 1910 年才最终完成。不过，这套四卷本的《陀螺理论》（*Über die Theorie des Kreisels*），洋洋千余页，作为处理刚体转动这个数学物理之难题的经典，任何人凭此一项成就即足以傲视物理学界。尤其值得一提的，关于克莱因发起的"数学科学百科全书"，索末菲承担了第五卷的编辑任务，一干就是 28 年（1898–1926），收录了许多数学物理名篇，其中就包括泡利（Wolfgang Pauli，1900–1958）的成名作《相对论》（*Die Relativitätstheorie*）。

索末菲自己是近代原子理论的主要贡献者。基于他自己的研究成果以及讲课内容，索末菲于 1919 年出版了《原子构造与谱线》（*Atombau und Spektrallinien*）一书，此书被誉为原子物理的圣经。1929 年，索末菲又针对这本书补充出版了《原子构造与谱线：波动力学补充》（*Atombau und Spektrallinien:*

Wellenmechanischer Ergänzungband)。1933 年，索末菲又出版了一本近300 页的《金属电子理论》(Elektronentheorie der Metalle)，这是索末菲在自己拓展了德鲁德（Paul Drude, 1863–1906）的自由电子理论的基础上撰写的。将这三本著作放到一起，能勾勒出一个原子理论和老量子论 [①] 构建者形象，也就能够理解为什么他的学生能成为量子力学和量子化学的奠基人。

　　索末菲在慕尼黑的讲课资料最后集成了六卷本《理论物理教程》(Vorlesungen über theoretische Physik)，其中第一卷《力学》(Mechanik) 出版于 1943 年，第二卷《形变介质的力学》(Mechanik der deformierbaren Medien) 出版于 1945 年，第三卷《电动力学》(Elektrodynamik) 出版于 1948 年，第四卷《光学》(Optik) 出版于 1950 年，第五卷《热力学与统计》(Thermodynamik und Statistik) 出版于 1952 年，第六卷《物理中的偏微分方程》(Partielle Differentialgleichungen der Physik) 出版于 1947 年。第五卷算是遗作，是由 Fritz Bopp 和 Josef Meixner 代为整理的。第六卷非常有名，早在 1949 年即有了英文版。其他各卷英译本出版于 1964 年。世界上有一些著名的数学、物理学系列教程，有必要在此介绍给我国的物理学爱好者。比索末菲早的、比较有名的有克莱因的各种数学教程，以及亥尔姆霍兹（Hermann von Helmholtz, 1821–1894）的系列物理教程（包括光的电磁理论、声学原理、分立质点动力学、连续介质动力学、热学）等。在索末菲之后，有他的学生泡利的九卷本《泡利物理学讲义》(Pauli Lectures on Physics)，是对泡利讲课材料的

－－－－－－－－－－

① 我不觉得 old quantum theory 是旧量子论。它不陈旧，它只是量子理论的早期形态而已

整理，算是对索末菲《理论物理教程》的风格继承。其他有名的理论物理教程有十卷本《朗道理论物理教程》，那是朗道（Лев Давидович Ландáу, 1908–1968）发起的、由多人完成的著作，除第八卷署名为 Landau, Lifshitz, Pitaevskii 外，其他各卷署名为 Landau, Lifshitz。就实验物理而言，八卷本的德语《实验物理教程》（*Bergmann-Schaefer Lehrbuch der Experimentalphysik*）对我国的物理学教育可能具有特别的意义，因为它提供了太多我们可能未加关注的实验细节。这是一套众多作者编纂的、开放式的教程，内容随时修订，多次再版。此外，还有卷数不等的、更多是针对学生设计的系列物理教程，各有千秋，恕不一一介绍。索末菲的六卷本《理论物理教程》是独特的，因为它包含着作者自己创造的学问，是作者自己在课堂上实际使用过的，最重要的是从作者培养出来的学生水平来看这套教程是卓有成效的。此外，值得强调的是，除了第五卷遗作借助他人之手才整理完成，这套系列教程是索末菲一人的功绩，没有掠他人之美。近时期加入自己的成果与思考来系统地表述物理学的，有美国物理学家赫斯特内斯（David Hestenes, 1933– ），但做不到如索末菲那样融会贯通且凭一己之力。不是凭一己之力完成的系列物理学教程，无法保证风格的统一只是小事，缺乏思想的整体内在一致性（integrity）才是最遗憾的地方。

索末菲在慕尼黑的理论物理研究所汇集了一批少年英才，他们在索末菲的指导和引领下几乎全部成为了在各自研究领域里扬名立万的人物。据说在 1928 年前后的那段时间里，德语国家的理论物理教授有三分之一都出自索末菲门下。索末菲的博士生中有四人获诺贝尔奖，分别是海森堡（Werner Heisenberg,

1901–1976）、德拜（Peter Debye, 1884–1966）、泡利和贝特（Hans Bethe, 1906–2005），博士后中有三人获得诺贝尔奖，鲍林（Linus Pauling, 1901–1994）、拉比（Isidor I. Rabi, 1898–1988）和劳厄（Max von Laue, 1879–1960），其中德拜获得的是化学奖，鲍林获得的是化学奖与和平奖。索末菲的作为大师之大导师（MacTutor of Maestros）的成就，仅英国的汤姆孙（J. J. Thomson, 1856–1940）可与之比肩。

我国古代哲人孟子认为"得天下英才而教育之"为君子之第三乐，此乃为人师之言也。索末菲是充分享受了这样的乐趣的。然而，对于所有求学者来说，人生的首要问题却是从什么人而受教的问题。我想说，"从合格之老师而受教，不亦大幸运乎？"然而，令人痛心的现实是，绝大部分的人材可能一生中都不会遇到一个合格的老师。倘若没有强大的自学能力与格外的机遇，难免早晚坠入"泯然众人矣"的遗憾。索末菲之所以有那么大的成就，与他求学的地方学术宗师云集有关；而索末菲的那些学生们之所以都各有所成就，又何尝不是因为遇到了索末菲这样的名师及其门下一众出类拔萃的同学们的缘故。笔者在博士毕业进入研究领域多年以后才悟到同这些学术巨擘之间学问的天渊之别究竟差在哪里。倘若一个人上过的学校中没有几个值得提起的老师或者同学，那你自己可要好好努力了。

值得庆幸的是，索末菲的《理论物理教程》英文版如今在我国出版了。愿这影响了无数物理学家成长的经典之作在中华大地上能收获更多的辉煌。

曹则贤

2022 年 7 月 27 日于中国科学院物理研究所

# 序 言

此一理论物理教程的作者，阿诺德·索末菲，是促成1910–1930 年间物理学所经历之变迁的关键人物之一。没有他的满怀雄心的、不懈的努力，关于原子的量子理论，不管是其狂飙式的进展还是广泛的传播，都不可能是我们所看到的那个样子。索末菲在慕尼黑的理论物理所[①] 里形成了一个学派，那里的原子论研习者们，德国人和外国人，初出茅庐的和小有成就的，源源不断地创作出研究论文。他的名著《原子构造与谱线》( *Atombau und Spektrallinien* )[②]，以及随后出版的《波动力学》( *Wellenmechanik* )[③]，在很长时间里是独有的关于此一基础领域之全面的、权威的论述；其后续版本令人印象深刻，乃是对自玻尔第一批文章之后原子理论的快速进展的概述。

索末菲所受的学术训练以及早期研究都根植于经典物理的数学方法，也因此他熟练掌握特别是在 1926 年薛定谔波动力学之后出现的那些新生的量子物理方法。自然地，也更因为其本人从经典理论的审美中所体会的愉悦，索末菲会给他的学生们以经典方法的系统训练。数学表述，还有它的物理诠释，与实验具体化之间的和谐在索末菲的教程中如浮雕一样被呈现出来，深刻地影

---

① 这里所说的 Institute，是大学里针对具体的某个教授设立的研究机构，规模很小。不是今天中文语境里的研究所

② 1919 年出版

③ 1929 年出版的 *Atombau und Spektrallinien* 第二卷，全名为《波动力学增补卷》( *Wellenmechanischer Ergänzungsband* )

响了他的学生们。

　　当索末菲整理他的教程准备将其付梓时，他已年过七十，结束了他长达四十年的学术培养生涯。他做这件事有二重意义：保全那些支撑物理取得伟大胜利的成果以度过危机，向年轻一代的物理学家传承那些在解决经典问题过程中形成的有价值的分析工具。自 1895 年撰写论物理中的任意函数的博士论文时起，索末菲在完善这些工具方面扮演了一个非常活跃的角色。在他早期的技艺高超的工作中，有一项是波在边缘上衍射问题的严格解的构造，为此他扩展了黎曼在函数理论中的方法，结果是用多值空间中的镜像法得到了衍射问题的一个解。读者会在第 5 卷（光学）中看到相关讨论。索末菲同哥廷恩的数学大家克莱因合著的论转动刚体理论的范本，即四卷本《陀螺理论》(*Theorie des Kreisels*)，跨越了其早年在哥廷恩的时期一直到其在慕尼黑的量子时代之初 [④]。该书的目的是通过将大量不同的数学内容，如函数理论、椭圆函数、四元数、Klein-Caley 参数等，应用于刚体动力学问题来展现纯数学与应用数学之间的密切联系。当其在亚珅工业专科学校做工业力学岗位教授时，即 1899–1905 年间，索末菲对工程问题产生了浓厚的兴趣。他关于润滑流体力学、工作于同一传输线上的发电机间的相互影响、火车刹车等问题的论文，无一不是采用了一般性处理方法而因此具有持久的价值。当无线电报出现时，索末菲及其弟子的系列论文开始研究无线电的发射与传播模式。这些都是在相应领域中索末菲堪为大师的那些数学方法的卓越案例。特别地，无线电绕地球的衍射问题被转化成了

---

④　出版时间跨度为 1897–1910

复积分的讨论，有了严格解（参见第 6 卷，第 6 章）[5]。

此处罗列一个详细的、索末菲因之丰富了物理理论的成就清单有点儿不合适。读者可以参阅文后列举的一些文章。但是，不妨就索末菲作为老师，以及如今要翻译出版的这个教程，多说几句。

慕尼黑的理论物理课程分为通识课和专业课两类。前者每周四小时，确切地说是分成 45–50 分钟一节的课，冬季学期持续 13 周，夏季学期持续 11 周。（索末菲的）那六门课就构成了当前六卷本的主题。每门课都是给选修了实验物理演示课程（任课教师先为伦琴，后来是维恩）的学生的导论。在实验物理课上，学生要求获得对物理现象的实地考察，以及基于本质上非数学的处理对考察结果的定量评价。而在理论物理课上，基础内容会被重修梳理一遍，但是目标瞄向了未来可用于高等问题的形成数学处理（能力）和构建全面理论（知识）。后者从一个系列的课程到另一个系列的课程是变动的，后期会纳入一些专题因而也会引起此前学过相关内容的高年级学生的兴致。除了讲座，还有每周两个小时的讨论课。

专业课是每周两小时的讲座，主题是那些在通识课上只能略加论述的主题，或者时令的话题。索末菲的讲座一般会和他当时的研究有关系，经常有部分内容很快会出现在他的原创论文中，这方面的例子有将洛伦兹变换诠释为四维空间里的转动（第 3 卷，第 26 节），从波动光学到几何光学的过渡（第 5 卷，第 35 节）和色散介质中信号速度的讨论（第 5 卷，第 22 节）。这些主题又

---

⑤ Complex integral，字面意思是复积分。但第 6 卷第 6 章那里似乎只是一些复杂的积分。

会被纳入通识课中，此前通识课中不那么有趣的部分会被拿掉。

　　除了讲座类课程，还有关于高级课题的讨论班、报告会。为此学生要总结分到手的课题、作报告，这常常意味着持续数星期的高强度学习。

　　在学生眼中，索末菲讲座最吸引人的地方是（条理）清晰：从物理一侧的处理方式，数学问题的提取，对所采用方法之简单但具有一般性的解释，重新用物理实验语汇对结果的充分讨论。他在黑板上有力的、布局得当的板书，清爽的图示，极大地帮助了学生在每一阶段之后钻研所涉及过的主题。此外，课程的标准足够高，能耗尽优秀学生的气力，还要求细心的合作。这在一个不点到、全靠自觉的大学系统里是重要的。在就一个问题从头讨论的练习中，初学者也会引起索末菲或者助教的注意，其会因自己的付出被赞赏而受到鼓舞。

　　无论学生岁数大小，索末菲对于真正的努力都具有特别的鉴赏力。这是德拜、泡利、海森堡（只提如今已获得诺奖的几位）这个级别的科学家早年求学生涯中投奔他的原因。但是，那些资质一般的学生也得到了很好的照料，会赋予较小的问题和较少的期待以锻炼其能力。懒散的学生不久就会自行离开。故此，索末菲的学生们形成了一个精英小团体，但人数也维持足够多，能够产生一股协助新手迅速独自扬帆起航的和风。祝愿索末菲教程的（英文）翻译广为散播这股和风，协助其他的群体去航行在发现的海洋上。

论及（教程）作者工作的其他文章

Anon., Current Biographies, 1950, pp. 537–538.

P. Kirkpatrick, *Am. J. Physics* (1949). **17**, 5, 312–316.

M. Born, *Proc. Roy. Soc., London. A.* (1952).

P. P. Ewald, *Nature* (1951). **168**, 364–366.

W. Heisenberg, *Naturwissenschaften*（自然科学）(1951). **38**, 337.

M. v. Laue, *Naturwissenschaften*（自然科学）(1951). **38**, 513–518.

埃瓦尔特 [6]

布鲁克林工业研究所，纽约

---

[6] Paul Peter Ewald (1888–1985)，有汉译埃瓦、埃瓦尔德等，比如 Ewald's diffraction sphere 就被译为埃瓦球、埃瓦尔德球。Ewald 是德国晶体学家，X-射线衍射研究晶体的先驱。1906–1907 年，Ewald 在哥廷恩大学学的数学。1907 年，Ewald 转到慕尼黑大学跟索末菲继续学习数学，1912 年以单晶中 X-射线传播定律的论文获得博士学位。博士毕业后，Ewald 接着在慕尼黑给索末菲做了一段时间的助手。

# PREFACE*

It is due in part to the encouragement of former students of mine that I have decided to publish my general course of lectures on theoretical physics which I gave regularly during thirty-two years at the University of Munich.

They were of an introductory nature, attended not only by the physics majors of the University and the Polytechnic Institute, but also by the many candidates for degrees in the teaching of mathematics and physics —usually in their fourth to eighth semesters—as well as by astronomers and some physical chemists. Classes were held four times a week and supplemented by a two-hour problem period. Special courses in modern physics which were given concurrently have not been included in this series of books; their subject matter has found its way into my papers and treatises. While it is true that quantum mechanics always hovers in the background, reference being made to it now and then, the actual substance of these lectures is classical physics.

The order of subjects, which is retained in their publication, was

1. Mechanics
2. Mechanics of Deformable Media
3. Electrodynamics
4. Optics
5. Thermodynamics and Statistical Mechanics
6. Partial Differential Equations in Physics

The lectures on mechanics were given in alternate semesters by myself and by my colleagues in mathematics. Concurrent courses in hydrodynamics, electrodynamics and thermodynamics were taught by younger members of the faculty. Vector analysis was given in a separate course so that its systematic development could be omitted from my lectures.

Here, as in my classes, I shall not detain myself with the mathematical

---

* From the forewords to Vol. I and Vol. II, first and second edition.

v

foundations, but proceed as rapidly as possible to the physical problems themselves. My aim is to give the reader a vivid picture of the vast and varied material that comes within the scope of theory when a reasonably elevated vantage-point is chosen. With this purpose in mind I shall not be too concerned if I have left some gaps in the systematic justification and axiomatic structure of the work. At any rate, I have avoided frightening the student with drawn-out investigations of a mathematical or logical nature and distracting his attention from that which is physically interesting. I think that this attitude has proved its worth in my lectures; it has therefore been retained in the printed text. And when Planck's lectures are impeccable in their systematic organization, then I may perhaps claim for my own lectures a greater variety of subject matter and a more flexible handling of the mathematical apparatus. However, I gladly refer the reader to the more complete and often more thorough treatment of Planck, especially for thermodynamics and statistical mechanics.

The problems collected at the end of each volume should be considered as supplements to the text. They were presented by the students during the problem periods after they had worked them out and handed them in in writing. Elementary numerical calculations, such as are found in great number in textbooks and collections of exercises, have, in general, not been included. The problems are numbered by chapter. Sections are numbered through in each volume, and equations in each section. Within each volume references to earlier equations can thus be made by merely giving the numbers of section and equation.

*In this second volume* a fairly complete development of certain mathematical methods had to be given. These methods are often taken up in a separate introductory course on theoretical physics; their incorporation in Vol. II accounts for its larger size. But the actual subject of this volume is the mechanics of systems with an infinite number of degrees of freedom. The place of ordinary differential equations (governing the mechanics of systems with a finite number of degrees of freedom) is here taken by partial differential equations, the place of vector algebra by vector analysis, which is briefly summarized in Chap. 1. Besides, it was necessary to develop the fundamentals of tensor analysis, being an indispensable tool in the theory of elastic solids and viscous fluids. This has been done for Cartesian coordinates and to some extent also for orthogonal curvilinear coordinates.

Some points may be mentioned here in which this presentation seems to be more complete than the one generally found in textbooks on the same level: In Chap. I.2 it is proved that the curl is an axial vector (or antisymmetric tensor). Following O. Reynolds we consider in II.10 *two* laws of similitude and *two* corresponding invariants, viz. a dimensionless number $S$ characterizing the pressure dependence in addition to the usual Reynolds number. In III.15 we discuss the quasi-elastic body (gyroscopic ether); its logical place is among the continuous media that are compatible with the fundamental theorem of kinematics in I.1. Rather than explain Maxwell's equations by a mechanical model, we want to show in this discussion the fundamental difference between electrodynamics and mechanics. In Chap. V, Sections 27 and 28 deal with the somewhat involved problems of circular waves and ship waves. Complete calculations are supplied using the method of stationary phase which is a simplified version of the method of steepest descent. In the plate and jet problems of Chap. VI the dimensions of plate and of orifice etc. are carried as parameters throughout the entire calculation. This form of analysis has perhaps a stronger physical appeal than the usual one which employs dimensionless quantities. Kármán's vortex street (32) is extended according to Maue to include the unsymmetrical case in which the flow is not parallel to the street. The hydrodynamic theory of journal bearings is briefly dealt with in VII.36. Riemann's theory of shock waves is discussed in 37 with a particular view to the results that Bechert obtained in certain elementary integrable cases. Sec. 38 is a report on the history and present situation of the difficult turbulence problem and includes also Burger's mathematical model of turbulence. In VIII.43 one finds, in the problem of the helical spring, an example of combined bending and torsion. In 44 the boundary conditions for an oscillating parallelepiped are discussed, and the foundations laid for the quantum-theoretical thermodynamics of the solid body.

It is obvious that not all the topics found in this volume could have been touched upon in the brief period of one term; several of the subjects mentioned before have in fact been added for the print.

The *second edition* of Vol. II has been supplemented by a representation of general tensor calculus limited to three dimensions and orthogonal line elements. Tensor calculus does not hold an advantage over the simpler vector analytic formulas for the cases considered here (cf. App. IV), but, because of its importance in the general theory of relativity, it cannot be entirely omitted if a fairly complete exposition of the mathematical methods of theoretical physics is the goal.

The discussion of the turbulence problem, which already presented great difficulties in the first edition, had to be revised on the basis of so-far unpublished work by C. F. von Weizsäcker and W. Heisenberg. My opinion of long standing that turbulence would finally be accounted for by integration of the Navier-Stokes equations in their complete, non-linear, form, proved wrong in the special case of "isotropic turbulence" investigated by the two authors; here, as in the kinetic theory of gases, statistical methods have shown their superiority. It was, of course, impossible to review the new results in detail, but the previous representation had to be corrected in many places in accordance with the new point of view.

Munich, July 1946                                    Arnold Sommerfeld


In this translation of the second volume of Professor Sommerfeld's "Vorlesungen über theoretische Physik", the text of the second edition (1947, D. V. B., Wiesbaden) has been followed. Disregarding a few non-technical remarks that hold interest only for German readers, nothing has been omitted, and no changes have been made apart from necessary adaptations of the notation. Several brief footnotes have been inserted, however, to bridge over certain differences, mostly of a terminological nature. Topical references to the first volume have been supplemented by corresponding references to well-known American and British texts with the aim of making this volume independent from the preceding one, and a new index has been prepared.

G.K.

# TABLE OF CONTENTS

# ERRATA

p. 9, 2nd line of the table: read $\frac{1}{2}e_{xy} = \frac{1}{2}e_{yx}$ , etc., instead of $e_{xy} = e_{yx}$ , etc.

p. 61, 3rd and 4th lines of table: read $Z_x$ instead of $Z$

p. 99, 13th line: read gr instead of gr-wt

p. 99, 15th line: read 1.293 instead of 1,293

p. 113, Eq. (1): read $\nabla^2$ instead of $\Delta$

p. 181, Fig. 39b is upside down

p. 233, last equation: read $\bar{z}_k = \bar{x}_k + i\bar{y}_k$ instead of $\bar{z}_k = \bar{x}_k = i\bar{y}_k$

p. 234, 6th line: read $z_k$ instead of $\bar{z}_k$

p. 234, 7th line: read $\dot{\bar{z}}_0$ instead of $\bar{z}_0$

p. 289, 2nd line of footnotes: read p. 9 instead of p. 000

p. 338, 22nd line: omit "and note that they are not equal"

p. 347, last term of 14th line: read $\dfrac{\partial A_\varphi}{\partial \varphi}$ instead of $\dfrac{\partial A_\varphi}{\partial \varphi}$

p. 348, third formula: read $\nabla^2 \mathbf{A}$ instead of $(\mathbf{A}\ \mathrm{grad})\mathbf{A}$ and add:

$$(\mathbf{A}\ \mathrm{grad})\mathbf{A} = \begin{cases} \mathbf{A}\cdot\mathrm{grad}\ A_r - \dfrac{1}{r}(A_\vartheta^2 + A_\varphi^2) \\[2mm] \mathbf{A}\cdot\mathrm{grad}\ A_\delta + \dfrac{1}{r}(A_r A_\vartheta - A_\varphi^2 \cot\delta) \\[2mm] \mathbf{A}\cdot\mathrm{grad}\ A_\varphi + \dfrac{1}{r}(A_r A_\varphi + A_\vartheta A_\varphi \cot\delta) \end{cases}$$

p. 354, line 5: read $|\,M_e\,| = |\,M_i\,| = 4\pi\mu U r_e r_i^2/(r_e^2 - r_i^2)$ instead of the values given for $M_e$ and $M_i$, and replace the first part of the next sentence by "The difference $-\,|\,M_e\,|/2\pi r_e + |\,M_i\,|/2\pi r_i$ equals $2\mu U/(1 + r_e/r_i)$, which, in the limit, becomes $\mu U$;"

p. 386, Eq. (1): read $p_{ik}$ instead of $_{ik}$

# KINEMATICS OF DEFORMABLE BODIES

## 1. A Fundamental Theorem of Kinematics

Helmholtz establishes the following theorem at the beginning of his paper on vortex motions:[1] The most general motion of a sufficiently small element of a deformable (i.e., not rigid) body can be represented as the sum of

1. a *translation*,
2. a *rotation*,
3. an *extension (contraction) in three mutually orthogonal directions*.

The proof is based on the Taylor expansion of the relative displacement of two neighboring points $P$ and $O$ in terms of their original coordinate differences.

Let $P$ be a point of the volume element under consideration and $x, y, z$ its coordinates in a rectangular system the origin of which, $O$, lies within the element. In a general motion of the body the points $P$ and $O$ will both experience changes of position, which we denote by $\xi, \eta, \zeta$ and $\xi_0, \eta_0, \zeta_0$ respectively, referring to the chosen space-fixed coordinate system. Taylor's formula then gives for the displacement of $P$

$$\xi = \xi_0 + \frac{\partial \xi}{\partial x} x + \frac{\partial \xi}{\partial y} y + \frac{\partial \xi}{\partial z} z + \cdots$$

(1)
$$\eta = \eta_0 + \frac{\partial \eta}{\partial x} x + \frac{\partial \eta}{\partial y} y + \frac{\partial \eta}{\partial z} z + \cdots$$

$$\zeta = \zeta_0 + \frac{\partial \zeta}{\partial x} x + \frac{\partial \zeta}{\partial y} y + \frac{\partial \zeta}{\partial z} z + \cdots .$$

Introducing for brevity

(2) $\quad \dfrac{\partial \xi}{\partial x} = a_{11} , \qquad \dfrac{\partial \xi}{\partial y} = a_{12} , \qquad \dfrac{\partial \eta}{\partial x} = a_{21} , \qquad \dfrac{\partial \eta}{\partial y} = a_{22} , \quad \cdots ,$

---

[1]Über Integrale der hydrodynamischen Gleichungen, welche den Wirbelbewegungen entsprechen.  Crelles J. 55, 25 (1858).

we write each of the quantities $a_{ik}$ as a sum of an "odd" and an "even" term

$$a_{ik} = \frac{a_{ik} - a_{ki}}{2} + \frac{a_{ik} + a_{ki}}{2} \; ;$$

"odd" and "even" refer, of course, to the commutation of the subscripts.

Now, (1) may be rewritten in the following form

$$\xi = \xi_0 + \left| 0 + \frac{a_{12} - a_{21}}{2} y + \frac{a_{13} - a_{31}}{2} z \right.$$

$$\left. + \left\| a_{11}x + \frac{a_{12} + a_{21}}{2} y + \frac{a_{13} + a_{31}}{2} z \right\| \right.$$

$$\eta = \eta_0 + \left| \frac{a_{21} - a_{12}}{2} x + 0 + \frac{a_{23} - a_{32}}{2} z \right.$$

(3)

$$\left. + \left\| \frac{a_{21} + a_{12}}{2} x + a_{22}y + \frac{a_{23} + a_{32}}{2} z \right\| \right.$$

$$\zeta = \zeta_0 + \left| \frac{a_{31} - a_{13}}{2} x + \frac{a_{32} - a_{23}}{2} y + 0 \right.$$

$$\left. + \left\| \frac{a_{31} + a_{13}}{2} x + \frac{a_{32} + a_{23}}{2} y + a_{33}z \right\| \right. ,$$

where terms higher than first order in $x$, $y$, $z$ have been omitted. Let us introduce the symbol $\mathbf{s}$ for the total change of position of $P$; the vertical separation lines in (3) indicate that the displacement $\mathbf{s}$ is compounded of three partial motions $\mathbf{s}_0$, $\mathbf{s}_1$, $\mathbf{s}_2$, or

(4)     $$\mathbf{s} = \mathbf{s}_0 + \mathbf{s}_1 + \mathbf{s}_2 .$$

The displacement $\mathbf{s}_0$, with the components $\xi_0$, $\eta_0$, $\zeta_0$, is the same for all points $P$ of a volume element and is therefore a *translation*.

The central portion of the set (3), $\mathbf{s}_1$, is a *rotation*. Upon introducing the vector $\boldsymbol{\phi}$, with the components

(5)     $$\varphi_x = \frac{a_{32} - a_{23}}{2} , \qquad \varphi_y = \frac{a_{13} - a_{31}}{2} , \qquad \varphi_z = \frac{a_{21} - a_{12}}{2} ,$$

and the position vector $OP = \mathbf{r}$, we obtain simply

(6)     $$\mathbf{s}_1 = \boldsymbol{\phi} \times \mathbf{r}.$$

This displacement is well known from rigid body kinematics (Vol. $I^2$, Eq. 22.3) and corresponds to the infinitesimal rotation $\phi$ (the appropriate notation would be $\delta\phi$) whose axis and magnitude are given by the components $\varphi_x$, $\varphi_y$, $\varphi_z$.[3]

The infinitesimal rotation is not a vector in the proper sense, such as a *polar* vector that characterizes a translatory displacement. One may, nevertheless, denote it by the usual vectorial symbolism, associating with it a polar vector $\phi$ that points in the direction of the *axis* of rotation and has a length equal to the (infinitesimal) angle of rotation. More about *axial* vectors is found in 2.

There is, of course, *no change in the length of the vector OP* due to the displacement $\mathbf{s}_1$. This is self-evident if we consider $\mathbf{s}_1$ from the point of view of rigid body kinematics, but will be shown here independently. The square of the distance $\overline{OP}$ is

$$\overline{OP}^2 = |\mathbf{r} + \mathbf{s}_1|^2 = |\mathbf{r}|^2 + 2\mathbf{r}\cdot\mathbf{s}_1 + |\mathbf{s}_1|^2,$$

but according to (6)

$$\mathbf{r}\cdot\mathbf{s}_1 = \mathbf{r}\cdot(\phi \times \mathbf{r}) = \phi\cdot(\mathbf{r} \times \mathbf{r}) = 0.$$

Disregarding second order terms as in (3), we find

$$|\mathbf{r} + \mathbf{s}_1|^2 = |\mathbf{r}|^2 = r^2$$

(It will be noticed that here and in all subsequent arguments of a similar nature, "no change" means "no change *in first approximation*.")

In order to pass from infinitesimal to finite quantities the displacements must be considered in their time dependence, which amounts to introducing velocities instead of displacements, viz.

(7)                    $$\mathbf{s} = \mathbf{v}\Delta t, \qquad \phi = \boldsymbol{\omega}\Delta t,$$

or in components

(8)
$$\xi = u\Delta t, \qquad \eta = v\Delta t, \qquad \zeta = w\Delta t,$$
$$\varphi_x = \omega_x\Delta t, \qquad \varphi_y = \omega_y\Delta t, \qquad \varphi_z = \omega_z\Delta t.$$

Here $\mathbf{v}$ is the velocity of the particle or volume element considered, and $\boldsymbol{\omega}$ its vortex vector; the physical importance of the latter quantity was first recognized by Helmholtz.

[2]A. Sommerfeld, Vorlesungen über Theoretische Physik, Akademische Verlagsgesellschaft, Leipzig, 1944, Bd. I, quoted henceforth as Vol. I. English translation in press (Academic Press Inc., New York).

[3]See, e.g., J. L. Synge and B. A. Griffith, Principles of Mechanics, McGraw-Hill, New York, 1942 Sec. 10.5.

The components of the vortex vector in the system $x$, $y$, $z$ are according to (8), (5), and (2)

$$(9) \quad \omega_x = \frac{1}{2}\left(\frac{\partial w}{\partial y} - \frac{\partial v}{\partial z}\right), \qquad \omega_y = \frac{1}{2}\left(\frac{\partial u}{\partial z} - \frac{\partial w}{\partial x}\right), \qquad \omega_z = \frac{1}{2}\left(\frac{\partial v}{\partial x} - \frac{\partial u}{\partial y}\right).$$

A more complete analysis of this fundamental definition belongs also to the review of vector analysis in the following article.

We turn now to the third partial motion $s_2$ indicated in the right hand part of our scheme (3). Here the deformability becomes significant, the two other partial motions having been recognized as the elements of the general motion of a rigid body.

The displacement $s_2$ is a *linear vector function* of the position vector $r$, in the sense of Vol. I, 22.[4] We denote the components of $s_2$ by $\xi_2$, $\eta_2$, $\zeta_2$, and introduce the notation

$$\xi_2 = \epsilon_{xx}x + \epsilon_{xy}y + \epsilon_{xz}z,$$

$$(10) \qquad \eta_2 = \epsilon_{yx}x + \epsilon_{yy}y + \epsilon_{yz}z,$$

$$\zeta_2 = \epsilon_{zx}x + \epsilon_{zy}y + \epsilon_{zz}z.$$

The meaning of the coefficients $\epsilon_{ik}$ follows from (2) and (3):

$$\epsilon_{xx} = \frac{\partial \xi}{\partial x}, \qquad \epsilon_{xy} = \epsilon_{yx} = \frac{1}{2}\left(\frac{\partial \xi}{\partial y} + \frac{\partial \eta}{\partial x}\right),$$

$$(11) \qquad \epsilon_{yy} = \frac{\partial \eta}{\partial y}, \qquad \epsilon_{yz} = \epsilon_{zy} = \frac{1}{2}\left(\frac{\partial \eta}{\partial z} + \frac{\partial \zeta}{\partial y}\right),$$

$$\epsilon_{zz} = \frac{\partial \zeta}{\partial z}, \qquad \epsilon_{zx} = \epsilon_{xz} = \frac{1}{2}\left(\frac{\partial \zeta}{\partial x} + \frac{\partial \xi}{\partial z}\right).$$

The quantities $\epsilon$ are the components of the *strain tensor*. The tensor itself may be symbolized in a similar way as the moment of inertia in Vol. I, 22.13b[5] by the quadratic array

$$(12) \qquad \epsilon = \begin{pmatrix} \epsilon_{xx} & \epsilon_{xy} & \epsilon_{xz} \\ \epsilon_{yx} & \epsilon_{yy} & \epsilon_{yz} \\ \epsilon_{zx} & \epsilon_{zy} & \epsilon_{zz} \end{pmatrix}$$

---

[4] See, e.g., L. Brand, Vector and Tensor Analysis, Wiley, New York, 1947, §§60 and 62.

[5] See, e.g., Synge and Griffith, *op. cit.*, Sec. 11.3.

In the present case the tensor is *symmetric* (i.e., the components are symmetric to the main diagonal that connects the upper left and lower right corners), a fact which is an immediate consequence of the definition of the third partial motion $s_3$ in (3). On the other hand, the scheme of coefficients of the partial motion $s_1$ represents an *antisymmetric tensor*. Its definition, given by Eq. (6) or Eq. (3) (the part between the vertical lines), yields the array

$$
(12a) \qquad \hat{\phi} = \begin{pmatrix} 0 & \varphi_{xy} & \varphi_{xz} \\ \varphi_{yx} & 0 & \varphi_{yz} \\ \varphi_{zx} & \varphi_{zy} & 0 \end{pmatrix}, \qquad \varphi_{ik} = -\varphi_{ki} ,
$$

and our change of notation $\varphi_z = \varphi_{xy} = -\varphi_{yx}$ etc. establishes accordance with the double subscript notation introduced in (10). Note here that the antisymmetric tensor can always be represented by a vector, which is, of course, not true for a symmetric tensor, as will be proved in 2.

In the analysis of the strain tensor we make use of the concept of the *tensor quadric* as in Vol. I, 22.15 in the analysis of the moment of inertia. Consider for this purpose the scalar product

$$
(13) \qquad s_2 \cdot r = \epsilon_{xx} x^2 + 2\epsilon_{xy} xy + \epsilon_{yy} y^2 + \cdots = f(x, y, z).
$$

By putting $f(x, y, z) = $ const, we obtain a surface of second order, the *strain* (or *deformation*) *quadric*, also called the *ellipsoid of strain* or *of dilatation*; this terminology, of course, does not exclude contraction, nor does it imply that the quadric is necessarily an ellipsoid as in the case of the inertia tensor. It may be any one of the surfaces of second order, e.g., a hyperboloid of one or two sheets or a degenerate surface such as a pair of planes.

In order to decide this, we refer the strain quadric to its principal axes. Upon introducing the corresponding rectangular coordinates $X_1$, $X_2$, $X_3$, Eq. (13) takes the form

$$
(14) \qquad F(X_1, X_2, X_3) = \epsilon_1 X_1^2 + \epsilon_2 X_2^2 + \epsilon_3 X_3^2 = \text{const.}
$$

The coefficients $\epsilon_1$, $\epsilon_2$, $\epsilon_3$ are called the *principal extensions* (or *contractions*, if negative).

If (13) is used, the linear vector function (10) may be rewritten in the form

$$
(15) \qquad \xi_2 = \frac{1}{2} \frac{\partial f}{\partial x}, \qquad \eta_2 = \frac{1}{2} \frac{\partial f}{\partial y}, \qquad \zeta_2 = \frac{1}{2} \frac{\partial f}{\partial z},
$$

and in terms of the principal coordinates we have the correspondingly simplified linear vector function

$$\Xi_1 = \frac{1}{2}\frac{\partial F}{\partial X_1} = \epsilon_1 X_1 , \qquad \Xi_2 = \frac{1}{2}\frac{\partial F}{\partial X_2} = \epsilon_2 X_2 ,$$

(16)

$$\Xi_3 = \frac{1}{2}\frac{\partial F}{\partial X_3} = \epsilon_3 X_3 .$$

The principal components of displacement $\Xi_i$ introduced here are the projections of the vector $\mathbf{s}_2$ upon the principal axes $X_i$ .

The proof of Helmholtz's theorem has been completed by the establishment of relations (16), which indicate that the third partial motion is composed of three extensions (contractions) in the mutually orthogonal directions of the principal axes of the deformation quadric. In fact, any point $P$ of our volume element with coordinates $X_i$ in the principal system is carried over, according to (16), into a point $P'$ with coordinates

(17)          $$X_i + \Xi_i = X_i(1 + \epsilon_i).$$

The tensor quadric has been introduced here by formal definition, but a direct physical interpretation is possible when coordinate differentials or the equivalent direction cosines are introduced in place of the coordinates $x$, $y$, $z$ [see Eq. (4.21)].

We now proceed from the linear extensions $\epsilon_i$ to the *cubical* extension or *dilatation* $\Theta$ which is defined as the specific change of volume (i.e. the volume change per unit of original volume)

(18)          $$\Theta = \frac{\Delta V' - \Delta V}{\Delta V} .$$

Here $\Delta V$ is the original and $\Delta V'$ the expanded (or compressed) volume of the element. The calculation is easy for a volume element that has the shape of a rectangular cell with sides parallel to the principal axes. Let one corner coincide with the point 0 and denote the sides before and after dilatation by $a_i$ and $a_i'$ . Then $\Delta V = a_1 a_2 a_3$ , $\Delta V' = a_1' a_2' a_3'$ . According to (17), $a_i' = a_i(1 + \epsilon_i)$, therefore $\Delta V' = \Delta V(1 + \epsilon_1)(1 + \epsilon_2)(1 + \epsilon_3)$ and, according to (18),

(19)          $$\Theta = (1 + \epsilon_1)(1 + \epsilon_2)(1 + \epsilon_3) - 1 = \epsilon_1 + \epsilon_2 + \epsilon_3 .$$

(The product terms have again been omitted.)

The representation of $\Theta$ by (19) is valid in any Cartesian system: take for example the original system $x$, $y$, $z$. We assert that there is always

(20)          $$\Theta = \epsilon_{xx} + \epsilon_{yy} + \epsilon_{zz} .$$

For the proof we have to use certain properties of the transformation by which the surface $f = $ const of Eq. (13) is transformed into $F = $ const of Eq. (14). Writing the transformation in the schematic form

(21)

| | $x$ | $y$ | $z$ |
|---|---|---|---|
| $X_1$ | $\alpha_1$ | $\beta_1$ | $\gamma_1$ |
| $X_2$ | $\alpha_2$ | $\beta_2$ | $\gamma_2$ |
| $X_3$ | $\alpha_3$ | $\beta_3$ | $\gamma_3$ |

$$\sum \alpha_i^2 = 1, \qquad \sum \alpha_i \beta_i = 0, \cdots$$

(21a) $\quad \alpha_1^2 + \beta_1^2 + \gamma_1^2 = 1, \cdots$

$$\alpha_1 \alpha_2 + \beta_1 \beta_2 + \gamma_1 \gamma_2 = 0, \cdots$$

and reading the scheme (21) from left to right (i.e., solving for the $X_i$), we obtain by substituting in (14)

$$F = \sum \epsilon_i X_i^2 = \sum \epsilon_i (\alpha_i x + \beta_i y + \gamma_i z)^2$$

$$= x^2 \sum \epsilon_i \alpha_i^2 + 2xy \sum \epsilon_i \alpha_i \beta_i + y^2 \sum \epsilon_i \beta_i^2 + \cdots$$

This expression must be identical with (13), therefore

(22)
$$\epsilon_{xx} = \alpha_1^2 \epsilon_1 + \alpha_2^2 \epsilon_2 + \alpha_3^2 \epsilon_3 ,$$

$$\epsilon_{yy} = \beta_1^2 \epsilon_1 + \beta_2^2 \epsilon_2 + \beta_3^2 \epsilon_3 ,$$

$$\epsilon_{zz} = \gamma_1^2 \epsilon_1 + \gamma_2^2 \epsilon_2 + \gamma_3^2 \epsilon_3 .$$

(22a)
$$\epsilon_{xy} = \alpha_1 \beta_1 \epsilon_1 + \alpha_2 \beta_2 \epsilon_2 + \alpha_3 \beta_3 \epsilon_3 ,$$

$$\epsilon_{yz} = \beta_1 \gamma_1 \epsilon_1 + \beta_2 \gamma_2 \epsilon_2 + \beta_3 \gamma_3 \epsilon_3 ,$$

$$\epsilon_{zx} = \gamma_1 \alpha_1 \epsilon_1 + \gamma_2 \alpha_2 \epsilon_2 + \gamma_3 \alpha_3 \epsilon_3 .$$

On adding Eqs. (22) and taking into account the orthogonality relations (21a), we finally obtain

(23) $\qquad\qquad \epsilon_{xx} + \epsilon_{yy} + \epsilon_{zz} = \epsilon_1 + \epsilon_2 + \epsilon_3 .$

The result is: The sum of the elements in the principal diagonal of the strain tensor is independent of the choice of the system of coordinates; it is an *invariant of the tensor*. In 4 this result will be supplemented by a search for all invariants of a tensor. Also the relations (22a) not fully exploited so far will reappear there.

By (19), (20), and (11) the dilatation $\Theta$ can be put in the simple form

(24) $\qquad\qquad \Theta = \dfrac{\partial \xi}{\partial x} + \dfrac{\partial \eta}{\partial y} + \dfrac{\partial \zeta}{\partial z} .$

Having established the geometrical significance of the *principal* extensions $\epsilon_i$ , our next task is the geometrical interpretation of the *general* strain components $\epsilon_{zz}$ and $\epsilon_{zy}$ . For the diagonal components $\epsilon_{zz}$ , $\cdots$ the case is not different from that of the quantities $\epsilon_1$ , $\cdots : \epsilon_{zz}$ *is the increment in length of an x-fiber per unit of original length.* If such a fiber is cut from the body between the points $x = 0$ and $x = \Delta l$, its length after straining $\Delta l' = \Delta l + \xi - \xi_0$ , whence the specific change of length is

$$\frac{\Delta l' - \Delta l}{\Delta l} = \frac{\xi - \xi_0}{\Delta l} .$$

But the last expression is seen to be equal to

$$\frac{\partial \xi}{\partial x} = \epsilon_{zz} ,$$

when we substitute the special values $x = \Delta l$, $y = z = 0$ in Eq. (1).

For the purpose of interpreting the components $\epsilon_{zy}$ we consider an "$x,y$-lamina" instead of an "x-fiber". Let 0123 denote an infinitesimal rectangular part of the lamina with sides $a$ and $b$ (Fig. 1). In the process

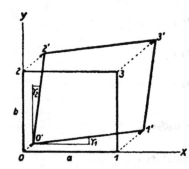

FIG. 1. Geometric interpretation of the strain component $\epsilon_{zy}$ as angular change in an originally orthogonal volume element. The diagram gives the projection in the $x,y$-plane.

of straining, the points 0, 1, 2 are displaced to 0', 1', 2', and the rectangle is deformed into a parallelogram. (Note that displacements in $z$-direction are disregarded; think of the strained figure as projected on the $x,y$-plane: the projection 3' of the strained point 3 will then coincide with the fourth corner of the parallelogram subtended by 0'1' and 0'2', if higher order terms are omitted. Note the analogous omission also in the case of the $\epsilon_{zz}$ .) The angles $\gamma_1$ and $\gamma_2$ indicated in Fig. 1 can now be calculated, again omitting higher order terms:

$$\gamma_1 = \operatorname{tg} \gamma_1 = \frac{\eta_1 - \eta_0}{a + \xi_1 - \xi_0} = \frac{\eta_1 - \eta_0}{a} = \frac{\partial \eta}{\partial x}$$

(25)
$$\gamma_2 = \operatorname{tg} \gamma_2 = \frac{\xi_2 - \xi_0}{b + \eta_2 - \eta_0} = \frac{\xi_2 - \xi_0}{b} = \frac{\partial \xi}{\partial y}$$

$$\gamma_1 + \gamma_2 = \frac{\partial \eta}{\partial x} + \frac{\partial \xi}{\partial y} = 2\epsilon_{zy} .$$

In this equation, subscripts on $\xi$, $\eta$ refer to the displaced points. The angle $\gamma = \gamma_1 + \gamma_2$ is equal to the change of the originally right angle at 0 due to the distortion of the rectangle 0123; $\epsilon_{xy}$, which is half of that angle, is the *shearing strain* or, simply the *shear*. The quantities $\epsilon_{yz}$ and $\epsilon_{zx}$ have the same significance relative to the $y,z$- and $z,x$-laminae.

A short compilation of the more common notations of the strain tensor follows.

| | | | | | | |
|---|---|---|---|---|---|---|
| This book . . . . . | $\epsilon_{xx}$ | $\epsilon_{yy}$ | $\epsilon_{zz}$ | $\epsilon_{xy}=\epsilon_{yx}$ | $\epsilon_{yz}=\epsilon_{zy}$ | $\epsilon_{zx}=\epsilon_{xz}$ |
| Love and some American authors . | $e_{xx}$ | $e_{yy}$ | $e_{zz}$ | $e_{xy}=e_{yx}$ | $e_{yz}=e_{zy}$ | $e_{zx}=e_{xz}$ |
| Kirchhoff and Planck | $x_x$ | $y_y$ | $z_z$ | $\frac{1}{2}x_y=\frac{1}{2}y_x$ | $\frac{1}{2}y_z=\frac{1}{2}z_y$ | $\frac{1}{2}z_x=\frac{1}{2}x_z$ |
| Some English authors | $e$ | $f$ | $g$ | $\frac{1}{2}a$ | $\frac{1}{2}b$ | $\frac{1}{2}c$ |
| Engineering usage . | $\epsilon_x$ | $\epsilon_y$ | $\epsilon_y$ | $\frac{1}{2}\gamma_{zy}=\frac{1}{2}\gamma_{yx}$ | $\frac{1}{2}\gamma_{yz}=\frac{1}{2}\gamma_{zy}$ | $\frac{1}{2}\gamma_{zx}=\frac{1}{2}\gamma_{zx}$ |

The inclusion of the factor $\frac{1}{2}$ in our definition of the $\epsilon_{xy}$, $\cdots$ is suggested by the system of notation of general tensor analysis, but we shall see in 40 that Kirchhoff's choice of the $x_y$, $\cdots$ which sets this factor in evidence has a certain advantage when simple expressions for energetic quantities are desired.

## 2. Review of Vector Analysis

Throughout this volume we shall make continual use of *vector analysis*, that is, calculus applied to vector quantities. Thus, while familiarity with *vector algebra* and with the basic concepts of *vector analysis* is assumed on the part of the reader, a clarification of some of the fundamental points seems appropriate.

First, we repeat here the definition of vector and scalar given in Vol. I, following 22.6. A quantity is a vector if, in an orthogonal transformation of the coordinate system, it follows the same transformation rules as the position vector $\mathbf{r} = x, y, z$; it is a scalar if it is invariant in such transformations.[6]

Let us put this definition to use in the proof that the quantity $\Theta$, derived from the vector $\mathbf{s} = x, y, z$ in (1.24), is a scalar. We replace $\mathbf{s}$ by

---

[6]For this definition compare H. and B. S. Jeffreys, Methods of Mathematical Physics, Cambridge University Press, Cambridge, 1946, Chap. 2.023.

an arbitrary vector $\mathbf{A} = A_x$ , $A_y$ , $A_z$[7] and consider instead of $\Theta$ more generally the quantity

(1)
$$\operatorname{div} \mathbf{A} = \frac{\partial A_x}{\partial x} + \frac{\partial A_y}{\partial y} + \frac{\partial A_z}{\partial z} \, ,$$

called the divergence of the vector field $\mathbf{A}$.

The invariance of div $\mathbf{A}$, which we now prove, amounts to the following equality:

$$\operatorname{div}' \mathbf{A} = \frac{\partial A_{x'}}{\partial x'} + \frac{\partial A_{y'}}{\partial y'} + \frac{\partial A_{z'}}{\partial z'} = \operatorname{div} \mathbf{A}$$

For the sake of brevity, the coordinates, primed or unprimed, will be numbered rather than lettered. Transformation (1.21) reads in the simplified notation

(2)

|        | $x_1$       | $x_2$       | $x_3$       |
| ------ | ----------- | ----------- | ----------- |
| $x_1'$ | $\alpha_{11}$ | $\alpha_{12}$ | $\alpha_{13}$ |
| $x_2'$ | $\alpha_{21}$ | $\alpha_{22}$ | $\alpha_{23}$ |
| $x_3'$ | $\alpha_{31}$ | $\alpha_{32}$ | $\alpha_{33}$ |

(2a)
$$\sum_k \alpha_{ik}\alpha_{jk} = \delta_{ij} \, ,$$
$$\sum_i \alpha_{ik}\alpha_{il} = \delta_{kl} \, .$$

The meaning of the symbol $\delta_{ij}$ (and correspondingly $\delta_{kl}$) is the usual one:

$$\delta_{ij} = \begin{cases} 1 & \text{for} \quad i = j \\ 0 & \text{for} \quad i \neq j \end{cases}$$

For an arbitrary function $U\,(x_1 \, , \, x_2 \, , \, x_3)$, which may at the same time be considered as a function of $x_1'$ , $x_2'$ , $x_3'$ because of (2), one has

(3)
$$\frac{\partial U}{\partial x_i'} = \sum_k \frac{\partial U}{\partial x_k} \frac{\partial x_k}{\partial x_i'} \, .$$

This can be rewritten as

(3a)
$$\frac{\partial U}{\partial x_i'} = \sum_k \alpha_{ik} \frac{\partial U}{\partial x_k}$$

if the scheme of coefficients (2) is read vertically downward. According to our vector definition we may apply the transformation rule for $\mathbf{r}$, that is

---

[7]In the original German text, vectors *and* their components are printed in the same Gothic types. The more usual way of making a typographical distinction between the two has been adopted for this translation.

the scheme (2), to the vector **A**. Thus, on reading (2) from left to right, it follows

(4) $$A_i' = \sum_l \alpha_{il} A_l .$$

We now identify the arbitrary function $U$ in (3a) with the vector component $A_i$. Doing this for $i = 1, 2, 3$ and adding leads us to

$$\sum_i \frac{\partial A_i'}{\partial x_i'} = \sum_i \sum_k \sum_l \alpha_{ik}\alpha_{il} \frac{\partial A_l}{\partial x_k} = \sum_k \sum_l \delta_{kl} \frac{\partial A_l}{\partial x_k} .$$

Here the first member is div' **A** and the last member div **A** if proper account is taken of the $\delta$-symbol, hence the invariance of the divergence has been proved. The cubical dilatation (1.24) can now be written as $\theta = \text{div } \mathbf{s}$.

The divergence operator derives a scalar from a vector. The gradient operator, on the other hand, is the simplest differential operator that derives a vector from a scalar. Denoting the scalar by $U$ (a scalar point function as before), we construct the vector

(5) $$\text{grad } U = \frac{\partial U}{\partial x_1}, \frac{\partial U}{\partial x_2}, \frac{\partial U}{\partial x_3} .$$

Its vector character follows at once from (3a) which can be written by means of the symbol "grad"

(5a) $$\text{grad}_i' U = \sum_k \alpha_{ik} \text{grad}_k U.$$

Thus the gradient transforms in the same way as the position vector.

We now turn to the differential operation introduced in (1.9) for which the symbol "curl" is used. Writing (1.9) with this symbol, we have

(6) $$\mathbf{B} = \text{curl } \mathbf{A}, \qquad \text{curl}_x \mathbf{A} = \frac{\partial A_z}{\partial y} - \frac{\partial A_y}{\partial z} , \qquad \text{etc.,}$$

where **B** stands for $2\boldsymbol{\omega}$[8] and **A** for $\mathbf{v} = u, v, w$. Eq. (1.6) already contains a definition of the vector **ω**. On setting as in (1.7)

$$\mathbf{s}_1 = \mathbf{v}_1 \Delta t, \qquad \boldsymbol{\varphi} = \boldsymbol{\omega}\Delta t,$$

(1.6) is carried over into

(7) $$\mathbf{v}_1 = \boldsymbol{\omega} \times \mathbf{r}.$$

---

[8]We shall later come back to this factor $\frac{1}{2}$ that is now included in the definition of **B**.

The developments in 1 make it evident that the definition of ω by (6) [or (1.9)] agrees with that by (7) [or (1.6)]; this will be checked by direct calculation in problem I.1.

Let us now write (7) with the same lettering as (6); we can again use **B** instead of 2ω, but we may use **A** also for $v_1$ , since in what follows we are only concerned with the vector character of this quantity and curl $v_1$ is identical with curl **v**. In this way (7) becomes

(7a)  $$2\mathbf{A} = \mathbf{B} \times \mathbf{r}.$$

Our aim is to show that the quantity **B** has, at least in a certain sense, vector character. Such a proof, which furnishes a better understanding of the concept "curl", is by no means superfluous although it is omitted in most textbooks.

The formal proof can be based directly on the definition (6), in which case it amounts to a rather cumbersome calculation (problem I, 2), but a clarification of the issue can also be obtained if one starts from Eq. (7a).[9] There the components of **B** appear as coefficients of a linear *vector function*, like the quantities $\epsilon_{ik}$ in (1.10). Introducing double subscripts and numbering the vector components (instead of lettering), we write

(8)
$$B_x = B_1 = B_{32} = -B_{23} , \qquad A_x = A_1 ,$$
$$B_y = B_2 = B_{13} = -B_{31} , \qquad A_y = A_2 ,$$
$$B_z = B_3 = B_{21} = -B_{12} , \qquad A_z = A_3 .$$

This notation for the components of **B** corresponds to the representation of $\phi$ by (1.12a); in both cases the diagonal terms ($B_{ii}$ , $\varphi_{ii}$) vanish and the "vector" **B** is really an *antisymmetric tensor* like $\varphi$.

Eq. (7a) now takes the simple form

(9)  $$2A_i = \sum_k B_{ik}x_k$$

which we can check, say, for $i = 1$; we find

$$2A_1 = 2A_x = B_{11}x_1 + B_{12}x_2 + B_{13}x_3 = -B_z y + B_y z = (\mathbf{B} \times \mathbf{r})_x$$

in agreement with (7a).

In order to define the quantities $B$ for the rotated coordinate system $x'$, we write Eqs. (9) in this system

(10)  $$2A_i' = \sum_k B_{ik}'x_k'$$

[9]Note, however, that Eq. (6) by no means implies Eq. (7a) in the case of a general vector field **A**. Observe the careful formulation of problem I.1.

and reintroduce $x$ and $A$ by

$$x'_k = \sum_l \alpha_{kl} x_l , \qquad A'_i = \sum_k \alpha_{ik} A_k$$

according to (2). This transforms (10) into

(11)
$$2 \sum_k \alpha_{ik} A_k = \sum_k \sum_l B'_{ik} \alpha_{kl} x_l .$$

On substituting for the components $A$ in the first member of this equation according to (9) which, obviously, can be written as $2A_k = \sum_l B_{kl} x_l$ by a mere relettering of the subscripts, Eq. (11) becomes

$$\sum_k \sum_l B_{kl} \alpha_{ik} x_l = \sum_k \sum_l B'_{ik} \alpha_{kl} x_l ,$$

which is an identity in the three independent variables $x_1$ , $x_2$ , $x_3$ . By comparing the coefficients, the following three equations between the $B$ and $B'$ are obtained

(12)
$$\sum_k B_{kl} \alpha_{ik} = \sum_k B'_{ik} \alpha_{kl} , \qquad l = 1, 2, 3,$$

from which the $B'$ must be determined. This is done by multiplying Eqs. (12) with $\alpha_{jl}$ and summing over $l$. The second members of (12) add up to

(12a)
$$\sum_l \sum_k B'_{ik} \alpha_{kl} \alpha_{jl} = \sum_k B'_{ik} \sum_l \alpha_{kl} \alpha_{jl} = \sum_k B'_{ik} \delta_{kj} = B'_{ij} .$$

Thus the new components $B'$ have been found, viz.

(13)
$$\sum_l \sum_k B_{kl} \alpha_{ik} \alpha_{jl} = B'_{ij} .$$

As it stands, each $B'$ is a linear function of all nine components $B$, but, with $B_{kl} = -B_{lk}$ and $B_{kk} = 0$, the number of terms reduces to three. Written more completely, the expressions for $B'$ are

(14)
$$B'_{ij} = B_{32}(\alpha_{i3}\alpha_{j2} - \alpha_{i2}\alpha_{j3}) + B_{13}(\alpha_{i1}\alpha_{j3} - \alpha_{i3}\alpha_{j1})$$
$$+ B_{21}(\alpha_{i2}\alpha_{j1} - \alpha_{i1}\alpha_{j2}).$$

Let us write out one of them, say, for $i = 3, j = 2$:

(15)
$$B'_{32} = B_{32}(\alpha_{33}\alpha_{22} - \alpha_{32}\alpha_{23}) + B_{13}(\alpha_{31}\alpha_{23} - \alpha_{33}\alpha_{21})$$
$$+ B_{21}(\alpha_{32}\alpha_{21} - \alpha_{31}\alpha_{22}).$$

The coefficients of the quantities $B$ in (15) are nothing else but the co-factors of the elements in the first row of the determinant of the transformation (2). Depending on whether this determinant is $+1$ or $-1$, the cofactors are *equal* or *opposite* to the elements to which they belong[10] (see, e.g., Vol. I, problem I.10.). Eq. (15) may therefore be rewritten as

$$\pm B_1' = \alpha_{11}B_1 + \alpha_{12}B_2 + \alpha_{13}B_3$$

where the single subscript notation has been reintroduced for the $B$. In the same way $B_2'$ and $B_3'$ are obtained. The result can be written in the form

(16) $$\pm B_i' = \sum_l \alpha_{il}B_l ,$$

which is exactly the vector transformation rule of Eq. (4) except for the ambiguity of the sign.

Two things can be learned from this result:

1. *The quantity* $\mathbf{B} = \text{curl } \mathbf{A}$ *behaves like an ordinary (polar) vector in a rotation, that is, in an orthogonal coordinate transformation with determinant* $+1$, hence the general use of the simple terminology "curl vector" which we shall also adopt.

2. *The curl is, strictly speaking, not a polar, but an axial vector* for the following reason: *In a transition from a dextral to a sinistral system or, more generally, in a rotation with a subsequent change of orientation (determinant* $-1$) *the new components of the curl are the negatives of the new components of the (polar) vector that represents the curl in the original system.*

We can now easily see why the use of double subscripts is an advantage in the foregoing proof: An axial vector is not appropriately represented by an arrow in the direction[11] of the axis; the correct symbol should be a circular arrow about the axis. In this way the plane of the circular arrow, say the $y$, $z$-plane, takes over the function of the $x$-axis, and a better way to write the left member of Eq. (6) would be

(17) $$\text{curl}_{yz} \mathbf{A} = \frac{\partial A_z}{\partial y} - \frac{\partial A_y}{\partial z} .$$

The notation $\text{curl}_x \mathbf{A}$ *represents* the axial vector by a polar vector. This is possible in pure rotation according to 1. and is indeed quite often an aid to the imagination.

In this connection a well-known and often applied rule may be men-

---

[10]Cf., e.g., Brand, *op. cit.*, §149.

[11]In fact, two orientations are possible and only one is chosen in accordance with the orientation of the coordinate system.

tioned that helps to keep in mind the structure of the expressions (6) or (17) and in particular the sequence of signs in those formulas. It reads:

$$(18) \qquad \text{curl } \mathbf{A} = \begin{vmatrix} \mathbf{i} & \mathbf{j} & \mathbf{k} \\ \dfrac{\partial}{\partial x} & \dfrac{\partial}{\partial y} & \dfrac{\partial}{\partial z} \\ A_x & A_y & A_z \end{vmatrix}.$$

The letters $\mathbf{i}$, $\mathbf{j}$, $\mathbf{k}$ denote here unit vectors[12] in the directions of the $x$-, $y$-, $z$-axes. Multiplication with $\partial/\partial x$, $\partial/\partial y$, $\partial/\partial z$ means differentiation with respect to the corresponding variable. The expansion of the determinant (18) yields

$$(18a) \qquad \text{curl } \mathbf{A} = \mathbf{i}\left(\frac{\partial A_z}{\partial y} - \frac{\partial A_y}{\partial z}\right) + \mathbf{j}\left(\frac{\partial A_x}{\partial z} - \frac{\partial A_z}{\partial x}\right) + \mathbf{k}\left(\frac{\partial A_y}{\partial x} - \frac{\partial A_x}{\partial y}\right),$$

which represents the correct way to form the curl components.

It should be noted, however, that the mapping of an axial upon a polar vector is possible only in *three-dimensional space*. This can be inferred from the number of components $\nu$ which is, of course, equal to the number of dimensions $n$ in the case of the polar vector, while for an axial vector in general

$$\nu = \frac{n(n-1)}{2}, \qquad \text{or} \begin{cases} \text{for} \quad n = 1, 2, 3, 4, \cdots \\ \nu = 0, 1, 3, 6, \cdots \end{cases}$$

The number $\nu$ is correlated with the double subscript notation of the axial vector: it is the number of combinations of the $n$ directions in space in groups of two. That $\nu$ coincides with $n$ for $n = 3$ is, one might say, a coincidence. The number of components $\nu = 6$ in four-dimensional space plays a decisive part in electrodynamics, where it coincides with the *number of components of the electromagnetic field* (3 electric + 3 magnetic components).

We return to our starting point, Eq. (19), which we write in vector form

$$(19) \qquad \qquad \omega = \frac{1}{2} \text{ curl } \mathbf{v}.$$

---

[12]The $i$, $j$, $k$ in the original German text have not been printed as vectors. The symbolic notation $\mathbf{V} = iV_x + jV_y + kV_z$ has then the character of a "higher complex number" and indicates more strongly the conceptual relationship between vectors, ordinary complex numbers, and Hamilton's quaternions.

It cannot be denied that the factor $\frac{1}{2}$ in this formula is a flaw. The meaning of $\omega$ is, as we know now, the whirl or rotation associated with the velocity distribution **v**. Our formal introduction of the symbol curl in (6) implies that the "physical rotation" is only half of the "mathematical rotation". It is not feasible to remove this paradox by absorbing the factor $\frac{1}{2}$ in the definition of the curl symbol; that would have the most uncomfortable consequences for general vector-analysis, in particular for electrodynamics. Nothing can be done but to point out the flaw and apologize.

The three operators grad, div, and curl have so far been defined by *differential* operations. There exists, however, a very convenient way of introducing div and curl by means of *integral* operations. Let us start with the *divergence*.

Visualize the vector field **A** as a field of flow, that is, let magnitude and direction of the flow be given everywhere by the magnitude and direction of **A**. Let the field point $P$ be surrounded by a closed surface and $\Delta\tau$ be the enclosed volume. If $n$ is the outward normal associated with the surface element $d\sigma$, $A_n$ represents the outflow per unit of surface, normal to the surface. On forming $(1/\Delta\tau) \int A_n \, d\sigma$, the limit for vanishing $\Delta\tau$ (or $\sigma$) becomes

$$(20) \qquad \operatorname{div} \mathbf{A} = \operatorname*{Lim}_{\Delta\tau \to 0} \frac{1}{\Delta\tau} \int A_n \, d\sigma.$$

The agreement between this definition and (1) is evident in the special case that $\Delta\tau$ is chosen as a rectangular cell with center $P$ and sides $\Delta x$, $\Delta y$, $\Delta z$ (Fig. 2). If $A_x$ at the point $x + \Delta x$ is expanded after Taylor and

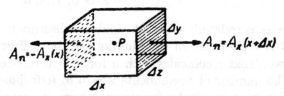

FIG. 2. Calculation of div A by an integral over the surface of an infinitesimal parallelepiped.

higher powers of $\Delta x$ are neglected as before, one finds for the two $x$-surfaces (that is, the surfaces normal to the $x$-axis)

$$\int A_n \, d\sigma = \int (A_x(x + \Delta x) - A_x(x)) \, dy \, dz = \int \frac{\partial A_x}{\partial x} \Delta x \, dy \, dz = \frac{\partial A_x}{\partial x} \Delta\tau$$

and two corresponding terms for the two pairs of $y$- and $z$-surfaces; altogether one obtains the expression for div **A**. In writing (20) mathematicians would add the proviso that the limit indicated therein actually exists. This we have omitted, since the restrictions as to continuity imposed on the vector field by such a proviso certainly would not be more severe than those which are inherent in definition (1). Thus, as a definition, (20) is certainly equivalent to (1); it should be preferred, however, as it is of a less formalistic nature.

The curl can be defined by the following construction in our field of flow. Let an arbitrarily chosen, oriented line $a$ pass through the point $P$; in the plane that contains $P$ and is normal to $a$, draw a closed curve $s$ surrounding $P$, and denote the enclosed area by $\Delta\sigma$. Let $A_s$ be the component of $\mathbf{A}$ in direction of the arc element $ds$, taken in that sense of direction which forms a right hand screw with the axis $a$. Consider now the line integral $\oint A_s \, ds$, for which Lord Kelvin introduced the appropriate name *circulation*. The limit of the ratio of the circulation to the area $\Delta\sigma$ is

$$(21) \qquad \operatorname{curl}_a \mathbf{A} = \operatorname*{Lim}_{\Delta\sigma \to 0} \frac{1}{\Delta\sigma} \oint A_s \, ds.$$

The agreement of this definition with the previous one, (17), can again be seen quite easily if the curve $s$ is appropriately chosen. Suppose one wants to verify (21) for the $y$, $z$-component of the curl. The line $a$ then coincides with the $+x$-direction, and $s$ may be chosen as a rectangle with center $P$ and sides $\Delta y$, $\Delta z$ (see Fig. 3). On expanding the contributions

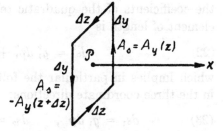

FIG. 3. Calculation of curl **A** by a line integral over an orthogonal circuit.

of each pair of sides as before, the first pair (which is parallel to the $y$-axis) yields

$$\int A_s \, ds = \int \left( -A_v(z + \Delta z) + A_v(z) \right) dy = - \frac{\partial A_v}{\partial z} \Delta y \Delta z,$$

and the second pair

$$\int A_s \, ds = \frac{\partial A_z}{\partial y} \Delta y \Delta z.$$

Altogether one obtains according to (21)

$$\operatorname{curl}_x \mathbf{A} = \operatorname{curl}_{yz} \mathbf{A} = \frac{\partial A_z}{\partial y} - \frac{\partial A_v}{\partial z}.$$

The other components come out if the line $a$ is made to coincide with the other coordinate axes, and the results are found by cyclic permutation ("rotation") of the letters.

The "integral" definitions put geometrical meaning into the operators div and curl previously introduced by formal analytic means, but a visualization of the operation grad $U$ is still lacking. Imagine for this purpose the level surfaces (that is, the surfaces $U = $ const) and their orthogonal trajectories marked in space. The direction of the gradient is everywhere tangential to the trajectory; its magnitude, $\partial U/\partial n$, expresses itself in the concentration of the level surfaces as they follow each other: for constant $\delta U$, $\partial U/\partial n$ is the larger, the smaller $\delta n$.

Compared with the original definitions (1), (5), and (17), the advantage of the geometrical definitions is not only their less formalistic nature, but also their validity in *general* coordinates, which permits in particular an easy transition to any required system of curvilinear *orthogonal* coordinates

We show this for the general case of a system of three mutually orthogonal families of surfaces, while the more elementary special cases such as polar coordinates will be treated in prob. I.3. Let $p_1$, $p_2$, $p_3$ be the parameters of the three families of surfaces. As a consequence of the mutual orthogonality, the analytic expression for the square of the distance of two neighbored points contains no mixed terms of the type $dp_i\, dp_k$. Following the notation $g_{ik}$ of the general theory of relativity, we denote the coefficients of the quadratic terms by $g_i^2$, so that the square of the element of length is

$$(22) \qquad ds^2 = g_1^2\, dp_1^2 + g_2^2\, dp_2^2 + g_3^2\, dp_3^2 ,$$

which implies in particular the following expressions for the differentials in the three coordinate directions:

$$(23) \qquad ds_1 = g_1\, dp_1 , \qquad ds_2 = g_2\, dp_2 , \qquad ds_3 = g_3\, dp_3 .$$

These expressions replace $dx$, $dy$, $dz$, if orthogonal curvilinear coordinates are used instead of the usual Cartesian coordinates. The components of the gradient, heretofore

$$\operatorname{grad} U = \frac{\partial U}{\partial x} , \frac{\partial U}{\partial y} , \frac{\partial U}{\partial z} ,$$

now become

$$(24) \qquad \operatorname{grad} U = \frac{1}{g_1}\frac{\partial U}{\partial p_1} , \frac{1}{g_2}\frac{\partial U}{\partial p_2} , \frac{1}{g_3}\frac{\partial U}{\partial p_3} .$$

The expression for the divergence in the new coordinates is obtained from (20) if the volume $\Delta\tau$ is chosen as an infinitesimal cell with (curvilinear) sides $\Delta s_1$, $\Delta s_2$, $\Delta s_3$ :

$$(25) \qquad \text{div } \mathbf{A} = \frac{1}{g_1 g_2 g_3} \left\{ \frac{\partial}{\partial p_1} (g_2 g_3 A_1) + \frac{\partial}{\partial p_2} (g_3 g_1 A_2) + \frac{\partial}{\partial p_3} (g_1 g_2 A_3) \right\}.$$

The expression for the 1-component of the curl is obtained by choosing the surface that contains the "rectangular" circuit orthogonal to the co-ordinate direction $p_1$ :

$$(26) \qquad \text{curl}_1 \mathbf{A} = \frac{1}{g_2 g_3} \left( \frac{\partial (g_3 A_3)}{\partial p_2} - \frac{\partial (g_2 A_2)}{\partial p_3} \right),$$

The 2- and 3-components follow by "rotating" the numbers 1, 2, 3.

### 3. The Theorems of Gauss, Stokes, and Green

It is not difficult to proceed from the integral definitions of divergence and curl in (2.20) and (2.21) to the integral theorems of Gauss and Stokes.

Let the vector field $\mathbf{A}$ be differentiable inside the closed surface $\sigma$ so as to establish the value of div $\mathbf{A}$ everywhere in the interior. The following integral relation then constitutes *Gauss's theorem*:

$$(1) \qquad \int \text{div } \mathbf{A} \, d\tau = \int A_n \, d\sigma.$$

Here $n$ is the outward[13] normal associated with the surface element $d\sigma$.

To prove this relation we divide the volume $\tau$ in a number of small cells $\Delta \tau_m$ ; in accordance with the mathematical definition of the definite integral, a set of such divisions with decreasing cell size may be used to write the first member of (1) as the limit

$$(1a) \qquad \text{Lim} \sum_{\substack{m=0}}^{m=M} \text{div } \mathbf{A} \, \Delta \tau_m,$$
$$\scriptstyle M \to \infty$$

where the value of div $\mathbf{A}$ may be taken at an arbitrary point of the cell $\Delta \tau_m$ . If the integral definition (2.20) is substituted for div $\mathbf{A}$ the sum in (1a) becomes

$$(1b) \qquad \sum_m \Delta \tau_m \lim_{\Delta \tau \to 0} \frac{1}{\Delta \tau} \int A_n \, d\sigma.$$

There is no objection against writing the factor $\Delta \tau_m$ , which is still finite, to the right of the limit sign; also the summation, which so far refers only

---

[13]This convention may cause some inconvenience, as in the case $\mathbf{A} = \text{grad } U$, where $\partial U / \partial n$ is then in general not defined outside of $\sigma$; we may, however, replace $\partial U / \partial n$ by $-\partial U / \partial n'$ where $n'$ is now the inward normal.

to a finite number of terms, may be interchanged with the limit sign. The expression (1b) then takes the form

$$\text{(1c)} \qquad \lim_{\Delta \tau \to 0} \sum_m \frac{\Delta \tau_m}{\Delta \tau} \int A_n \, d\sigma.$$

Let it now be agreed upon that the cell $\Delta \tau$ which serves to define the divergence be *identical* with the cell $\Delta \tau_m$ that belongs to the definite integral.[14] At the same time let us write $d\sigma_m$ instead of $d\sigma$, to indicate that $d\sigma_m$ is now an element of the surface of $\Delta \tau_m$ . After these changes (1c) becomes

$$\text{(1d)} \qquad \lim_{\Delta \tau_m \to 0} \sum_m \int A_n \, d\sigma_m \, .$$

In carrying out the summation over $m$ the parts of the $\sigma_m$-surfaces that lie *inside* $\sigma$ cancel each other in pairs because of the opposite signs of the normals $n$. Only the contributions of the surface elements that are adjacent to the $\sigma$-surface are left over; hence the sum in (1d) may simply be written

$$\text{(1e)} \qquad \int A_n \, d\sigma,$$

where the integral refers to the *surface of the originally given region in space* as in the integral (1). Since this integral is independent of the division into cells $\Delta \tau_m$ , the limiting process required in (1d) does not change anything, neither does the summation required in (1a). Expression (1e), as it stands, is therefore the value of the expression (1a) which, in turn, was only another way of writing the volume integral on the left side of (1). Since (1e) is identical with the right member of (1), everything is proved.

The formal proof of Gauss's theorem usually found in textbooks is based on a partial integration in rectangular coordinates; it can be replaced, as we have seen, by an argument that uses only the general principles of integral calculus and vector analysis. The same is true for Stokes's theorem, for which we shall outline only the general idea of the proof, referring to the foregoing for the formal details.

In terms of fluid flow, the second member of (1) means the outflow through the boundary surface (inflow, if negative). The first member of (1) represents the algebraic sum of the strengths of all sources or sinks, continuously distributed throughout the interior. The equality of the two members is thus intuitively evident.

---

[14]To be sure, this means that the limiting processes $\Delta \tau \to 0$ and $\Delta \tau_m \to 0$ (or $m \to \infty$) are carried out simultaneously, whereas the two processes should be carried out independently according to (1b).

In *Stokes's theorem* we consider an arbitrary, in general curved, surface $\sigma$ bounded by a closed, oriented curve $s$ (circuit). At each surface element $d\sigma$ construct the normal $n$ and let it point in the direction that forms a right hand screw with the orientation of $s$. The content of Stokes's theorem is then: For any vector field $\mathbf{A}$, continuous in the neighborhood of $\sigma$,

$$(2) \qquad \int \operatorname{curl}_n \mathbf{A}\, d\sigma = \oint \mathbf{A}\cdot ds.$$

For a proof divide $\sigma$ into small surface elements that can be considered as parts of the associated tangential planes, and let their boundaries $s_m$ be oriented in accordance with the orientation of $s$. Write down equation (2.21) for each of the elements $\Delta\sigma_m$, multiply with $\Delta\sigma_m$ and sum over $m$. This leads to

$$(3) \qquad \operatorname*{Lim}_{m\to\infty} \sum_m \operatorname{curl}_n \mathbf{A}\, \Delta\sigma_m = \operatorname*{Lim}_{m\to\infty} \sum_m \oint \operatorname*{Lim}_{\Delta\sigma\to0} \frac{\Delta\sigma_m}{\Delta\sigma} \mathbf{A}\cdot ds_m .$$

On making $\Delta\sigma = \Delta\sigma_m$ and proceeding as in Eqs. (1a)-(1d), the contributions of all internal boundaries cancel each other, since each segment belongs to two adjacent surface elements and is therefore traversed twice in opposite directions. Only the contributions of the external boundaries remain and sum up to the line integral on the right hand side of (2). In terms of fluid flow, Eq. (2) states: the circulation around the boundary curve $s$ equals the flux of the vortex vector through an *arbitrary* surface $\sigma$ subtended by $s$.

The integral taken over the boundary curve vanishes if the surface integral in Stokes's theorem refers to a *closed* surface. We indicate this as follows:

$$(4) \qquad \oint \operatorname{curl}_n \mathbf{A}\, d\sigma = 0$$

where we assume, of course, the continuity of curl $\mathbf{A}$, a point which has been emphasized before.

If the vector $\mathbf{A}$ is, in particular, chosen as a gradient, $\mathbf{A} = \operatorname{grad} U$, say, then the differential expression

$$(5) \qquad \mathbf{A}\cdot ds = A_x\, dx + A_y\, dy + A_z\, dz$$

becomes a total differential $dU$. In this case

$$(6) \qquad \oint \mathbf{A}\cdot ds = 0$$

for any boundary curve. The first member of (2) must then vanish for any surface $\sigma$ and any normal direction $n$. This implies that quite generally

$$(7) \qquad \operatorname{curl\,grad} U = 0,$$

a formula which can also be read from the differential definition of the curl, Eq. (2.6).

The equation curl $\mathbf{A} = 0$, written in rectangular coordinates, is equivalent to the three conditions for the differential expression $A_x\, dx + A_y\, dy + A_z\, dz$ to be a total differential [see Eq. (5)]. These same conditions already appeared in Vol. I, Eq. (18.17) for the existence of a potential in a force field; they can now be written in the comprehensive form curl $\mathbf{A} = 0$.

Note, by the way, that the operation curl grad has a well defined meaning, but grad curl has not, since the gradient, by its definition, cannot be applied to a vector.

When Gauss's theorem is applied to a curl vector, that is, when curl $\mathbf{B}$ is substituted for $\mathbf{A}$ in (1), the right member of (1) becomes an integral of the form (4), hence it vanishes. Consequently the expression on the left side of (1) is zero for any closed surface $\sigma$ in $\tau$. This means that generally

$$\text{(8)} \qquad\qquad \text{div curl } \mathbf{B} = 0,$$

an identity which can also be read directly from the differential definition (2.1) and (2.6). Again the converse symbol curl div $\mathbf{B}$ carries no meaning since the operation curl cannot be applied to a scalar.

Let us also have a look at the symbols div grad and grad div. The first applies to a scalar, say $U$. We write

$$\text{(9)} \qquad\qquad \text{div grad } U = \nabla^2 U.$$

In Cartesian coordinates the operator $\nabla^2$ is

$$\text{(9a)} \qquad\qquad \nabla^2 = \frac{\partial^2}{\partial x^2} + \frac{\partial^2}{\partial y^2} + \frac{\partial^2}{\partial z^2}\ .$$

In orthogonal curvilinear coordinates $p_1$, $p_2$, $p_3$ the relations (2.24) and (2.25) yield

$$\text{(9b)} \quad \nabla^2 U = \frac{1}{g_1 g_2 g_3}\left\{\frac{\partial}{\partial p_1}\left(\frac{g_2 g_3}{g_1}\frac{\partial U}{\partial p_1}\right) + \frac{\partial}{\partial p_2}\left(\frac{g_3 g_1}{g_2}\frac{\partial U}{\partial p_2}\right) + \frac{\partial}{\partial p_3}\left(\frac{g_1 g_2}{g_3}\frac{\partial U}{\partial p_3}\right)\right\}.$$

The symbol $\nabla^2$ is called the *Laplace Operator* or *Laplacian*, also the *second differential parameter*. The *first differential parameter* is defined by

$$DU = (\nabla U)^2 = \left(\frac{\partial U}{\partial x}\right)^2 + \left(\frac{\partial U}{\partial y}\right)^2 + \left(\frac{\partial U}{\partial z}\right)^2$$

$$\text{(9c)}$$

$$= \frac{1}{g_1^2}\left(\frac{\partial U}{\partial p_1}\right)^2 + \frac{1}{g_2^2}\left(\frac{\partial U}{\partial p_2}\right)^2 + \frac{1}{g_3^2}\left(\frac{\partial U}{\partial p_3}\right)^2.$$

Both symbols operate on scalar point functions. Other notations for the

Laplacian are $\Delta^2$ and $\Delta$, the latter mainly in French and German literature.[15]

The converse symbol, grad div, applies to vectors. Let it here be
sufficient to mention that it appears in the useful formula

(10)                     curl curl $\mathbf{A}$ = grad div $\mathbf{A}$ $-$ $\nabla^2\mathbf{A}$.

Caution is necessary in the application of (10) since in this formula the
symbol $\nabla^2$ = div grad operates on a vector, contrary to its definition.
Actually, the formula refers to the *Cartesian components* of the vector $\mathbf{A}$,
and $\nabla^2$ operates on the individual component of $\mathbf{A}$, considered as a
scalar. Accordingly, the proof of (10) must be carried out in Cartesian
coordinates and consists in checking the equality of the two members by
an easy calculation; the operator $\nabla$ can also be used.

Formula (10), conversely, serves to transform into general coordinates
the quantity $\nabla^2\mathbf{A}$, which so far is only known for Cartesian components.
We define

(10a)                     $\nabla^2\mathbf{A}$ = grad div $\mathbf{A}$ $-$ curl curl $\mathbf{A}$.

The right member is known in arbitrary orthogonal coordinates according to (2.24)-(2.26) so as to give a definite meaning to the left member.
We shall see this procedure applied in 9, after Eq. (18). In problem I.4
we discuss and interpret the difference between the correctly and in-
correctly calculated components of $\nabla^2\mathbf{A}$ in cylindrical coordinates.

We turn now to *Green's theorem*[16] which occupies a *unique place* among

---

[15]The sign $\nabla$ (read "del" or "nabla", the latter being the name of a Hebrew musical
instrument of inverted triangular shape) is symbolically defined by

$$\nabla = \mathbf{i}\frac{\partial}{\partial x} + \mathbf{j}\frac{\partial}{\partial y} + \mathbf{k}\frac{\partial}{\partial z}.$$

$\nabla^2$ is meant as a scalar product:

$$\nabla^2 = \nabla\cdot\nabla = \frac{\partial^2}{\partial x^2} + \frac{\partial^2}{\partial y^2} + \frac{\partial^2}{\partial z^2}.$$

The product of $\nabla$ with a scalar $\varphi$ is the gradient of $\varphi$: The scalar product $\nabla\cdot\mathbf{A}$, if
evaluated according to the formal rules of vector algebra, gives the divergence of $\mathbf{A}$:

$$\nabla\cdot\mathbf{A} = \frac{\partial A_x}{\partial x} + \frac{\partial A_y}{\partial y} + \frac{\partial A_z}{\partial z} = \text{div }\mathbf{A}.$$

The vector product $\nabla \times \mathbf{A}$ means, accordingly, the curl of $\mathbf{A}$; one obtains, with
$\mathbf{i}\times\mathbf{j} = -\mathbf{j}\times\mathbf{i} = \mathbf{k}$, $\mathbf{i}\cdot\mathbf{i} = 0$, etc., as in (2.18b),

$$\nabla\times\mathbf{A} = \mathbf{i}\left(\frac{\partial A_z}{\partial y} - \frac{\partial A_y}{\partial z}\right) + \mathbf{j}\left(\frac{\partial A_x}{\partial z} - \frac{\partial A_z}{\partial x}\right) + \mathbf{k}\left(\frac{\partial A_y}{\partial x} - \frac{\partial A_x}{\partial y}\right) = \text{curl }\mathbf{A}.$$

[16]George Green, *An Essay on the Application of Mathematical Analysis to the Theories
of Electricity and Magnetism*, published in Nottingham, 1828. The theorem is an-
nounced and established in full generality in art. 3 of this paper, the calculations are,
of course, carried out in (Cartesian) coordinates, not in vectors. The notations $U$, $V$,
used in (11), are Green's.

the vector-analytic integral theorems. Its applications in mathematical physics are very numerous. It is of fundamental importance in potential theory for which Green first devised it; indispensable in hydrodynamics, electrodynamics, and optics, it is also highly useful in pure mathematics. Riemann based his theory of complex functions upon the two-dimensional form of Green's theorem; it is likely to turn up at any time in the calculus of variations and in the theory of eigenfunctions and integral equations. Green proved his theorem by partial integrations as it is often done at present. We shall derive it more simply from Gauss's theorem which, apparently, was not known to Green. Let $U$ and $V$ be two scalar point functions satisfying the necessary continuity requirements (existence of the second and continuity of the first partial derivatives) and substitute for $\mathbf{A}$ in (1)

$$(11) \qquad \mathbf{A} = U \operatorname{grad} V - V \operatorname{grad} U,$$

and, consequently, for $A_n$

$$(12) \qquad A_n = U \frac{\partial V}{\partial n} - V \frac{\partial U}{\partial n}.$$

We calculate[17]

$$(13a) \qquad \operatorname{div}(U \operatorname{grad} V) = \operatorname{grad} U \cdot \operatorname{grad} V + U \operatorname{div} \operatorname{grad} V,$$

$$(13b) \qquad \operatorname{div}(V \operatorname{grad} U) = \operatorname{grad} V \cdot \operatorname{grad} U + V \operatorname{div} \operatorname{grad} U$$

and find for div $\mathbf{A}$ from (11)

$$(14) \qquad \operatorname{div} \mathbf{A} = U \operatorname{div} \operatorname{grad} V - V \operatorname{div} \operatorname{grad} U = U \nabla^2 V - V \nabla^2 U.$$

On using (14) and (12), Gauss's theorem (1) becomes

$$(15) \qquad \int (U\nabla^2 V - V\nabla^2 U)\, d\tau = \int \left( U \frac{\partial V}{\partial n} - V \frac{\partial U}{\partial n} \right) d\sigma,$$

which constitutes *Green's theorem in its first form.* (Unfortunately, the nomenclature is not uniform. Many authors refer to (15) as *Green's second formula* which seems to be the preferred usage in current American literature.)

Green's theorem is correct for any two functions $U$ and $V$ and any boundary $\sigma$ of the volume $\tau$. Note that the positive direction of $n$ points outward.

It was assumed so far that the volume $\tau$ has only *one* exterior boundary surface. If an inner boundary exists also as in the case of a body with

---

[17]For an arbitrary scalar $\psi$ and an arbitrary vector $\mathbf{A}$ one easily verifies the formula div $\psi \mathbf{A} = \psi \operatorname{div} \mathbf{A} + \mathbf{A} \cdot \operatorname{grad} \psi$.

one or more cavities, the integration on the right side of (15) must include the inner boundary. This happens in particular if the interior contains points where one of the functions $U$ or $V$ becomes infinite or otherwise singular, so that these points must be excluded from the integration by inner boundaries $\sigma_i$. The corresponding integrals then appear in addition on the right side of (15).

If one substitutes $\mathbf{A} = U \, \text{grad} \, V$, that is, $A_n = U \partial V / \partial n$ in Gauss's theorem, (1) becomes on account of (13a)

$$(16) \qquad \int \text{grad} \, U \cdot \text{grad} \, V \, d\tau + \int U \nabla^2 V \, d\tau = \int U \frac{\partial V}{\partial n} \, d\sigma$$

which constitutes the *second form of Green's theorem* (or *Green's first formula* if the other terminology is used). On identifying $U$ and $V$ one obtains the form

$$(16a) \qquad \int (\nabla U)^2 \, d\tau + \int U \nabla^2 U \, d\tau = \int U \frac{\partial U}{\partial n} \, d\sigma,$$

which contains both differential parameters (9b, c).

The last identity leads directly to conclusions of fundamental importance in potential theory, which is, mathematically speaking, nothing else but the theory of the differential equation

$$(17) \qquad\qquad\qquad \nabla^2 U = 0$$

the solutions of which are known as *harmonic* or *potential functions*. From formula (16a) one derives at once the following lemma: *A potential function that vanishes on a closed surface $\sigma$ and is regular inside, vanishes everywhere in the interior of $\sigma$.* In fact, Eq. (16a) implies

$$(17a) \qquad\qquad \int (\nabla U)^2 \, d\tau = 0 \qquad \text{inside } \sigma$$

because of (17) and the boundary condition $U = 0$; but (17a) requires $\nabla U = 0$, the integrand being non-negative. This means $U = \text{const}$, and since $U$ vanishes on $\sigma$, $U = 0$.

The *uniqueness of solution of the boundary value problem* of potential theory can also be proved by means of (16a). We formulate this problem as follows: required an integral $U$ of (17) that assumes given values on a given boundary surface and is regular everywhere inside. If there were two solutions $U_1$ and $U_2$ satisfying the conditions of the problem, the difference $U = U_1 - U_2$ would satisfy the conditions of the foregoing lemma; thus $U = 0$ inside $\sigma$ as before, and $U_1 = U_2$.

The result holds, with a slight modification, also in the case that instead of $U$ the normal derivative $\partial U / \partial n$ is prescribed on the boundary (see problem I.5).

These few remarks should give a general idea of the scope of Green's theorem; 20 will bring a characteristic example of how it is put to work.

## 4. Some Remarks on Tensor Analysis

One distinguishes between *symmetric, antisymmetric,* and *general* or *asymmetric* tensors. A general tensor can be considered as the sum of a symmetric and an antisymmetric tensor in exactly the same way as the coefficients $a_{ik}$ in 1. The antisymmetric tensor, in turn, is representable by an (axial) vector as in (1.12a), hence we have only to deal with *symmetric* tensors in this article. As we shall limit our discussion to second order tensors (i.e. tensors whose components carry not more than two subscripts), the representation (1.12) of the strain tensor may be taken as a starting point. Our remarks will be formulated so as to refer to the strain tensor, but they are of course valid for any symmetric tensor of second order, in particular for the stress and friction tensors of Chapter II.

The behavior of a tensor in a transformation of coordinates, which is so closely connected with the general significance of the tensor concept, can be studied either by means of its representation through the scheme of coefficients of a linear vector function as in (1.10), or by means of its relation to the tensor quadric (1.13). In rewriting these two equations we replace $x, y, z$ by $x_1$, $x_2$, $x_3$ ; $\epsilon_{xx}$, $\epsilon_{xy}$, $\cdots$ by $\epsilon_{ik}$ ; the special vector $\xi_2$ , $\eta_2$ , $\zeta_2$ by the general vector symbol $\mathbf{A}$ with components $A_1$ , $A_2$ , $A_3$ , and obtain:

(1)     $A_i = \sum_k \epsilon_{ik} x_k$ ,          (2)     $\sum_i \sum_k \epsilon_{ik} x_i x_k = \text{const.}$

We compare these with the transformed equations

(1')     $A_i' = \sum_k \epsilon_{ik}' x_k'$ ,          (2')     $\sum_l \sum_m \epsilon_{lm}' x_l' x_m' = \text{const,}$

in order to find the relations that connect the tensor components $\epsilon$ and $\epsilon'$. We shall again use the relations between the vector components $x, A$ and $x', A'$ known from (2.2), viz.

(3)     $A_i' = \sum_k \alpha_{ik} A_k$ ,     $A_i = \sum_k \alpha_{ki} A_k'$ ,

(4)     $x_i = \sum_l \alpha_{li} x_l'$ ,     $x_k = \sum_m \alpha_{mk} x_m'$ .

The comparison of (1) with (1') need not be carried out here, since the procedure is formally identical with the transformation of the anti-symmetric tensor, carried out in (2.9)-(2.13); one has only to write $\epsilon_{ik}$ instead of $B_{ik}$, $\mathbf{A}$ instead of $2\mathbf{A}$, and to observe that the antisymmetry of the $B$ is not made use of until (2.14). Then the result (2.13) reads in the present notation

(5) $$\epsilon'_{ii} = \sum_l \sum_k \alpha_{ik}\alpha_{il}\epsilon_{kl} .$$

This same result can also be obtained by comparison of (2) with (2'): Substitute, according to (4), $x_i$ and $x_k$ in (2) and reverse the order of summation. The expression $\sum_l \sum_m \sum_i \sum_k \epsilon_{ik}\alpha_{li}\alpha_{mk}x'_l x'_m = \text{const}$ must be identical with (2'), therefore

(6) $$\epsilon'_{lm} = \sum_i \sum_k \epsilon_{ik}\alpha_{li}\alpha_{mk} ,$$

which is indeed the same as (5), except for the lettering of the subscripts.

The second method to establish the transformation rules is, of course, much simpler, but could not have been applied in the case of the anti-symmetric tensor of rotation. The attempt to associate a quadric with such a tensor fails as the quadric vanishes identically: the coefficients of the squares vanish ($\varphi_{ii} = 0$), the product terms cancel in pairs because of $\varphi_{ik} = -\varphi_{ki}$ .

Our result (6) may be expressed in form of the following proposition: *The components of the symmetric tensor transform as the products and squares of the coordinates*, a statement which, conversely, can be considered as the definition of a tensor. The corresponding statement for a polar vector would be: *The components transform as the coordinates themselves.*

Eqs. (6) are generalizations of our previous relations (1.22) and (1.22a); they are sums of six terms, while (1.22) and (1.22a) are sums of three terms. The reason is evidently that the strain tensor was assumed in "diagonal form", that is, referred to its principal axes. But a rectangular cell cut parallel to the principal axes remains rectangular after straining; its angular changes and, therefore, its shears vanish. Thus three of the six tensor components in (6) are zero whenever the $\epsilon$ refer to principal axes.

We shall now discuss the *invariants of a tensor*, supplementing what was learned in Eq. (1.23) about the invariance of the diagonal sum, and prove first that the determinant of the coefficients of the quadric (2) is invariant in orthogonal coordinate transformations.

Denoting the determinant by $D$ we have to show

(7) $$D = \begin{vmatrix} \epsilon_{11} & \epsilon_{12} & \epsilon_{13} \\ \epsilon_{21} & \epsilon_{22} & \epsilon_{23} \\ \epsilon_{31} & \epsilon_{32} & \epsilon_{33} \end{vmatrix} = D',$$

where $D'$ stands for the determinant of the $\epsilon'$ in Eq. (6). The proof is based on the multiplication theorem for determinants, which we apply here as follows: when the determinant of the $\epsilon$ is multiplied with the determinant of the transformation coefficients $\alpha$, the result can again be written as a determinant whose elements $\beta_{kl}$ are $\sum_i \epsilon_{ik}\alpha_{li}$. On multiplying the determinant of the $\beta$ once more with the determinant of the $\alpha$, the elements of the resulting determinant are

$$\sum_k \beta_{kl}\alpha_{mk} = \sum_k \sum_i \epsilon_{ik}\alpha_{li}\alpha_{mk} .$$

According to (6) this is identical with the transformed tensor component $\epsilon'_{ik}$, the determinant of which was denoted by $D'$. Thus we have found

(7a) $$D' = D \cdot |\alpha_{li}| \, |\alpha_{mk}|,$$

but this does not differ from assertion (7) as the determinant of the $\alpha$ equals unity.

When the surface $f = \sum \sum \epsilon_{ik} \, x_i x_k = $ const is orthogonally transformed into $f' = \sum \sum \epsilon'_{ik} x'_i x'_k = $ const, the quadratic form $\varphi = \sum x_i^2$ is transformed into $\varphi' = \sum x_i'^2$. (The "circle at infinity" $\varphi = 0$ which is the gauge quadric of Euclidean metric, transforms into itself in orthogonal transformations.) Consequently a specified quadric of the family $f + \lambda\varphi = $ const transforms into $f' + \lambda\varphi' = $ const and the parameter $\lambda$ has the same value in these two equations of the same surface of second order. But since the determinant of the coefficients of *this* surface is also invariant, it follows that

(8) $$D_\lambda = \begin{vmatrix} \epsilon_{11} + \lambda & \epsilon_{12} & \epsilon_{13} \\ \epsilon_{21} & \epsilon_{22} + \lambda & \epsilon_{23} \\ \epsilon_{31} & \epsilon_{32} & \epsilon_{33} + \lambda \end{vmatrix} = D'_\lambda ,$$

and this must hold for any choice of the arbitrary parameter $\lambda$. Thus a generalization of (7) has been proved.

The determinant $D_\lambda$ may now be expanded in powers of $\lambda$,

(8a) $$D_\lambda = D + \lambda\Delta + \lambda^2\Theta + \lambda^3.$$

Here $\Theta$ is our previous diagonal sum of (1.23) and $\Delta$ the quadratic expression

(9) $$\Delta = \epsilon_{11}\epsilon_{22} + \epsilon_{22}\epsilon_{33} + \epsilon_{33}\epsilon_{11} - \epsilon_{12}\epsilon_{21} - \epsilon_{23}\epsilon_{32} - \epsilon_{31}\epsilon_{13} .$$

On doing the same with $D'_\lambda$ and denoting the coefficients by $D'$, $\Delta'$, $\Theta'$ in analogy to (8a), we obtain because of (8)

$$D + \lambda\Delta + \lambda^2\Theta + \lambda^3 = D' + \lambda\Delta' + \lambda^2\Theta' + \lambda^3.$$

*This must hold for any value of* $\lambda$, therefore

(10) $$D = D', \qquad \Delta = \Delta', \qquad \Theta = \Theta'.$$

*The symmetrical tensor possesses three invariants, that is, three characteristic scalars that are independent of the incidental way in which the tensor is represented by its components.* The geometrical relations between the invariants and the tensor quadric are the object of problem I.6. The invariant $\Theta$ is, in mechanical interpretation, equivalent to the cubical dilatation, as previously shown. In general, $\Theta$ is known as the *first scalar* or *spur* of the tensor.

The preceding argument relates to quadratic forms in a general way and must be valid for the antisymmetric as well as for the symmetric tensor. Hence, also the antisymmetric tensor ought to have three invariants, but the invariant $\Theta$ is identically zero (the diagonal terms vanish) and the determinant $D$ vanishes (so does any antisymmetric determinant of odd order). There remains only the invariant $\Delta$ for which (9) supplies the special form

(11) $$\Delta = \epsilon_{23}^2 + \epsilon_{31}^2 + \epsilon_{12}^2 ,$$

which is equal to the sum of squares of the components of the associated axial vector. Hence in the case of the axial vector only *one quantity exists that is independent of the choice of the coordinate system*, exactly as in the case of the polar vector. For the polar vector this is the length and for the axial vector the magnitude of the "twist", which equals the length of the polar vector by which the axial vector is represented.

As in the case of the moment of inertia[18] (Vol. I, Fig. 40 *a*, *b*, *c*), we shall now discuss some particular cases of the tensor quadric. The significance of a *spherical* tensor quadric is evident. In terms of strain it means *uniform extension or contraction in all directions*: no shears occur since every axis may be considered as principal axis; all right angles are preserved.

*Unidirectional extension*, e.g., in *x*-direction with vanishing extensions in all directions normal to $x$ is associated with the quadric $x^2 = $ const, which is a *pair of planes*. Note, however, that this is not the state of strain caused by unidirectional stress in *x*-direction, since an additional contraction normal to $x$ is produced in that case. The strain quadric is then a *hyperboloid of one sheet*, axisymmetric with respect to the *x*-axis.

A state of strain that presents some interest arises when the points of the body are displaced along concentric circles about an invariable axis in such a way that the displacements are inversely proportional to the

---

[18]Cf. Synge and Griffith, *op. cit.*; Sec. 11.3 contains a table.

distance from the axis, $r$. It prompts itself to introduce cylindrical coordinates and to define the displacement vector by

$$(12) \qquad s_r = 0, \qquad s_\varphi = \frac{A}{r}, \qquad s_z = 0.$$

The strain components in the same coordinates are determined on the
basis of formulas (26) and (28) to be derived presently; they are correct
for arbitrary orthogonal curvilinear coordinates and yield in the present
case

$$(13) \qquad \epsilon_{r\varphi} = -\frac{A}{r^2}, \qquad \epsilon_{rr} = \epsilon_{\varphi\varphi} = \epsilon_{zz} = \epsilon_{\varphi z} = \epsilon_{zr} = 0.$$

We now write the tensor quadric for a point $r$, $\varphi$, $z$ in (local) coordinates
$\xi$, $\eta$, $\zeta$ oriented parallel to the increments $dr$, $d\varphi$, $dz$:

$$(14) \qquad 2\epsilon_{r\varphi}\xi\eta = \text{const.}$$

The coefficient $\epsilon_{r\varphi}$ varies with the distance $r$, but has a fixed value in
Eq. (14), which refers to a certain point of the body. We may then include
the factor $2\epsilon_{r\varphi}$ in the constant, and have instead of (14)

$$(15) \qquad \xi\eta = \text{const,}$$

an equilateral hyperbola. Hence the tensor quadric is quite generally an
*equilateral hyperbolic cylinder*. The axes of the hyperbola (15) form an
angle of 45° with the asymptotes $\xi = 0$, $\eta = 0$.

A volume element cut parallel to $dr$, $d\varphi$, $dz$ is after straining in a state

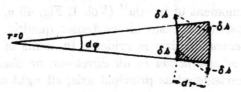

Fig. 4a. In the state of pure shear an originally
rectangular volume element is transformed into a
rhomboid.

Fig. 4b. The princi-
pal strains form an
angle of 45° with the
directions of the shear
and are equal and op-
posite to each other.

of *pure shear*. The face that originally had four right angles is distorted
into a rhomboid-like shape (see Fig. 4a). By setting

$$s(r) = s_0 + \delta s, \qquad s(r + dr) = s_0 - \delta s, \qquad (\varphi = \text{const})$$

it is seen that only $\delta s$ contributes to the distortion, while the effect of $s_0$
is a mere translation (not indicated in the figure). On the other hand, the
volume element drawn in Fig. 4b which is oriented parallel to the *principal*

*axes* of the tensor quadric is not only *free of shear strain* (as required by the definition of the principal axes) but also *free of dilatation*.[19]

This is a consequence of the "equilateral shape" of the tensor quadric: from (13) it is seen that the spur $\Theta$ vanishes; hence it must also vanish for the volume element cut parallel to the principal axes (invariance of the spur). Since $\Theta = \epsilon_1 + \epsilon_2 + \epsilon_3$ and $\epsilon_3 = \epsilon_{zz} = 0$, we have $\epsilon_2 = -\epsilon_1$. The state of *pure shear* is characterized by *opposite principal extensions*.

In conclusion we derive the formulas that express *the strain tensor in curvilinear coordinates* (we have already used them to obtain (13)). These formulas cannot be written down in such a direct way as those for grad, div, and curl in Eqs. (2.24)-(2.26), because there exists no definition of the tensor that is intuitively meaningful or carries self-evident geometrical invariance such as the definitions of grad, div, curl in 2. The following analysis[20] should make up for that and contribute to a deeper understanding of the tensor quadric. It is somewhat laborious, but will prove a great help later, e.g. in the torsion problem, 42.

Let $ds$ be the distance of two neighboring points $P$ and $P'$ of the medium under consideration, let $dx$, $dy$, $dz$ be the differences of their Cartesian and $dp_1$, $dp_2$, $dp_3$ the differences of their curvilinear coordinates. According to (2.22) one has

$$(16) \qquad ds^2 = dx^2 + dy^2 + dz^2 = g_1^2\, dp_1^2 + g_2^2\, dp_2^2 + g_3^2\, dp_3^2.$$

In the deformation of the medium all points suffer displacements given by (1.10), which we shall now denote by $\delta\mathbf{q}$. By $d\mathbf{q}$ we denote, on the other

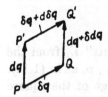

Fig. 5. To illustrate the transformation of the strain tensor in general orthogonal coordinates: a line element $d\mathbf{q}$ is shifted along the displacement vector $\delta\mathbf{q}$.

hand, the relative position vector of two neighboring points before the deformation, so that $|\,d\mathbf{q}\,| = ds$. The symbol $d\delta\mathbf{q}$ denotes the change of the displacement $\delta\mathbf{q}$ caused by the transition $d$ from one field point to a neighboring field point. With the notation of Fig. 5, this is the difference of the displacement vectors $P'Q'$ and $PQ$. The symbol $\delta d\mathbf{q}$ means, again in terms of Fig. 5, the change of the position vector $PP'$ caused by the

---

[19]Note also that this state of strain is formally the same as in a bar of circular cross section subjected to torsion, but the generators of the tensor quadric are perpendicular to the bar axis in the latter case [compare (4.13) with (42.3)].

[20]Cf. E. Beltrami, Annali di Mathematica 10, 188 (1881). Opere Mathematiche, III, 383.

displacement operation $\delta$, which carries $P$ into $Q$ and $P'$ into $Q'$; $\delta dq$ is thus the difference of the position vectors $QQ'$ and $PP'$, after and before deformation. But the quadrilateral $PQQ'P'$ supplies the vector relation $PP' + P'Q' = PQ + QQ'$ or $P'Q' - PQ = QQ' - PP'$, which is equivalent to the commutative relation

(17) $$d\delta q = \delta dq.$$

We return now to (1.10) and have in Cartesian coordinates

(18) $$d\delta q = \begin{cases} d\delta x = \epsilon_{xx}\, dx + \epsilon_{xy}\, dy + \epsilon_{xz}\, dz, \\ d\delta y = \epsilon_{yx}\, dx + \epsilon_{yy}\, dy + \epsilon_{yz}\, dz, \\ d\delta z = \epsilon_{zx}\, dx + \epsilon_{zy}\, dy + \epsilon_{zz}\, dz. \end{cases}$$

These relations give the change of the displacement $d\delta q$ as a linear vector function of the Cartesian components of the vector $dq$. If we decompose $dq$ into its curvilinear components given by (2.23), viz.

(19) $$ds_1 = g_1 dp_1, \qquad ds_2 = g_2 dp_2, \qquad ds_3 = g_3 dp_3$$

we obtain again a linear vector function for $d\delta q$ the coefficients of which are the tensor components we look for:

(20) $$d\delta q = \begin{cases} d\delta s_1 = \epsilon_{11}\, ds_1 + \epsilon_{12}\, ds_2 + \epsilon_{13}\, ds_3, \\ d\delta s_2 = \epsilon_{21}\, ds_1 + \epsilon_{22}\, ds_2 + \epsilon_{23}\, ds_3, \\ d\delta s_3 = \epsilon_{31}\, ds_1 + \epsilon_{32}\, ds_2 + \epsilon_{33}\, ds_3. \end{cases}$$

Our task is to calculate the "curvilinear tensor components" $\epsilon_{ik}$ from the displacement components $\delta q_1$, $\delta q_2$, $\delta q_3$ in directions $p_1$, $p_2$, $p_3$ as in (1.11).

We first determine in Cartesian coordinates the change of the square of the element of length $ds^2$ produced by the displacement $\delta q$. From (16), (17), and (18) we infer

$$\tfrac{1}{2}\delta\, ds^2 = dx\,\delta dx + dy\,\delta dy + dz\,\delta dz$$

(21) $$= \epsilon_{xx}\, dx^2 + \epsilon_{yy}\, dy^2 + \epsilon_{zz}\, dz^2$$

$$+ 2\epsilon_{xy}\, dx\, dy + 2\epsilon_{yz}\, dy\, dz + 2\epsilon_{zx}\, dz\, dx.$$

The last member is identical with the polynomial of the tensor quadric except for the notation of the variables $dx$, $dy$, $dz$ that replace the former $x$, $y$, $z$. The first member of (21) may be written in the form $ds\,\delta ds$. On dividing by $ds^2$ we obtain $\delta ds/ds$ which is the *extension of the ds-fiber per unit of length*; the variables are now the quantities $dx/ds$, $dy/ds$, $dz/ds$. Hence the *tensor quadric gives* directly the *specific extension* of the $ds$-fiber *in terms of the direction cosines* $dx/ds$, $dy/ds$, $dz/ds$ and thus lends intuitive physical meaning to the concept of the symmetrical tensor.

The procedure leading to Eq. (21) can just as well be carried out in curvilinear coordinates $p_1$, $p_2$, $p_3$. Using (19) we write the length element in the form $ds^2 = ds_1^2 + ds_2^2 + ds_3^2$. Eqs. (20) and (17) then yield

$$\tfrac{1}{2}\delta \, ds^2 = ds_1 \, \delta ds_1 + ds_2 \, \delta ds_2 + ds_3 \, \delta ds_3$$

(22)
$$= \epsilon_{11} \, ds_1^2 + \epsilon_{22} \, ds_2^2 + \epsilon_{33} \, ds_3^2$$

$$+ 2\epsilon_{12} \, ds_1 \, ds_2 + 2\epsilon_{23} \, ds_2 \, ds_3 + 2\epsilon_{31} \, ds_3 \, ds_1 \, .$$

On dividing this equation by $ds^2$ the extension of the $ds$-fiber is obtained as before.

The required relations between the $\epsilon_{ik}$ and the displacement components $\delta q_i$ are now obtained by the direct calculation of $\tfrac{1}{2}\delta ds^2$ in terms of the $ds_i$ and subsequent comparison with the right member of (22). Writing (16) in condensed form, we obtain first

(22a)
$$\tfrac{1}{2} \, \delta ds^2 = \sum_i g_i \, \delta g_i \, dp_i^2 + \sum_i g_i^2 \, dp_i \, \delta dp_i \, .$$

Here we have to express the variations, imposed on the coefficients $g_i$ and on the coordinate differentials $dp_i$ by the displacement $\delta \mathbf{q}$, in terms of the variations and differentials of the coordinates themselves; in other words, we have to substitute in (22a)

$$\delta g_i = \sum_k \frac{\partial g_i}{\partial p_k} \, \delta p_k \, , \qquad \delta dp_i = d \delta p_i = \sum_k \frac{\partial \delta p_i}{\partial p_k} \, dp_k$$

and obtain

(23)
$$\tfrac{1}{2} \, \delta ds^2 = \sum_i \sum_k g_i \frac{\partial g_i}{\partial p_k} \, \delta p_k \, dp_i^2 + \sum_i \sum_k g_i^2 \frac{\partial \delta p_i}{\partial p_k} \, dp_i \, dp_k \, .$$

Now we are ready to introduce the $ds_i$. According to (19), we write $ds_i/g_i$ for the $dp_i$. The "curvilinear" displacement components $\delta p_i$ are obviously connected with the components $\delta q_i$ by

$$\delta q_i = g_i \delta p_i \, , \qquad \text{or} \qquad \delta q_k = g_k \delta p_k \, .$$

In this way the first of the two sums on the right side of (23) becomes

(23a)
$$\sum_i \sum_k \frac{\delta q_k}{g_i g_k} \frac{\partial g_i}{\partial p_k} \, ds_i^2 \, ;$$

the second sum becomes

$$\sum_i \sum_k \frac{g_i}{g_k} \frac{\partial}{\partial p_k} \left(\frac{\delta q_i}{g_i}\right) ds_i \, ds_k$$

or, on carrying out the differentiations,

(23b)    $$\sum_i \sum_k \frac{1}{g_k} \frac{\partial \delta q_i}{\partial p_k} ds_i \, ds_k - \sum_i \sum_k \frac{\delta q_i}{g_i g_k} \frac{\partial g_i}{\partial p_k} ds_i \, ds_k .$$

Now we can compare the coefficients of $ds_i \, ds_k$ in (22) with those in the sum of (23a) and (23b). Let us do this first for the diagonal terms: the coefficient of a *specified* $ds_i^2$ in (23a) is

(24a)    $$\sum_k \frac{\delta q_k}{g_i g_k} \frac{\partial g_i}{\partial p_k} ;$$

in (23b), $k$ must be identified with the given $i$. The coefficient is then simply

(24b)    $$\frac{1}{g_i} \frac{\partial \delta q_i}{\partial p_i} - \frac{\delta q_i}{g_i^2} \frac{\partial g_i}{\partial p_i} .$$

Upon adding (24a) and (24b), two terms cancel and one obtains

(25)    $$\frac{1}{g_i} \left(\frac{\partial \delta q_i}{\partial p_i} + \sum_{k \neq i} \frac{\delta q_k}{g_k} \frac{\partial g_i}{\partial p_k}\right).$$

This, then, is the value of $\epsilon_{ii}$ . Written in extenso, (25) means

$$\epsilon_{11} = \frac{1}{g_1} \left(\frac{\partial \delta q_1}{\partial p_1} + \frac{\delta q_2}{g_2} \frac{\partial g_1}{\partial p_2} + \frac{\delta q_3}{g_3} \frac{\partial g_1}{\partial p_3}\right)$$

(26)    $$\epsilon_{22} = \frac{1}{g_2} \left(\frac{\partial \delta q_2}{\partial p_2} + \frac{\delta q_3}{g_3} \frac{\partial g_2}{\partial p_3} + \frac{\delta q_1}{g_1} \frac{\partial g_2}{\partial p_1}\right)$$

$$\epsilon_{33} = \frac{1}{g_3} \left(\frac{\partial \delta q_3}{\partial p_3} + \frac{\delta q_1}{g_1} \frac{\partial g_3}{\partial p_1} + \frac{\delta q_2}{g_2} \frac{\partial g_3}{\partial p_2}\right).$$

In the determination of the shears $\epsilon_{ik}$ , we note first that (23a) gives no contribution. In (23b), either sum yields two terms if $i$ and $k$ are specified, viz. one term $\cdots ds_i \, ds_k$ and another $\cdots ds_k \, ds_i$ . This amounts to the four term expression

(27)    $$\frac{1}{g_k} \frac{\partial \delta q_i}{\partial p_k} + \frac{1}{g_i} \frac{\partial \delta q_k}{\partial p_i} - \frac{1}{g_i g_k} \left(\delta q_i \frac{\partial g_i}{\partial p_k} + \delta q_k \frac{\partial g_k}{\partial p_i}\right)$$

which, on the other hand, equals $2\epsilon_{ik}$ . If we write (27) in extenso, we obtain

$$\epsilon_{12} = \frac{1}{2}\left(\frac{1}{g_2}\frac{\partial \delta q_1}{\partial p_2} + \frac{1}{g_1}\frac{\partial \delta q_2}{\partial p_1}\right) - \frac{1}{2g_1g_2}\left(\delta q_1 \frac{\partial g_1}{\partial p_2} + \delta q_2 \frac{\partial g_2}{\partial p_1}\right)$$

$$(28) \quad \epsilon_{23} = \frac{1}{2}\left(\frac{1}{g_3}\frac{\partial \delta q_2}{\partial p_3} + \frac{1}{g_2}\frac{\partial \delta q_3}{\partial p_2}\right) - \frac{1}{2g_2g_3}\left(\delta q_2 \frac{\partial g_2}{\partial p_3} + \delta q_3 \frac{\partial g_3}{\partial p_2}\right)$$

$$\epsilon_{31} = \frac{1}{2}\left(\frac{1}{g_1}\frac{\partial \delta q_3}{\partial p_1} + \frac{1}{g_3}\frac{\partial \delta q_1}{\partial p_3}\right) - \frac{1}{2g_3g_1}\left(\delta q_3 \frac{\partial g_3}{\partial p_1} + \delta q_1 \frac{\partial g_1}{\partial p_3}\right).$$

Formulas (26) and (28) can be cast in a much more comprehensive form if the notation of general tensor calculus is employed as in Appendix I.

For the present we apply these formulas only to verify our Eqs. (13). If we specify the $p_i$ as cylindrical coordinates by setting $p_1 = r$, $p_2 = \varphi$, $p_3 = z$, it follows that $g_1 = g_3 = 1$, $g_2 = r$, and the displacement (12) reads

$$\delta q_1 = \delta q_3 = 0, \qquad \delta q_2 = \frac{A}{r}.$$

As a consequence, only those terms in (26) and (28) contribute something that contain $\delta q_2$ or $\partial \delta q_2/\partial p_1$ and do not have a vanishing coefficient $\partial g_i/\partial p_k$ (note that all these derivatives vanish except $\partial g_2/\partial p_1$ which equals unity). Only $\epsilon_{12}$ contains such terms, viz.

$$\frac{1}{2g_1}\frac{\partial \delta q_2}{\partial p_1} = -\frac{1}{2}\frac{A}{r^2} \quad \text{and} \quad -\frac{1}{2g_1g_2}\delta q_2 \frac{\partial g_2}{\partial p_1} = -\frac{1}{2}\frac{A}{r^2}.$$

Thus the system (26) and (28) reduces in our case to: $\epsilon_{12} = -A/r^2$, and all other $\epsilon$ vanish, in agreement with (13).

Let us finally interpret the displacement field (12) as a velocity field by setting

$$(29) \qquad v_r = 0, \qquad v_\varphi = \frac{A}{r}, \qquad v_z = 0.$$

This motion is of fundamental importance in the theory of vortices. We shall see that it is *irrotational* wherever $r > 0$, that is, at all points outside the axis. This is true in spite of our inclination to "see" (29) as a whirl about the axis $r = 0$. The steady motion characterized by the velocity distribution (29) is known as a *circulating motion*, also as a (straight) *line vortex* or *simple vortex*. The latter terminology makes it evident that the points of the axis have a particular significance: they form a vortex line, that is, a concentrated vortex filament. This will be discussed more fully in 19, p. 140, where the irrotationality of the vortex motion will also be shown; the latter point, however, can be verified right now by means of the tabulated answer to problem I.3. One has only to identify our vector (29) with the vector **A** in the formulas giving the curl components in cylindrical coordinates.

The tensor concept has been developed in this article by studying the behavior of a tensor in the transition from one Cartesian coordinate system to another. If more general transformations are to be considered, the methods of general tensor analysis must be employed of which a short account is found in Appendix I-IV.

# CHAPTER II

# STATICS OF DEFORMABLE BODIES

## 5. Concept of Stress; General Classification of Deformable Bodies

The term statics will be used throughout this book in the same sense as in the mechanics of a finite number of degrees of freedom, where it was applied according to the following definition: Statics treats of the forces to which matter is subjected regardless of the motion caused by them (cf. Vol. I, 5).

One distinguishes between external and internal forces. Once a mechanical system has been delimited, the *internal* forces of the system are those which obey Newton's third law (equality of action and reaction). Among the *external* forces, gravity and centrifugal force[1] are commonly encountered in mechanics of deformable bodies. Capillary forces occur also; they act between internal and surface molecules or between a solid boundary and the molecules of the surrounding fluid. Examples of internal forces are the reactions (tension and pressure forces, surface forces) that have their origin in the distribution of matter over a limited space. They play the most important part[2] in the mechanics of deformable bodies.

In solids the reactions measured per unit of area are known as *stresses*; in the case of fluids we usually speak of *pressures* (negative stresses); both have the dimension of a force per unit area, dyne/cm$^2$ in the CGS system.

To analyze the stress concept in the case of an elastic solid, we imagine a plane surface cutting through the body and consider the interaction between the two parts of the body across this surface. In Fig. 6 such a surface of separation has been drawn normal to the $x$-axis, and the two parts of the body have been disjoined for the sake of clarity in the drawing. On the part to the left a "positive $x$-surface" has been marked by cross hatching. (A surface with an outward normal pointing in the positive $x$-direction is called positive $x$-surface for brevity.) We assume the $x$-surface

---

[1]The proper place for the centrifugal force, which is an inertia effect, is in dynamics (Chap. III), but it may be considered in statics since it is equivalent to an external force for an observer moving with the body.

[2]The distinction between external and internal forces in a specified system does not in general coincide with the distinction between forces of physical origin (impressed forces) and forces of geometrical origin (due to spatial constraints to which the system is subjected).

in this figure bounded in the shape of a rectangle with the sides $\Delta y$, $\Delta z$. The $x$-surface bounding the right part of the body, which in reality coincides with the positive $x$-surface, has also been cross hatched. It is a negative $x$-surface since its outward normal points in the negative $x$-direction.

Consider now *the forces which the negative x-surface exerts upon the positive x-surface*. When these forces are parallel to the *positive* $x$-direction we speak of a positive *normal stress* $\sigma_{xx}$ ; its magnitude equals the sum of the forces divided by the area $\Delta y \Delta z$. In general, however, the total force will by no means be perpendicular to the $x$-surface. We then resolve the force and the associated stress into components in the positive $x$-, $y$-, and $z$-directions. The $y$ and $z$ components are known as *tangential* or *shear* stresses and denoted by $\sigma_{xy}$ , $\sigma_{xz}$ . This notation indicates the position of the surface by the first subscript and the direction of the stress by the second.

The principle of action and reaction requires us to draw for the right part in Fig. 6 the stress arrows that act across the negative $x$-surface in

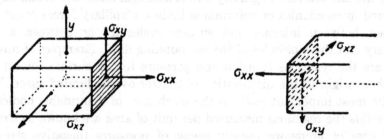

Fig. 6. Normal and shear stresses acting on a volume element of an elastic body. Positive and negative $x$-surfaces.

opposite direction. A positive $\sigma_{xk}$ means here a stress in the negative $k$-direction. The same is true for the negative $x$-surface of the body element in the left part of the figure which is obtained by a mere displacement (through $-\Delta x$) of the negative $x$-surface considered before. In speaking of the entirety of quantities $\sigma_{ik}$ characteristic for a volume element, we use the term *stress tensor* although the proof of the tensor character will be given in 8 only, by showing the symmetry of the $\sigma_{ik}$ and their transformation rules. The term stress tensor is, by the way, redundant since the word tensor in itself implies tension.

The forces exerted by the surrounding elements of the body may be directed either toward the exterior or the interior of the body element under consideration. In the latter case (compression) it is more appropriate to speak of pressures than of stresses. We shall, however, retain the uniform terminology "stress" in the case of an elastic *solid*, and consider pressures as negative stresses.

Not so in the case of *fluids*. The cohesive forces in a fluid are weak, its "ultimate strength" under ordinary circumstances is exceedingly small if compared with that of a solid. In other words, the resistance of fluids against tension is so small that positive stresses practically never occur; we then prefer to deal in our equations with *pressures* $p$ rather than stresses $\sigma$. In a sketch such as Fig. 6, the positive $p_{xx}$ would have to be drawn as an arrow directed inward.[3]

The totality of pressures $p_{ik}$ acting upon a fluid element will again be called a *tensor*. There is, however, an essential difference between the normal pressures $p_{ii}$ and the tangential pressures $p_{ik}$, at least in the state of rest or of slow motion. A fluid layer may be moved along the adjacent layer without expenditure of a noticeable amount of energy provided the motion is slow; this means that the tangential pressures $p_{ik}$ are quite small if the two layers are slowly displaced parallel to each other. By extension, we must assume that the tangential pressures vanish altogether in the state of rest. In this way a fundamental simplification of the pressure tensor $p$ is accomplished which will be applied in 6 (hydrostatics).

For larger relative velocities of the fluid particles the complete pressure tensor must be studied and its dependence on the velocity gradient investigated. This will be done in 10 (viscous flow).

The concept of the fluid state is meant to cover also *gases* and *vapors*, the latter two being characterized by larger compressibility. Fluids in the restricted sense, or liquids, are essentially incompressible. Of course, they yield to a certain extent if subjected to strong external forces, but their compressibility is very small in comparison with gases or vapors and can be neglected in a great number of problems; so is water often considered as *incompressible*.

The *condition for incompressibility* reads $\Theta = 0$ according to (1.24); on passing from the displacement vector **s** to the velocity vector **v** as in (1.7), we obtain the relation

$$(1) \qquad\qquad \text{div } \mathbf{v} = 0,$$

expressing the fact that the balance of outflow and inflow for a given volume element $\Delta\tau$ is zero at any time. If $\Delta m$ denotes the mass contained in $\Delta\tau$, we have

$$(2) \qquad\qquad \Delta m = \rho\Delta\tau$$

where $\rho$ stands for the *density*. For homogeneous incompressible fluids $\rho$ is constant in time as well as in space.

---

[3]It is only consequent to invert the direction of the arrows also for $p_{xy}$, $p_{xz}$, etc. so that $p_{xy}$ is counted *positive* if it points in the *negative* $y$-direction. This rule will be followed from now on.

What corresponds to Eq. (1) in the case of a *compressible fluid* where not the volume but the mass is preserved? The mass $\Delta m$ cannot decrease unless part of it passes through the boundary of the volume element $\Delta \tau$. The mass flow at any point of its boundary is represented by the normal component of the vector $\rho \mathbf{v}$. The total outflow of mass during the time element $\Delta t$ is therefore [cf. e.g. (2.20)]

$$(3) \qquad \int \rho v_n \, d\sigma \, \Delta t = \operatorname{div}(\rho \mathbf{v}) \Delta \tau \Delta t.$$

On the other hand, the loss of mass during the same time is found from (2) as

$$-\frac{\partial \rho}{\partial t} \Delta t \Delta \tau.$$

Comparison with (3) yields the following *condition for the conservation of mass*

$$(4) \qquad \frac{\partial \rho}{\partial t} + \operatorname{div}(\rho \mathbf{v}) = 0,$$

also known under the name of *equation of continuity*. It may be written in the form:

$$(4a) \qquad \frac{d\rho}{dt} + \rho \operatorname{div} \mathbf{v} = 0$$

The first term in this equation is the *material differential quotient* which will be discussed more fully at the beginning of 11. Its definition is

$$(4b) \qquad \frac{d\rho}{dt} = \frac{\partial \rho}{\partial t} + \mathbf{v} \cdot \operatorname{grad} \rho.$$

If $\rho$ is constant in time *and* space, Eq. (4) takes again the form of the incompressibility condition (1). In an incompressible inhomogeneous fluid ($\rho$ variable in space) we have by (4b)

$$(4c) \qquad \frac{d\rho}{dt} = 0 \qquad \text{that is} \qquad \frac{\partial \rho}{\partial t} = -\mathbf{v} \cdot \operatorname{grad} \rho,$$

which describes the temporal fluctuations of $\rho$ at a fixed point of observation, while the material rate of change of $\rho$ vanishes.

We add here Eqs. (1), (4), and (4a) spelled out in Cartesian coordinates:

$$(1') \qquad \frac{\partial u}{\partial x} + \frac{\partial v}{\partial y} + \frac{\partial w}{\partial z} = 0,$$

(4')
$$\frac{\partial \rho}{\partial t} + \frac{\partial(\rho u)}{\partial x} + \frac{\partial(\rho v)}{\partial y} + \frac{\partial(\rho w)}{\partial z} = 0,$$

(4a')
$$\frac{d\rho}{dt} + \rho\left(\frac{\partial u}{\partial x} + \frac{\partial v}{\partial y} + \frac{\partial w}{\partial z}\right) = 0.$$

In problems involving capillary forces we are concerned with the fluid layers that are closest to the surface (surface film). The *surface of separation* introduced previously (see Fig. 6) is now to be replaced by a *linear cut* through the film. The object is again to study the reaction forces transferred from one bank of the cut to the other. Referring the forces to the unit length of the cut we speak of *surface tension*; its dimension is dyne/cm.

## 6. *Equilibrium of Incompressible Fluids* (*Hydrostatics*).

The subject of this section is, of course, not only water ("hydor" in Greek) but any liquid in equilibrium since differences in the rheological characteristics have no bearing on the *state of rest*: shear stresses, as already observed, are zero in equilibrium, or

(1)                    $p_{ik} = 0, \quad i \neq k.$

This equation actually amounts to a *definition of the fluid* in the state of rest—in contrast to the state of solidification: glass which has no crystalline structure must be characterized physically as a *fluid in the solid state* for which Eq. (1) is, of course, incorrect.

A fundamental conclusion from (1) is that the hydrostatic pressure is associated with a *spherical* tensor quadric. Any diameter of the quadric may be taken as a principal axis, because (1) must be true in any coordinate system. The complete symmetry of the tensor of hydrostatic pressure can be expressed as follows. *At a given point the pressure acts on any surface element in the direction of its normal i, its magnitude being the same for all directions i*; $p_{11} = p_{22} = p_{33}$ for any three orthogonal directions. Pascal[4], it appears, was the first to perceive this law.

We can summarize this result in the statement that the hydrostatic pressure is a *scalar quantity*. It will be denoted by $p$, as is usually done with omission of the subscripts that are now abundant. Note that there is no preferential loading of the *horizontal* cross sections if a *vertical* force such as gravity acts on the fluid; the same pressure $p$ is transferred

---

[4]Blaise Pascal (1624-1662), the great geometer (Pascal's theorem) and mathematician (Pascal's triangle, foundation of the calculus of probabilities), a great writer and theologian too; he studied the laws of air pressure and at his instigation the first barometrical determination of an altitude was carried out.

through any cross-section (also a vertical one). *The unidirectional character of the force does not interfere with the symmetry of the pressure.*

In order to calculate the pressure if magnitude and direction of the external force are given, let $\mathbf{F}$ denote the external force per unit of fluid *volume*, so that the dimension of $\mathbf{F}$ is dyne/cm$^3$. Then the volume element $\Delta\tau$ is acted upon by the force $\mathbf{F}\Delta\tau$. On delimiting the volume element as rectangular cell ($\Delta\tau = \Delta x \Delta y \Delta z$) and marking the negative $x$-surface by the value $x$ of the $x$-coordinate and, consequently, the positive $x$-surface by $x + \Delta x$, the forces in positive $x$-direction are:

on the neg. $x$-surface    on the pos. $x$-surface    on the vol. element

(2)      $p(x)\Delta y \Delta z$      $-p(x + \Delta x)\Delta y \Delta z$      $F_x \Delta\tau$

The first and second contributions add up to

(3a) $$-\frac{\partial p}{\partial x} \Delta x \Delta y \Delta z = -\frac{\partial p}{\partial x} \Delta\tau$$

and yield after division by $\Delta\tau$

(3b) $$-\frac{\partial p}{\partial x} + F_x = 0;$$

this can be written in the form

(4) $$\operatorname{grad} p = \mathbf{F},$$

which is correct for any direction at any point of the liquid.

The relations (3b) or (4) for the force components obviously constitute a necessary condition for the equilibrium. From statics of rigid bodies a second condition is known requiring that the resultant moment be zero. This condition must also apply to non-rigid bodies, for, if we remove or relax the internal constraints of a rigid mechanical system, we obtain a non-rigid system with a vastly increased number of degrees of freedom. Hence, *quite generally, any equilibrium condition for a rigid body must be valid a fortiori for a deformable body.*

It is, however, easily seen that the second or moment equilibrium condition is automatically fulfilled in the case of a liquid. The normal pressure cannot produce a moment about any axis. Moments about the $x$-axis, for instance, could be produced only by the shearing stresses $p_{yz}$ and $p_{zy}$, but they vanish. The external force $\mathbf{F}$ cannot produce a moment either; this is certainly true when we assume a continuous distribution of lines of force; such a distribution can be always considered as parallel in any infinitesimal region (the field of gravity is a simple example for what is meant). If, on the other hand, the lines of the field $\mathbf{F}$ have a singular

point at our volume element (a point where the field lines intersect), then
$\mathbf{F}$ must needs vanish there, and so does the moment of $\mathbf{F}$. Eq. (4) is in
fact not only a *necessary* but also a *sufficient* condition of equilibrium.

Eq. (4) includes a very remarkable theorem: *equilibrium is only possible
if the external force has a potential*, that is, if $\mathbf{F}$ can be represented as the
gradient of a scalar function. In this case we may write

$$(5) \qquad \mathbf{F} = - \operatorname{grad} U$$

where the minus sign is prompted by the relation to the potential energy,
cf. Vol. I, appendix to 18.[5] The *existence* of the potential function $U$ is
not sufficient, $U$ must also be *single valued within the space occupied by
the liquid*. Only under this assumption is it possible to calculate $p$ from
(4) as a single-valued scalar point function in the form

$$(6) \qquad p + U = \text{const.}$$

The integration constant in (6) must be determined by special conventions
concerning the normalization of the potential and the pressure.

Postponing the further discussion of this theorem to the end of this
article, we deal for the present with the regular case in which the force
$\mathbf{F}$ has a single-valued potential and take gravity as an example. Let
$z = 0$ be the surface of the liquid (call it "water" for simplicity) and orient
the positive $z$-axis vertically downward. Then

$$(7a) \qquad \mathbf{F} = F_z \mathbf{k}; \; F_z = \rho g = \gamma,$$

$$(7b) \qquad U = - \rho g z = - \gamma z.$$

Here $\gamma (= \rho g)$ denotes the *specific weight* of the water as distinguished
from the *specific mass* or density $\rho$. In writing (7b) the arbitrary constant
in $U$ has been adjusted so as to give zero potential at the surface. We
obtain now from (6)

$$(8) \qquad p = \gamma z + \text{const},$$

where *const* is the pressure value at the surface $z = 0$, equal to the atmospheric pressure; however, $p$ can also be given the meaning of the
overpressure relative to the atmosphere (gauge pressure), which is more
convenient for what follows. Let $p$ have this meaning, then (8) becomes

$$(8a) \qquad p = \gamma z.$$

This equation contains the well known elementary rules about com-
municating vessels, viz. equal pressure at equal depth in each of the

---

[5]See e.g., T. C. Slater and N. H. Frank, *Mechanics*, McGraw-Hill, New York, 1947,
Sec. III,2.

interconnected vessels regardless of shape. The same equation may also be obtained directly from the differential formula (4). One follows the shape of the vessel along a broken line consisting of a series of vertical and horizontal segments inside the water and integrates. Along the horizontal segments $dp = 0$, along the vertical ones $dp = \gamma \, dz$. Summation of the vertical contributions gives immediately $p = \gamma z$ at the depth $z$.

Let us insert here a few basic numerical constants: normal atmospheric pressure $= 76$ cm Hg, density of Hg $13.596$ gr/cm$^3$; hence from (8a), $p_{\text{atm}} = 76 \times 13.596 \times g$ gr/cm$^2 = 1033\, g \times$ gr/cm$^2 = 1.033$ kg-wt./cm$^2$. One cm$^2$ of a water surface carries a pressure load of about 1 kg-weight which makes 10 metric tons per m$^2$. The height of a water column corresponding to 76 cm Hg would be 10.33m, since for this height $p_{\text{atm}} = 1033 \times 1 \times g$ gr/cm$^2 = 1.033$ kg-wt./cm$^2$ where $\rho = 1$ has been used.

We now consider an example that combines centrifugal force and gravity: a liquid in a drum (centrifuge) rotates with constant angular velocity $\omega$ about a vertical axis. The centrifugal force per unit of volume is

(9)    $\mathbf{F} = F_r \mathbf{r}; \; F_r = \rho r \omega^2; \; \mathbf{r}$ is a unit vector $\perp$ to the axis.

Its potential is given by

(9a)    $$U = -\frac{1}{2}\,\rho r^2 \omega^2.$$

The total potential of gravity *and* centrifugal force is (except for a constant) given by

$$U = -\rho g z - \frac{1}{2}\,\rho r^2 \omega^2 = -\gamma\left(z + \frac{r^2 \omega^2}{2g}\right).$$

From (6), the pressure in the rotating liquid is found as

(10)    $$p = \gamma\left(z + \frac{r^2 \omega^2}{2g}\right) + \text{const.}$$

To determine the constant in (10) let us first mark the water-level in the axis of the drum, i.e., at $r = 0$, by $z_0$. If $p$ is again the overpressure, we have for $z = z_0$ and $r = 0$ from (10)

$$0 = \gamma z_0 + \text{const}, \qquad \text{const} = -\gamma z_0$$

and in general

(11)    $$p = \gamma\left(z - z_0 + \frac{r^2 \omega^2}{2g}\right).$$

The free surface, characterized by $p = 0$, has the equation

(12)    $$z_0 - z = \frac{r^2 \omega^2}{2g}.$$

It is the well known paraboloid of rotation indicated in Fig. 7. Since $z$ is positive downward, $z_0 - z$ is the ordinate of the parabolic meridian curve counted positive upward. On denoting by $h$ the difference of the water

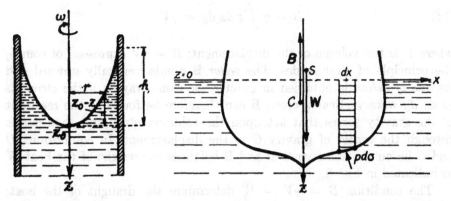

FIG. 7. The paraboloid surface of a fluid mass in rotation. The rise along the wall can be considered as velocity head.

FIG. 8. The equilibrium of a prismatic hull. The buoyancy is the resultant of all pressure forces acting on the surface.

levels at the circumference and the center, and by $v$ the circumferential velocity, $h$ appears in the well known form of the velocity head,

$$(13) \qquad h = \frac{v^2}{2g}.$$

The surfaces of constant pressure are congruent paraboloids, obtained by displacing the free surface vertically downward.

Next we turn to the problem of the *equilibrium of a boat* where again gravity is the only external force present. Fig. 8 represents a cross section normal to the longitudinal axis of the boat (positive $y$-axis forward). Let us think of the hull as a cylindrical surface with generatrices parallel to the $y$-axis. $S$ is the center of mass of the boat, $C$ the center of mass of the displaced water or of the "displacement", when the boat is upright.

Defining the buoyancy of the boat as the resultant of all pressure forces acting upon the wetted surface, we can show that this resultant passes through $C$. The point $C$ is thus called the center of buoyancy.

Let $d\sigma$ be a surface element of the hull. According to the simplified form of the hull, $d\sigma = ds\, dy$ where $ds$ is the line element of the cross section. The pressure $\gamma z$ acts in line with the surface normal of $d\sigma$, thus the pressure force is $\gamma z\, d\sigma$ and its vertical component is $\gamma z\, d\sigma \cos(n, z)$, where $n$ is the inward normal of $d\sigma$. With $d\sigma \cos(n, z)$ being numerically equal to the projection of $d\sigma$ upon the horizontal, the vertical component of the force element can be written $\gamma z\, dx\, dy$. The buoyancy is then found as

$$(14) \qquad\qquad B = \gamma \int z\, dx\, dy = \gamma V,$$

where $V$ is the volume of the displacement; $B = \gamma V$ expresses, of course, the principle of Archimedes. The *vector* **B** points vertically upward and its *line of action* is obtained in exactly the same way as if the elements $\gamma z\, dx\, dy$ were vertical vectors; **B** can therefore be found as the resultant of the gravity forces that act upon the columns $z\,dx\,dy$ and thus passes through the center of gravity $C$ of the displacement. In equilibrium, $C$ and $S$ lie on the same vertical and **B** balances the weight of the boat **W** as indicated in Fig. 8.

The condition $B = \gamma V = W$ determines the draught of the boat; naturally, the draught increases with the ship load. Note that also the horizontal components of the pressure forces acting on two opposite elements $d\sigma_1$ and $d\sigma_2$ at the same depth $z$ balance each other. This remains true, even if the boat is unsymmetrically loaded, and follows immediately from the equality of the two projections upon the vertical plane $\cos (n_1 , x)\, d\sigma_1 = \cos (n_2 , x)\, d\sigma_2$ .

We now want to learn something about the *oscillations of the boat* if the equilibrium position is disturbed. This is still a question well within the bounds of hydrostatics; for, in good approximation to reality, we may assume that the pressure distribution at any phase of the motion does not differ greatly from the hydrostatic pressure distribution that would prevail if this phase were a possible rest position.

Let us apply this idea to the rolling motion of the boat (rotation about the longitudinal or $y$-axis). The symmetry plane of the boat now subtends a variable angle with the vertical which will be denoted by $\vartheta$. During the roll the shape of the displacement and the position of $C$ will change. Assume a roll of small amplitude and let $C'$ be the center of mass of the displacement at the end of the roll, $C''$ the corresponding point when the roll has gone in the opposite direction. Let $\mathfrak{M}$ be the center of curvature of the curve $C'C\ C''$ at the point     (In Fig. 9, $\mathfrak{M}$ has been constructed in an approximate way by intersecting the normals at the points $C'$ and $C''$.) This point $\mathfrak{M}$ is known as the *metacenter*. Our figure illustrates the instant of the motion when the buoyancy **B** passes through $C'$. As it is, **B** is to the right of the weight vector **W** which always passes through $S$. The moment **M** of the couple formed by **B** and **W** is indicated in the figure by a circular arrow that acts so as to decrease the angle of roll $\vartheta$. This is the necessary and sufficient condition for the *stability of* *the boat.* It is fulfilled whenever the metacenter $\mathfrak{M}$ is *above* the center of mass $S$, (so drawn in the figure). If $\mathfrak{M}$ were below $S$, the moment **M** would tend to increase the disturbance and finally make the boat capsize.

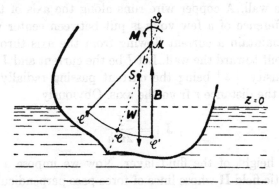

FIG. 9. Metacenter and restoring moment for the tilted hull.

The distance between $S$ and $\mathfrak{M}$ is called the metacentric height $h$. According to the figure one has for small $\vartheta$

$$|\mathbf{M}| = M_\nu = hW \sin \vartheta \sim hW\vartheta.$$

The differential equation of the rolling motion is therefore the same as that of a compound pendulum (see e.g. Vol. I, 16).[6] With $\Theta$ denoting the moment of inertia of the boat about its longitudinal axis,

$$(15) \qquad\qquad \Theta\ddot{\vartheta} = M_\nu = -hW\vartheta,$$

and the circular frequency of the oscillations

$$(16) \qquad\qquad \omega = \sqrt{\frac{hW}{\Theta}}.$$

Practical values of $h$ are about 3 to 5 times the distance $CS$.

The validity of the foregoing is limited, strictly speaking, to infinitesimal amplitudes, where only one metacenter $\mathfrak{M}$ has to be considered. $\mathfrak{M}$ is actually the limiting position of the metacenter for $\vartheta \rightarrow 0$. In naval architecture one has to investigate the stability also for finite $\vartheta$ values, and takes into account that the point $\mathfrak{M}$ moves in reference to the boat when $\vartheta$ changes.

We now turn to a discussion of what happens in a field of force that has *no single-valued potential*. Instead of an exposition in general terms

[6]Cf. Synge and Griffith, *op. cit.*, Sec. 7.2.

let us consider a simple example which can be easily shown in an actual experiment. A weakly conducting fluid such as a solution of $CuSO_4$ is placed in a shallow cylindrical container with insulating bottom and conducting side wall. A copper wire runs along the axis of the container. A potential difference of a few volts is put between center wire and side wall so as to maintain a current flowing from the axis through the fluid and spreading out toward the wall. Let $I$ be the current and $\mathbf{J}$ the vector of the current density, $|\mathbf{J}|$ being the current passing radially through the unit of area at the distance $r$ from the axis. Obviously

$$(17) \qquad\qquad |\mathbf{J}| = \frac{I}{2\pi r h}$$

where $h$ is the height of the fluid layer. Now we impose a fairly homogeneous magnetic field $\mathbf{H}$ whose lines of force pass perpendicularly through the fluid layer. The electromagnetic force acting upon a fluid element of unit volume is then given by

$$(18) \qquad\qquad \mathbf{F} = \mathbf{J} \times \mathbf{H},$$

a vector that is normal to $\mathbf{J}$ and $\mathbf{H}$ and points therefore in the tangential direction. On introducing the polar angle $\varphi$ and considering (17) and (18), we have

$$(19) \qquad |\mathbf{F}| = F_\varphi = \frac{A}{r}, \qquad A = \frac{I \cdot H}{2\pi h}.$$

This force has the potential

$$(20) \qquad\qquad U = -A\varphi.$$

The negative gradient of $U$ taken with respect to the cylindrical coordinates $r$, $z$ is indeed zero and for the $\varphi$-direction

$$- \operatorname{grad}_\varphi U = -\frac{1}{r}\frac{\partial U}{\partial \varphi} = \frac{A}{r} = F_\varphi .$$

Yet this potential is obviously not single-valued in the (doubly connected!) domain of the fluid, since it changes its value by the amount $2\pi A$ for each revolution of the variable $\varphi$ about the axis $r = 0$.

Since the hydrostatic pressure must be a single-valued point function, a pressure distribution that balances this potential in the sense of Eq. (6) cannot exist. Hence the fluid yields to the force $F_\varphi$ and begins to move. The velocity pattern of the motion is the single-vortex motion described in (4.29); the circumferential velocity is in our case

$$|\mathbf{v}| = v_\varphi = \frac{A}{r}\frac{t}{\rho}$$

where $t$ is the time through which the magnetic field has been acting and $\rho$ the density of the fluid. The velocity would grow continually if there were no friction.

The mathematician who studies a multi-valued function tries to arrange its set of values in a number of different branches. He leads an appropriate cut through the domain of definition of the function to restrict the possible types of curves along which the independent variable may be changed, and thereby establishes a single-valued branch of the function. We can do the analogous thing in our experiment by putting a separating wall (wood or cardboard) into the fluid between the axis and the side wall. The fluid builds up on one side and recedes from the other, and the potential is now a single-valued function in the domain of the fluid; a hydrostatic pressure $p = A\varphi$ is built up so as to balance the potential $U$, and the motion stops. The pressure distribution can be inferred from the shape of the surface; on one side of the separating wall, the level is by the amount $2\pi A$ higher than on the other side. The level difference is thus equal to the modulus of periodicity of the potential.

## 7. Statics of Compressible Fluids

The density of compressible gases and vapors depends on the pressure. Let the relation between the two have the form $p = C\rho^n$ which can be set up in reference to a normal state $p_0$, $\rho_0$ as follows:

$$\text{(1)} \qquad \frac{p}{p_0} = \left(\frac{\rho}{\rho_0}\right)^n.$$

The curve corresponding to this equation is known in technical and astrophysical applications as a *polytrope*; $n$ is called the *polytropic exponent*.

In the isothermal state (constant temperature throughout the fluid), $n$ equals 1. This is an immediate consequence of the equation of state

$$\text{(2)} \qquad p = \rho \frac{R}{\mu} T,$$

where $R$ = universal gas constant, $\mu$ = molar weight[7] of the gas (e.g.,

---

[7]As is well known the number of grams in a *mole* equals the sum of the atomic weights of the gas. Instead of *molar* weight the term *molecular* weight is frequently used, but we wish to save this term in order to apply it to the actual weight (or more properly, the mass) of the single molecule. As R. W. Pohl has pointed out (Z.f.Phys., 121, 543, 1943) the definition of the mole depends on the definition of the unit of mass. If gr is replaced by kg, the mole changes into the "kilomole". The molar volume and Loschmidt's number per mole (Avogadro's number) change correspondingly (see later).

$\mu = 32$ gr for $O_2$), $T$ = absolute temperature. In the *adiabatic* state (convective equilibrium such as results from a vigorous mixing process without heat exchange) $n = 1.4$ for diatomic gases.* In aeronautics a mean value of $n = 1.2$ is sometimes assumed (polytropic exponent of the atmosphere, see below).

It is known that the exponent $n = 1.4$ of the adiabatic state is equal to the ratio $c_p/c_v$ of the specific heats at constant pressure and constant volume. This ratio follows from the first law of thermodynamics by means of an application of classical statistics (equipartition theorem of energy). The theoretical value is obtained as $c_p/c_v = 1 + 2/f$, where $f$ is the number of degrees of freedom of the gas molecule. For a diatomic molecule such as $N_2$ or $O_2$ this number equals 5 (three degrees of freedom of translation and two of rotation; the rotation about the line connecting the two atoms cannot be excited for quantum-theoretical reasons), hence the value $n = 1 + 2/5 = 1.4$.

In the case of a compressible fluid, it is convenient to follow a different convention in measuring the external force. Previously referred to the *unit of volume* and denoted by $\mathbf{F}$, we refer the external force now to the *unit of mass* and denote it by $\mathbf{P}$. Thus

(3a) $$\mathbf{F}\Delta\tau = \mathbf{P}\Delta m, \quad \text{hence } \mathbf{F} = \rho\mathbf{P}.$$

In the case of gravity where $|\mathbf{F}| = \rho g$ we have simply

(3b) $$\pm|\mathbf{P}| = P_z = \pm g,$$

the sign depending on the choice of the orientation of the $z$-axis; in the case of the centrifugal force where $|\mathbf{F}| = \rho r\omega^2$ as in (6.9) we have

(4) $$|\mathbf{P}| = r\omega^2.$$

The advantage of the new notation becomes apparent if we consider the

---

In CGS units the gas constant $R = 8.31 \times 10^7$ erg/centigrade: on the other hand, if one uses Meter-Kilogram-Second (MKS) units as Pohl does, the powers of ten drop out. In the same way as the unit of force in the MKS system is defined by 1 Dyne = $10^5$ dyne, we can introduce as unit of energy in the MKS system

$$1 \text{ Erg} = 1 \text{ M}^2\text{KS}^{-2} = 1 \text{ M} \cdot 1 \text{ Dyn} = 10^7 \text{ erg} = 1 \text{ Joule.}$$

The gas constant is then $R = 8.31$ Erg/centigrade. Cf. the remarks about MKS-units in Vol. I, end of 1, also Slater and Frank, *op. cit.*, Sec. I.5.

*The pressure distribution within the gaseous mass is such that an arbitrary adiabatic displacement of a particle does not disturb its temperature equilibrium with the neighboring particles, cf. problem II,1a.

equilibrium condition (6.4) which is correct for both compressible and incompressible fluids. It reads now

$$(5) \qquad \frac{1}{\rho} \operatorname{grad} p = \mathbf{P},$$

where the second member is a known quantity and the first member can be expressed in terms of either $p$ or $\rho$, if a $p$, $\rho$-relation is known.

For the $p$,$\rho$-relation (1), Eq. 5 takes the form

$$(6) \quad \operatorname{grad} \mathcal{P} = \mathbf{P}, \qquad \text{where} \qquad (6a) \quad \mathcal{P} = \frac{p_0^{1/n}}{\rho_0} \frac{p^{1-(1/n)}}{1 - (1/n)}.$$

The same Eq. (6) is correct for any non-polytropic $p$,$\rho$-relation provided we explain the quantity $\mathcal{P}$ by

$$(6b) \qquad \mathcal{P} = \int_A^B \frac{dp}{\rho}.$$

Here $A$ is a fixed point, $B$ is the variable field point with coordinates $x, y, z$.

In the same way as (6.4), Eq. (6) implies that *in a compressible fluid, too, equilibrium is only possible when the external* force, that is in the present case, the external force per unit of mass $\mathbf{P}$, *has a potential*. On denoting the potential by $V$, we draw from (6)

$$(7) \qquad \operatorname{grad} (\mathcal{P} + V) = 0, \qquad \mathcal{P} + V = \text{const.}$$

For the field of gravity $V = gz$, where $z$ is counted positive upward. When the polytropic relation (1) is used, $\mathbf{P}$ is given by (6), and the second relation (7) takes the form

$$(8) \qquad \frac{p_0^{1/n}}{\rho_0} p^{1-(1/n)} = \frac{n-1}{n} (\text{const} - gz).$$

At $z = 0$, $p = p_0$ (reference pressure). The constant in (8) is found from $p_0/\rho_0 = \text{const} (n-1)/n$. Eq. (8) can then be written as

$$(9) \qquad p = p_0 \left( 1 - \frac{n-1}{n} \frac{\rho_0}{p_0} gz \right)^{n/(n-1)}.$$

On expressing $p$ by $\rho$ we obtain

$$(9a) \qquad \rho = \rho_0 \left( 1 - \frac{n-1}{n} \frac{\rho_0}{p_0} gz \right)^{1/(n-1)}$$

Formulas (9) and (9a) represent the distribution of pressure and density in a polytropic atmosphere; its altitude $h$ can be determined by setting $p = 0$ or $\rho = 0$:

$$(9b) \qquad h = \frac{n}{n-1} \frac{p_0}{\rho_0 g}.$$

By reintroducing $h$ in (9) and (9a) these equations can now be simplified and read

$$(10) \qquad p = p_0 \left(1 - \frac{z}{h}\right)^{n/(n-1)}, \qquad \rho = \rho_0 \left(1 - \frac{z}{h}\right)^{1/(n-1)}.$$

The distribution of the temperature can be found when the ratio $p/\rho$ is calculated from (10) and the equation of state (2) is applied:

$$(11) \qquad RT = \mu \frac{p_0}{\rho_0} \left(1 - \frac{z}{h}\right).$$

According to this, the temperature distribution in the polytropic atmosphere is a linear function of the altitude, dropping from $T_0 = \mu p_0/\rho_0 R$ to $T = 0$ between $z = 0$ and $z = h$.

Numerical results for the altitude of the polytropic atmosphere are found from (9b): in the adiabatic case, $n = 1.4$, the altitude is only 28 km; in the polytropic case, when one chooses $n = 1.2$, one obtains 48 km; the altitude of the isothermal atmosphere is, of course, *infinite*.

The assumption $T = $ const is not permissible for the entire atmosphere. In the troposphere, the lowest part of the atmosphere, there is a very noticeable temperature lapse as one goes up; in the stratosphere, the temperature first remains constant over a considerable interval and increases later on. The boundary between troposphere and stratosphere is located at an average height of 12 km, at the poles it is lower than at the equator. For minor altitude differences the assumption of constant temperature may be maintained. The pressure and density distribution resulting in this case is well known under the name of the *barometric formula*. Since this formula is of general interest, it will be derived directly from the fundamental Eq. (5).

On setting $P_z = -g$ and writing (2) in the form

$$(12) \qquad \rho = Cp, \qquad C = \frac{\mu}{RT},$$

we obtain from (5)

$$\frac{1}{p} \frac{dp}{dz} = -Cg.$$

This gives on integration

(13) $$\log p = -Cgz + \log p_0 ,$$

where $p_0$ is the pressure at the ground $(z = 0)$; on solving for $p$, one has

(14) $$p = p_0 e^{-Cgz} = p_0 e^{-\mu gz/RT} .$$

The same formula holds for the distribution of the density [comp. (12) or $\rho/\rho_0 = p/p_0$]:

(15) $$\rho = \rho_0 e^{-\mu gz/RT} .$$

According to (13), the altitude may be read directly from a barometer if the instrument is furnished with an appropriate scale.[8]

One can interpret Eq. (15) in the sense of a sedimentation equilibrium of the air masses. It was Jean Perrin who successfully applied this idea to make a laboratory model of the atmosphere using a homogeneous emulsion of mastic droplets in water. His aim was to determine in this way a fundamental constant of atomic theory. That this can be done is evident as soon as $\mu$ in Eq. (15) is replaced by $Lm$, where $L$ is Loschmidt's[9] (or Avogadro's) number, i.e. the number of molecules per mole, and $m$ is the mass of the single molecule; $mg$ would therefore be the molecular weight in the proper sense of the word. Eq. (15) then becomes

(15a) $$\rho = \rho_0 e^{-Lmgz/RT} .$$

While it is impossible to measure directly $m$ of an actual molecule, Perrin's model gas permits of a direct determination of *its* $m$ (e.g. by weighing a sufficiently large number of mastic droplets). Eq. (15a) can then be

---

[8]Eq. (14) or (15) can be derived from the general Eq. (9) or (9a) by making $n$ approach the limit 1 and using

$$e^z = \operatorname*{Lim}_{m \to \infty} \left(1 + \frac{x}{m}\right)^m .$$

[9]Sometimes one uses the term "Loschmidt's number" in the sense of the number of molecules per $cm^3$ and distinguishes it from "Avogadro's number" = number of molecules per mole, a definition of Loschmidt's number that was proposed by Boltzmann in a memorial address for Loschmidt (L. Boltzmann, Populäre Schriften, Leipzig 1905, p. 243). It is true that Loschmidt had focused his attention only on the first of the two numbers which he was able to estimate by several gas-kinetic methods, but he could have easily derived the second from the first number on the basis of Avogadro's law. Yet one must admit that not even the crudest estimate of the order of magnitude of either one of the numbers was within Avogadro's reach, if one considers the general state of knowledge of his time. Since the number of molecules *per mole* is a fundamental constant of all atomic physics, it appears reasonable to give a characteristic name to this number alone. Should it not be the name of the man who was the first to devise methods for its determination?

used for a determination of $L$ if the densities $\rho$ and $\rho_0$ have been observed. The value obtained by Perrin in this way is $68 \times 10^{22}$, which is remarkably close to the value $L = 60.2 \times 10^{22}$ accepted today.

When we introduce in (15a) instead of $R$ the *gas constant per molecule*[10]

$$(15b) \qquad k = \frac{R}{L},$$

the exponential in (15a) becomes

$$(15c) \qquad e^{-mgz/kT} = e^{-V/kT}.$$

Note that $V$ stands here for the potential energy of a mass point $m$ that has been elevated to the altitude $z$ in the field of gravity. The right member of (15c) is called the *Boltzmann factor*; it involves a specified energy $V$, but its physical meaning can be extended to the energy states of any physical system and represents, generally speaking, the relative probability of such a state in reference to a standard state of equal temperature.

The Boltzmann factor is fundamental in all statistical investigations; the purpose of this short digression into atomic theory has been to point to its connection with the elementary barometric formula.

Since the molar weight $\mu$ occurs in formula (15), a possibility arises to separate heavy from light molecules in a mixture of gases by sedimentation. Practically, this cannot be done in the field of gravity, but a centrifugal field may be effectively used for that purpose.

The potential of the centrifugal force (4) per unit of mass is

$$(16) \qquad V = -\frac{1}{2} r^2 \omega^2.$$

According to (7) we have for the quantity $\mathcal{P}$

$$(16a) \qquad \mathcal{P} = \text{const} + \frac{1}{2} r^2 \omega^2.$$

Here we assume $\omega$ to be constant [as in (6.9a)] all over the inside of the centrifuge. On the other hand, the integral $\mathcal{P}$ for an isothermal gas or mixture of gases is found from (6b) and (12)

$$(17) \qquad \mathcal{P} = \frac{RT}{\mu} \int \frac{dp}{p} = \frac{RT}{\mu} \log \frac{p}{p_0}.$$

Here $p_0$ is the pressure at the axis of the centrifuge $'r = 0$), and the con-

---

[10]Usually called Boltzmann's constant although Planck, following Boltzmann's ideas, was the first to introduce it definitely.

stant in (16a) vanishes. From (16a) and (17) the pressure distribution inside the centrifuge is obtained:

$$(18) \qquad \log \frac{p}{p_0} = \frac{\mu}{RT} \cdot \frac{1}{2} r^2 \omega^2, \qquad p = p_0 \exp \left( \frac{\mu}{2} \frac{r^2 \omega^2}{RT} \right).$$

The pressure is seen to increase exponentially with increasing $r$.

Consider now a mixture of two gases with molar weights $\mu_1$ and $\mu_2$. Eq. (18) gives two different partial pressures $p_1$ and $p_2$ with two different constants $p_{01}$, $p_{02}$ that depend on the mixing ratio. The total pressure is $p = p_1 + p_2$. According to Eq. (18), the partial pressure that belongs to the larger molar weight increases more rapidly than that belonging to the smaller $\mu$, when $r$ increases. Since for a given temperature the pressures and densities change proportionally, the heavier component of the mixture is concentrated in the proximity of the circumference of the centrifuge. The numerical magnitude—or rather minuteness—of this effect for air (21% $O_2$, 79% $N_2$) will be the object of problem II.2.

As in the incompressible case, we may consider gravity in addition to the centrifugal force; the *gravity potential* $-gz$ is then to be added in Eq. (16). Instead of (18) we have

$$(19) \qquad p = p_0 \exp \left\{ \frac{\mu}{RT} \left( -gz + \frac{1}{2} r^2 \omega^2 \right) \right\}.$$

The surfaces of constant pressure (hence constant density) are again paraboloids, but $p_0$ is now the pressure at the particular paraboloid $gz = r^2 \omega^2 / 2$, from which the other isobaric surfaces are obtained by an upward or downward translation. Note that there does not exist a free surface toward vacuum ($p = 0$) since the exponential function does not become zero for finite values of the argument.

Design and construction of centrifuges of enormous efficiency have become possible through the researches of Th. Svedberg. The speed of these devices which now are commercially available under the name of ultracentrifuges goes up to 100,000 r.p.m., the strength of the centrifugal field reaches 750,000g. One of the purposes of these machines is to sort protein molecules (molar weight about 30,000). Their principal field of application is in colloidal chemistry and biology where the objects of investigation are approximately incompressible mixtures of liquids rather than compressible gases.

## 8. The State of Stress of an Elastic Solid

The mathematical characterization of the stresses in a solid body is complete when the stress is recognized as a symmetrical tensor of second order. In other words, it must be shown that the array of stresses

$$\sigma = \begin{pmatrix} \sigma_{xx} & \sigma_{xy} & \sigma_{xz} \\ \sigma_{yx} & \sigma_{yy} & \sigma_{yz} \\ \sigma_{zx} & \sigma_{zy} & \sigma_{zz} \end{pmatrix}$$

is symmetric about the principal diagonal,

(1) $\qquad\qquad\qquad \sigma_{xy} = \sigma_{yx}$ , etc.,

and that the components transform in the same way as the squares and products of the coordinates in passing to another Cartesian system [cf. Eq. (4.6)]

The symmetry can be easily inferred from the condition of moment equilibrium. Let us first set up the moment about the $z$-axis which acts on the volume element with the sides $\Delta x$, $\Delta y$, $\Delta z$, cf. Fig. 10. The normal

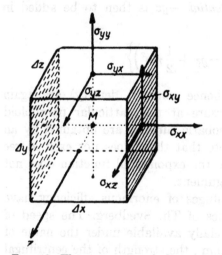

FIG. 10. The symmetry of the strain tensor. The resulting moment of the shear forces must vanish.

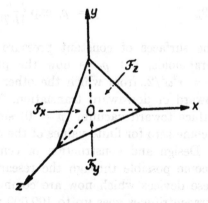

FIG. 11. Illustrating the equilibrium of forces acting on a tetrahedral volume element. F denotes the *area*.

stresses do not contribute anything to this moment since their vectors intersect with the $z$-axis. Consider now the tangential stresses $\sigma_{zy}$ and $\sigma_{yz}$ indicated in the figure. The corresponding forces transmitted to the volume element and the arms of these forces are

$$\sigma_{zy}\Delta y\Delta z, \qquad \frac{1}{2}\Delta x \qquad \text{and} \qquad \sigma_{yz}\Delta z\Delta x, \qquad \frac{1}{2}\Delta y.$$

For the signs of the resulting moments refer to Fig. 10, observing the convention that the direction of a stress component be indicated by its second index, when it acts on a "positive" surface element (cf. p. 38). The sum of the moments due to $\sigma_{zy}$ and $\sigma_{yz}$ then becomes

(2) $$\frac{\Delta\tau}{2}(\sigma_{zy} - \sigma_{yz}).$$

This does not take care of the contributions of the shear stresses acting across the "negative" $x$- and $y$-surfaces of our volume element. Let $\sigma'_{zy}$ and $\sigma'_{yz}$ be these stresses; on expanding the first in terms of $\Delta x$, the second in terms of $\Delta y$, one sees immediately that $\sigma'_{zy} = -\sigma_{zy}$ and $\sigma'_{yz} = -\sigma_{yz}$ if terms of the order $\Delta x$ and $\Delta y$ are neglected. The contributions of these stresses to the moment about the $z$-axis is therefore once more the quantity (2) if only terms of the order of magnitude $\Delta\tau$ are retained. An additional external force $\mathbf{F}$ (per unit of volume) will in general also be present; its moment is $|\mathbf{F}|\,\Delta\tau l$ where the arm $l$ is of a smaller order of magnitude than the side length [see 6, after Eq. (4)]: it cannot possibly balance the reactive moments. Thus the equilibrium condition requires that the expression (2) vanishes by itself, but this is identical with the first symmetry condition in (1). The other two follow in the same way by considering moments about the $x$- and $y$-axes.

The proof of our second point, the *tensor character of the stress*, is more involved. It must be based on the *vector character* of the *reactive forces* that are transmitted across a surface element, just as the corresponding proof for the strain tensor was based on the vector character of the displacement.

We first set up the conditions for the equilibrium of forces acting on a tetrahedron, three faces of which coincide with the coordinate planes while the fourth is formed by a plane of arbitrary normal direction $n$ (Fig. 11). Let the area of the oblique face be $F_n$, and $F_x$, $F_y$, $F_z$ the areas of the faces lying in the planes $x = 0$, $y = 0$, $z = 0$ and consider these areas as projections of $F_n$:

(3)     $F_x = F_n \cos(n, x),\qquad F_y = F_n \cos(n, y),\qquad F_z = F_n \cos(n, z).$

The equilibrium condition for the $x$-direction is then[11]

(4) $$F_n\sigma_{nx} = F_x\sigma_{xx} + F_y\sigma_{yx} + F_z\sigma_{zx},$$

which because of (3) may be written

(5) $$\sigma_{nx} = \alpha_1\sigma_{xx} + \beta_1\sigma_{yx} + \gamma_1\sigma_{zx}.$$

---

[11]One checks with the aid of Fig. 11 that $F_x$, $F_y$, $F_z$ are *negative* $x$-, $y$-, $z$-surfaces; hence, in the total $x$-force, the terms with factors $\sigma_{xx}$, $\sigma_{yx}$, $\sigma_{zx}$ are originally negative. In Eq. (4) they have been transferred to the right side with the *positive* sign.

Here $\alpha_1$, $\beta_1$, $\gamma_1$ have been written for the direction cosines in (3) as indicated in the following scheme:

|     | $n$        | $t$        | $t'$       |
|-----|------------|------------|------------|
| $x$ | $\alpha_1$ | $\alpha_2$ | $\alpha_3$ |
| $y$ | $\beta_1$  | $\beta_2$  | $\beta_3$  |
| $z$ | $\gamma_1$ | $\gamma_2$ | $\gamma_3$ |

Eq. (5) expresses the $x$-component of the stress acting across the $n$-surface. The $y$- and $z$-components can be determined analogously by setting up the equilibrium conditions for the $y$- and $z$-directions. One obtains

$$\sigma_{ny} = \alpha_1 \sigma_{zy} + \beta_1 \sigma_{yy} + \gamma_1 \sigma_{zy} , \tag{6}$$

$$\sigma_{nz} = \alpha_1 \sigma_{zz} + \beta_1 \sigma_{yz} + \gamma_1 \sigma_{zz} . \tag{7}$$

The *normal stress* on the $n$-surface results now from (5), (6) and (7) by projecting the stress vector

$$\sigma_{nz} \mathbf{i} + \sigma_{ny} \mathbf{k} + \sigma_{nz} \mathbf{l} \tag{8a}$$

on the $n$-direction. The projection equals

$$
\begin{aligned}
\sigma_{nn} &= \alpha_1 \sigma_{nz} + \beta_1 \sigma_{ny} + \gamma_1 \sigma_{nz} \\
&= \alpha_1^2 \sigma_{zz} + \beta_1^2 \sigma_{yy} + \gamma_1^2 \sigma_{zz} + 2\alpha_1 \beta_1 \sigma_{zy} + 2\beta_1 \gamma_1 \sigma_{yz} + 2\gamma_1 \alpha_1 \sigma_{zz} .
\end{aligned}
\tag{8}
$$

The directions $t$ and $t'$ already introduced may be realized as the legs of a right angle lying in the $n$-surface in an otherwise arbitrary position. We now project the stress vector (8a) also upon the directions $t$ and $t'$ and obtain the *tangential* or *shear* stress across the $n$-surface, resolved in a $t$- and $t'$-component:

$$
\begin{aligned}
\sigma_{nt} &= \alpha_2 \sigma_{nz} + \beta_2 \sigma_{ny} + \gamma_2 \sigma_{nz} \\
&= \alpha_1 \alpha_2 \sigma_{zz} + \beta_1 \beta_2 \sigma_{yy} + \gamma_1 \gamma_2 \sigma_{zz} \\
&\quad + (\beta_1 \alpha_2 + \alpha_1 \beta_2) \sigma_{zy} + (\gamma_1 \beta_2 + \beta_1 \gamma_2) \sigma_{yz} + (\alpha_1 \gamma_2 + \alpha_2 \gamma_1) \sigma_{zz}
\end{aligned}
\tag{9}
$$

(for $\sigma_{nt'}$, change subscript 2 into 3). When numerals are introduced instead of the subscripts $x$, $y$, $z$ and $n$, $t$, $t'$ and $\alpha_{ik}$ is written for the nine direction cosines, (8) and (9) and the corresponding relation for $\sigma_{nt'}$ can be written in the condensed form

$$(10) \qquad \sigma'_{ik} = \sum_l \sum_m \alpha_{il}\alpha_{km}\sigma_{lm} \, .$$

This equation contains no reference to the size of the tetrahedron and gives, therefore, normal and tangential stresses acting across an arbitrary $n$-plane through $O$ in Fig. 11. Its full significance is obtained if we disregard the particular roles of $n$, $t$, and $t'$. Eq. (10) is then seen to express the stresses acting on a cell parallel to the arbitrary orthogonal triad $n$, $t$, $t'$ by the stresses on a cell parallel to $x$, $y$, $z$.

Since (10) has the form of the tensor transformation formulas as derived for the strain tensor in (4.5), we may from now on refer to the *stress tensor* and visualize the state of stress in the neighborhood of a point by means of a stress quadric (elipsoid, hyperboloid, etc.) Magnitude and direction of the principal stresses are determined by the principal axes of the quadric. Any state of stress can be represented for a sufficiently small neighborhood of the field point as the sum of three mutually "orthogonal" tensors of simplest structure, viz. unidirectional tensions or compressions in correspondence with Helmholtz's theorem for the strains in 1 Eq. (16).

In setting up the equilibrium condition for the infinitesimal *tetrahedron* in (4), only the reactive forces transmitted through the surfaces, but no body forces have been considered. This is justified since the contributions of the latter are proportional to the *volume* while the stresses have the order of magnitude of the areas of the *faces*. The situation is different if we now turn to the equilibrium of external and internal forces for a small *rectangular cell.** We want to establish a relation that is analogous to the fundamental equation (6.4) of hydrostatics. To secure equilibrium in $x$-direction as in (6.3) we have to consider the contributions of the normal stresses acting on the positive and negative $x$-surfaces and, *in addition*, those due to the tangential $x$-stresses across the positive and negative $y$- and $z$-surfaces; the latter do not occur in hydrostatics. The sum must be balanced against the contribution of the body forces represented, per unit of volume, by the vector $\mathbf{F}$. This yields, after cancellation of the factor $\Delta\tau = \Delta x \Delta y \Delta z$, the equilibrium condition

$$(11) \qquad \frac{\partial\sigma_{xx}}{\partial x} + \frac{\partial\sigma_{yx}}{\partial y} + \frac{\partial\sigma_{zx}}{\partial z} + F_z = 0 .$$

The corresponding equations for the $y$- and $z$-directions are obtained by "rotating" the letters. Note that in (11) the $\mathbf{F}$ term has the same sign as

---

*Since the area terms due to opposite faces now cancel [cf. the remark following Eq. (2)].

the stress derivatives while in (6.4) the signs were opposite. The reason is obviously that the positive sign now denotes a tension, while it denoted a pressure previously.

The triplet of equations that corresponds to (11) reads in condensed form:

$$(12) \qquad \text{Div } \sigma + \mathbf{F} = 0.$$

The symbol Div stands for "vector divergence"; it derives a vector from a tensor while the operation div, formally explained in the same way, derives a scalar from a vector. In the general tensor analysis, where tensors of arbitrary order are considered, certain operations that decrease the order of a tensor are studied. The present case occurs there under the heading "differentiation with subsequent contraction".[12]

The definition of the symbol Div contained in (12) is obviously limited to Cartesian coordinates. It can be generalized for arbitrary orthogonal systems by the introduction of the limit of a surface integral as in (2.20). Enclose the field point, at which the $p$-component of the divergence vector associated with the tensor field $\sigma$ is to be calculated, in a closed surface $f$ with outward normal $n$. The generalized definition of Div is then

$$(13) \qquad \text{Div}_p \, \sigma = \lim_{\Delta\tau \to 0} \frac{1}{\Delta\tau} \int \sigma_{np} \, df$$

where $\Delta\tau$ is the volume enclosed by $f$, and $\sigma_{np}$ the component of the stress that acts across the $n$-surface element, taken in the direction $p$.

In the particular case of Cartesian coordinates we take for $\Delta\tau$ the rectangular cell $\Delta x$, $\Delta y$, $\Delta z$, and verify the new definition (13) in the same way as in the argument leading from Eq. (2.20) to (2.20a). In the general case of orthogonal coordinates [cf. Eq. (2.22)], the surface $f$ is again chosen as the boundary of the curvilinear rectangular cell $g_1\Delta p_1$, $g_2\Delta p_2$, $g_3\Delta p_3$. The component of the vector divergence in the direction $dp_k$ follows from (13) as

$$(14) \qquad \text{Div}_k \, \sigma = \frac{1}{g_1 g_2 g_3} \left\{ \frac{\partial}{\partial p_1} (g_2 g_3 \sigma_{1k}) + \frac{\partial}{\partial p_2} (g_3 g_1 \sigma_{2k}) + \frac{\partial}{\partial p_3} (g_1 g_2 \sigma_{3k}) \right\}.$$

The meaning of $\sigma_{ik}$ is, of course, the component of the stress in the direction of $dp_k$ acting across a surface element that is perpendicular to the direction of $dp_i$.

To obtain the $k^{th}$ component of the equilibrium condition the expression (14) must be augmented by the component of $\mathbf{F}$ acting in the direction

---

[12]Cf. Appendix I-IV.  General tensor analysis has been developed to serve as the adequate mathematical language of the general theory of relativity.

of the coordinate $p_k$ . This sum replaces the left member of Eq. (12) in the case of general coordinates.

But the conditions of equilibrium are not sufficient to determine the tensor field $\sigma$ from the vector field $\mathbf{F}$, since the tensor has six independent components while Eq. (12) is only equivalent to three scalar equations. The three missing equations are not supplied by the conditions of the moment equilibrium as one might be tempted to believe, since that condition has already been exploited in establishing the *symmetry* of the stress tensor. For the *determination* of the stress tensor the strain-stress relations must be used as will be done in the following 9.

In this connection one may ask for supplementary conditions which together with (12) would ensure that the six quantities $\sigma$ represent a physically possible state of stress. The answer is found in the *conditions of compatibility*. The same question could have been asked in the case of the strain field, since the six functions $\epsilon$ depend on the three displacement components $\xi$, $\eta$, $\zeta$ and, of course, may not be chosen arbitrarily.[13]

In conclusion we have again compiled the more frequent notations for the stress tensor:

| | | | | | | |
|---|---|---|---|---|---|---|
| This book | $\sigma_{xx}$ | $\sigma_{yy}$ | $\sigma_{zz}$ | $\sigma_{xy} = \sigma_{yx}$ | $\sigma_{yz} = \sigma_{zy}$ | $\sigma_{zx} = \sigma_{xz}$ |
| Many American authors | $\tau_{xx}$ | $\tau_{yy}$ | $\tau_{zz}$ | $\tau_{xy} = \tau_{yx}$ | $\tau_{yz} = \tau_{zy}$ | $\tau_{zx} = \tau_{xz}$ |
| Love and many English authors | $X_x$ | $Y_y$ | $Z_z$ | $Y_x = X_y$ | $Z_y = Y_z$ | $X_z = Z_x$ |
| Kirchhoff and Planck | $-X_x$ | $-Y_y$ | $-Z_z$ | $-Y_x = -X_y$ | $-Z_y = -Y_z$ | $-X_z = -Z_x$ |
| Some English authors | $P$ | $Q$ | $R$ | $S$ | $T$ | $U$ |
| Engineering usage | $\sigma_x$ | $\sigma_y$ | $\sigma_z$ | $\tau_{xy} = \tau_{yx}$ | $\tau_{yz} = \tau_{zy}$ | $\tau_{zx} = \tau_{xz}$ |

## 9. Strain-Stress Relations, Elastic Constants, Elastic Potential

In the present section the loads are supposed to be sufficiently small so that second and higher powers of the stress and strain components may be neglected. Then the general form of the relations between stress

---

[13]Cf. Love, Mathematical Theory of Elasticity, 4th ed., Cambridge, 1927, pp. 49 and 135.

and strain components becomes *linear*. Hooke, a contemporary of Newton, was the first to state this "law", although in simpler formulation. About the limits of the linear law (proportional and elastic limits) some remarks will be found in Chap. VIII, 39.

We further assume that the elastic body is isotropic: its physical properties should be the same in all directions. Now it is well known that the structure of most solids such as minerals, metals and engineering materials in general is crystalline or microcrystalline. The following analysis refers therefore to "volume elements" that are large compared to the microstructure of the material, that is, what we call a volume element contains a great number of irregularly oriented microcrystals. The elastic properties of macroscopic crystals will be discussed in Chap. VIII.

A volume element cut parallel to the principal axes of the stress ellipsoid makes a convenient starting point for our analysis. The three pairs of faces are subjected to the principal stresses $\sigma_1$ , $\sigma_2$ , $\sigma_3$ while shear stresses $\sigma_{ik}$ do not occur. It is then a first consequence of the assumed isotropy, that the shape of the volume element after loading is still rectangular. Thus there are no angular changes, the shear strains $\epsilon_{ik}$ are zero, and the state of strain consists in pure extensions along the principal axes cf the stress ellipsoid. *Hence the principal axes of the stress quadric coincide with the principal axes of the strain quadric.* The principal extensions will be denoted by $\epsilon_1$ , $\epsilon_2$ , $\epsilon_3$ , as in (1.14).

Now by the proposed linearity of the stress-strain relations the following relation between a stress component and the strain components is implied:

(1) $$\sigma_1 = a\epsilon_1 + b\epsilon_2 + c\epsilon_3 .$$

Since no axis is preferred, two further equations must be obtainable simply by rotation of the subscripts:

(1a) $$\sigma_2 = a\epsilon_2 + b\epsilon_3 + c\epsilon_1 , \qquad \sigma_3 = a\epsilon_3 + b\epsilon_1 + c\epsilon_2 .$$

Again for reasons of isotropy, $c$ must equal $b$, in Eq. (1), otherwise the directions (2) and (3) would not be equivalent with regard to the stress $\sigma_1$ . It is convenient to rewrite (1) by adding $\pm b\epsilon_1$ , viz.

$$\sigma_1 = (a - b)\epsilon_1 + b(\epsilon_1 + \epsilon_2 + \epsilon_3),$$

or, with a change in notation[14]

(2) $$\sigma_1 = 2\mu\epsilon_1 + \lambda(\epsilon_1 + \epsilon_2 + \epsilon_3).$$

---

[14]Replacing the factor of $\epsilon_1$ in Eq. (2) by $2\mu$ is an advantage, as will be seen below [Eqs. (16) and (18)].

On introducing the same notation in Eqs. (1a) the triplet (1) and (1a) reads in condensed writing

$$(3) \qquad \sigma_i = 2\mu\epsilon_i + \lambda(\epsilon_1 + \epsilon_2 + \epsilon_3).$$

To find the *general* form of (3) (i.e., for a volume element of arbitrary orientation) is our next task, and now the transformation rules (1.22) and (1.22a) are needed. They are general tensor equations and valid for the strain as well as for the stress tensor:

$$(4a) \qquad \epsilon_{xx} = \alpha_1^2\epsilon_1 + \alpha_2^2\epsilon_2 + \alpha_3^2\epsilon_3 , \qquad \epsilon_{xy} = \alpha_1\beta_1\epsilon_1 + \alpha_2\beta_2\epsilon_2 + \alpha_3\beta_3\epsilon_3 ,$$

$$(4b) \qquad \sigma_{xx} = \alpha_1^2\sigma_1 + \alpha_2^2\sigma_2 + \alpha_3^2\sigma_3 , \qquad \sigma_{xy} = \alpha_1\beta_1\sigma_1 + \alpha_2\beta_2\sigma_2 + \alpha_3\beta_3\sigma_3 .$$

In (4b) we substitute for $\sigma_i$ according to (3) and obtain with the aid of (4a)

$$(5) \qquad \begin{aligned} \sigma_{xx} &= 2\mu\epsilon_{xx} + \lambda(\alpha_1^2 + \alpha_2^2 + \alpha_3^2)(\epsilon_1 + \epsilon_2 + \epsilon_3), \\ \sigma_{xy} &= 2\mu\epsilon_{xy} + \lambda(\alpha_1\beta_1 + \alpha_2\beta_2 + \alpha_3\beta_3)(\epsilon_1 + \epsilon_2 + \epsilon_3). \end{aligned}$$

On observing that $\sum \alpha^2 = 1$, $\sum \alpha\beta = 0$, and $\Theta = \epsilon_1 + \epsilon_2 + \epsilon_3 = \epsilon_{xx} + \epsilon_{yy} + \epsilon_{zz}$ , (5) may be transformed into

$$\sigma_{xx} = 2\mu\epsilon_{xx} + \lambda(\epsilon_{xx} + \epsilon_{yy} + \epsilon_{zz}),$$

$$(6)$$

$$\sigma_{xy} = 2\mu\epsilon_{xy}$$

or, in condensed writing,[15]

$$(7) \qquad \sigma_{ik} = 2\mu\epsilon_{ik} + \lambda\delta_{ik}\Theta$$

Eq. (7) is the general form of the stress-strain relation for an arbitrary volume element.

The converse relations are easily found: Summing the first triplet (6) gives

$$\Sigma = \sigma_{xx} + \sigma_{yy} + \sigma_{zz} = (2\mu + 3\lambda)(\epsilon_{xx} + \epsilon_{yy} + \epsilon_{zz}).$$

This permits one to express the first scalar $\Theta$ by the first scalar $\Sigma$. On reintroducing $\Sigma$ in (6), one has at once

---

[15]For $\delta_{ik}$ see Eq. (22a).

$$\epsilon_{xx} = \frac{1}{2\mu}\left(\sigma_{xx} - \frac{\lambda}{2\mu + 3\lambda}\Sigma\right) = 2\mu'\sigma_{xx} + \lambda'\,\Sigma$$

. . . . . . . . . . . . . . . . .

(6a)

$$\epsilon_{xy} = \frac{1}{2\mu}\sigma_{xy} = 2\mu'\sigma_{xy}.$$

. . . . . . . . . . .

or, condensed,

(7a)                $$\epsilon_{ik} = 2\mu'\sigma_{ik} + \lambda'\delta_{ik}\Sigma.$$

These are the general strain-stress relations and the constants appearing in (7a) are, in terms of $\mu$ and $\lambda$,

(7b)          $$2\mu' = \frac{1}{2\mu}, \qquad \lambda' = -\frac{1}{2\mu}\frac{\lambda}{2\mu + 3\lambda}.$$

Eqs. (6) and (6a) show that the isotropic elastic body has *two elastic constants* which are physical characteristics of the material of the body and depend on the temperature. An anisotropic crystal has more than two elastic parameters; depending on the character of its asymmetry there may be up to 21 elastic constants, as will be seen in Chap. VIII. The quantities $\lambda$ and $\mu$, introduced by Lamé, are known as *Lamé's constants*, or better, *Lamé's moduli*.[16] For general investigations they are the most convenient characteristics of the elastic behavior, but their physical meaning is not immediately obvious. The quantities $\lambda$ and $\mu$ will now be expressed by two other constants that refer in a straightforward way to the simplest type of tension or compression experiment.

A vertical prismatic bar with cross-section $F$, the upper end of which is rigidly fixed, is subjected to a load $P$ that acts at the lower end. Let the load $P$ be uniformly distributed over the end cross-section. The tension across the end section (and across any other parallel section) is $\sigma = P/F$; $\sigma$ is a *principal stress* since no side forces, and therefore no shear stresses, are active. The associated *principal extension* $\epsilon$ equals $\Delta l/l$, where $l$ is the length of the bar and $\Delta l$ the displacement of the end cross-section. It turns out that the ratio of stress to extension is a *constant of the material*, that is, it is independent of $P$, $F$, and $l$, as long as the loading is not excessive. This ratio is denoted by $E$ and known as *Young's modulus*[17]

---

[16]It stands to reason to have two different words for the coefficients of the stress-strain and those of the strain-stress relations. Following Voigt, the former, like ($\lambda$, $\mu$), could be called *moduli*, the latter, like ($\lambda'$, $\mu'$), *constants* of elasticity [cf. 40].

[17]Introduced 1807 by Thomas Young, the same who discovered the interference of light.

or *modulus of elasticity*. Thus, for a bar under tension, we have

$$(8) \qquad E = \frac{\sigma}{\epsilon},$$

where $E$ has the same dimension [force/area] as $\sigma$ ($\epsilon$ is dimensionless). For steel and wrought iron an average value of $E$ is $2 \times 10^6$ kg.-wt./cm².

Elongation, however, is not the only response to the loading that can be observed in the tension experiment; actually, it is always accompanied by a contraction of the cross-section. It is this same contraction which we notice on a much larger scale when a rubber band is stretched, and we take it as the natural adaptation of the material to the loading. Denoting the contraction, which is the same for all fibers of the cross-section, by $-\epsilon'$, in this case a positive number, we define quite generally

$$(9) \qquad -\frac{\epsilon'}{\epsilon} = \nu \qquad \text{or, using (8),} \qquad \epsilon' = -\frac{\nu}{E}\sigma.$$

The quantity $\nu$ is also a *constant of the material*. It is called *Poisson's ratio* of transverse contraction to longitudinal extension.[18]

Poisson not only introduced this number, but actually determined its value as $\frac{1}{4}$, on the basis of a rather limited molecular theory. Modern lattice theory of the solid state has again taken up this problem and improved on Poisson's calculation; one of the results is that there is no universal theoretical value of $\nu$. In engineering applications the accepted value for iron is $\nu = 1/m = 0.3$.

When the bar is under compression instead of tension, longitudinal contraction is accompanied by transverse dilatation, but $E$ and $\nu$ remain the same in spite of the changes of sign that have occurred, provided, of course, that the loading is within the elastic limit (see also 39).

We now apply what we have learned in the simple tension experiment to analyze the general state of stress, which we consider as a superposition of three unidirectional stresses in the principal directions; the effect of each stress is the same as in the bar problem. Because of the linearity of all occurring relations we may superimpose the stresses as well as the associated strains. However, the extension along the principal axis 1 is not solely determined by the stress $\sigma_1$ but in addition (if considered as a transverse contraction) by the stresses acting in the principal directions 2 and 3. In this way one obtains from (8) and (9)

$$\epsilon_1 = \frac{\sigma_1}{E} - \frac{\nu}{E}(\sigma_2 + \sigma_3) = \frac{1+\nu}{E}\sigma_1 - \frac{\nu}{E}(\sigma_1 + \sigma_2 + \sigma_3)$$

---

[18]The preferred letter for Poisson's ratio in American and English literature seems to be $\sigma$.

or, in condensed writing,

$$(10) \qquad \epsilon_i = \frac{1 + \nu}{E}\, \sigma_i - \frac{\nu}{E}\, \Sigma$$

On summing the relations (10) for $i = 1, 2, 3$, one has

$$(10a) \qquad \Theta = \frac{1 - 2\nu}{E}\, \Sigma, \qquad \Sigma = \frac{E}{1 - 2\nu}\, \Theta,$$

and by substituting $\Sigma$ in (10) the result

$$(10b) \qquad \sigma_i = \frac{E}{1 + \nu}\left(\epsilon_i + \frac{\nu}{1 - 2\nu}\, \Theta\right)$$

is obtained. The relations which we wish to find result from a comparison between (10b) and (3). This leads to

$$(11) \qquad 2\mu = \frac{E}{1 + \nu}, \qquad \lambda = \frac{\nu E}{(1 + \nu)(1 - 2\nu)},$$

which express Lamé's parameters by $E$ and $\nu$ whose physical significance is more immediate.

Let us now take up another particularly simple case of a stress field, viz. *equal pressure in all directions* as it is realized, e.g., inside the cylinder of a hydraulic press. Both the stress and the strain ellipsoid are now spheres [see also 4 after (11)]:

$$\sigma_1 = \sigma_2 = \sigma_3 = -p, \qquad \Sigma = -3p, \qquad \epsilon_1 = \epsilon_2 = \epsilon_3 = -\frac{1}{3}\,|\,\Theta\,|.$$

Eq. (10b) gives in this case

$$(12) \qquad p = \frac{E\,|\,\Theta\,|}{3(1 - 2\nu)}.$$

We define now in analogy to (8) a *modulus of compression* $K$ by the equation

$$(13) \qquad K = \frac{p}{|\,\Theta\,|}$$

and infer from (12) and (13) at once

$$(14) \qquad K = \frac{1}{3}\frac{E}{1 - 2\nu}.$$

The *incompressible case* $\Theta = 0$ corresponds to $K = \infty$ or $\nu = \frac{1}{2}$. This value represents the upper limit for the possible values of Poisson's ratio. For $\nu > \frac{1}{2}$ the behavior of the body becomes *unstable*: its response to an external pressure would be an increase of volume ($\Theta > 0$) instead of a decrease.

We conclude this discussion with another particular stress field, that of pure shearing stress. Take a rectangular parallelepiped, the z-surfaces of which are free of stress while the shearing stress $\sigma_{zv} = \sigma_{vz} = \tau$ acts across the x- and y-surfaces. Our example is one of plain stress, that is, the stress field is independent of z and can be represented by simply drawing a cross-section parallel to the x-y-plane as in Fig. 12a. The

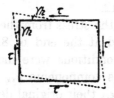

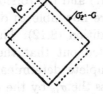

FIG. 12a. Change of shape in pure shear loading.

FIG. 12b. The principal stresses in pure shear loading are equal and opposite.

deformation due to $\tau$ changes the original rectangle into a rhomboid, one pair of right angles being diminished, the other pair increased by the angle $\gamma$. The stress quadric is, like the strain quadric in pure shear (cf. 4 Fig. 4a, b), a cylinder parallel to the z-axis the basis of which is an equilateral hyperbola. The principal axes form an angle of 45° with the x- and y-directions and the two principal stresses are numerically equal and of opposite sign: $\sigma_2 = -\sigma_1$ ; $\sigma_3$ is, of course, zero. This is a consequence of the invariance of the first scalar of the stress tensor, $\Sigma$: for the element considered originally (Fig. 12a), $\sigma_{xx} = \sigma_{vv} = \sigma_{zz} = 0$ by hypothesis, hence an element cut parallel to the principal axes must take tension and compression in equal amounts to make again $\Sigma = \sigma_1 + \sigma_2 = 0$.

The state of pure shear stress serves to define the *shear modulus* which we provisionally denote by $G$, a notation generally accepted in engineering practice. In partial analogy to (8) and (13) we set

$$(15) \qquad\qquad G = \frac{\tau}{\gamma} .$$

The analogy is not complete: the denominators in (8) and (13) represent the *extension* and (cubical) *compression* that correspond to the stresses in the numerators, but in the present case, (15), the denominator is the *angular change* produced by the stress that stands in the numerator, and this change is *twice* the strain component obtained according to the definition (1.25).

To find $G$, that is, to express it by the elastic constants already introduced, we go back to the general relation (6), $\sigma_{xy}/2\epsilon_{xy} = \mu$, which reads in our present notation $\tau/\gamma = \mu$. Thus (15) and (11) yield at once

$$(16) \qquad G = \mu = \frac{1}{2}\frac{E}{1+\nu}.$$

The shear modulus $G$ is identical with Lamé's modulus $\mu$ so that the notation $G$ is no longer needed. Also the name torsion modulus is in use for this constant because of its significance for the problem of torsional waves (see p. 107) and the torsion problem of a bar, 42.

We are now able to determine completely the stress from the equilibrium condition (8.12). This problem was posed at the end of 8, and it was pointed out that the three equilibrium conditions were insufficient for a complete determination of the six stress components $\sigma_{ik}$. But if we express the $\sigma_{ik}$ by the $\epsilon_{ik}$ and write for the $\epsilon_{ik}$ their original definitions (1.11) in terms of the displacements, we obtain three equations just sufficient to determine the displacement vector.

That is an easy job if Cartesian coordinates are used. The $x$-component of the vector divergence in (8.12) transforms, according to the stress-strain relation (6), into

$$\text{Div}_x\ \sigma = 2\mu\left(\frac{\partial\epsilon_{xx}}{\partial x} + \frac{\partial\epsilon_{yx}}{\partial y} + \frac{\partial\epsilon_{zx}}{\partial z}\right) + \lambda\frac{\partial\Theta}{\partial x},$$

which becomes, on going back to displacements by means of (1.11),

$$\text{Div}_x\ \sigma = \mu\left(2\frac{\partial^2\xi}{\partial x^2} + \frac{\partial^2\xi}{\partial y^2} + \frac{\partial^2\eta}{\partial y\,\partial x} + \frac{\partial^2\xi}{\partial z^2} + \frac{\partial^2\zeta}{\partial z\,\partial x}\right) + \lambda\frac{\partial\Theta}{\partial x}.$$

This is easily rearranged into

$$(17) \qquad \text{Div}_x\ \sigma = \mu\nabla^2\xi + (\mu+\lambda)\frac{\partial\Theta}{\partial x}.$$

On substituting (17) in the equilibrium conditions (8.12) and rotating the letters we obtain finally

$$\mu\nabla^2\xi + (\mu+\lambda)\frac{\partial\Theta}{\partial x} + F_x = 0,$$

$$(18) \qquad \mu\nabla^2\eta + (\mu+\lambda)\frac{\partial\Theta}{\partial y} + F_y = 0,$$

$$\mu\nabla^2\zeta + (\mu+\lambda)\frac{\partial\Theta}{\partial z} + F_z = 0.$$

*These three fundamental equations determine the displacement vector* $\xi$, $\eta$, $\zeta$ *everywhere in the interior of the body* if its behavior along the body surface is determined by appropriate boundary conditions. When the $\xi$, $\eta$, $\zeta$ have been found, the $\epsilon$ follow by differentiation and the $\sigma$ by way of the stress-strain relations. Thus Eqs. (18) are the definitive differential equations of elastic equilibrium. The quantity $\mu$ occurs in (18) without the factor 2, which again indicates that the choice of notation in Eq. (9.2) was to the purpose.

Eqs. (18) are limited to Cartesian coordinates. If we wish to write them in general coordinates, we have to interpret the symbol $\nabla^2$ as grad div $-$ curl curl according to (3.10a). On denoting the displacement vector $\xi$, $\eta$, $\zeta$ by $\delta\mathbf{q}$ as in (4.17)-(4.28), the legitimate vectorial form of Eq. (18) is obtained:

$$\mu \text{ grad div } \delta\mathbf{q} - \mu \text{ curl curl } \delta\mathbf{q} + (\mu + \lambda) \text{ grad } \Theta + \mathbf{F} = 0,$$

Instead of div $\delta\mathbf{q}$ we may write $\Theta$ and obtain

$$(19) \qquad (2\mu + \lambda) \text{ grad } \Theta - \mu \text{ curl curl } \delta\mathbf{q} + \mathbf{F} = 0.$$

The transition to general coordinates $p_1$, $p_2$, $p_3$ is now governed by the general expressions (2.24)-(2.26) for grad, etc. Setting $\mathbf{P} = \text{curl } \delta\mathbf{q}$ for brevity, one has immediately

$$(2\mu + \lambda) \frac{1}{g_1} \frac{\partial \Theta}{\partial p_1} - \frac{\mu}{g_2 g_3} \left( \frac{\partial g_3 P_3}{\partial p_2} - \frac{\partial g_2 P_2}{\partial p_3} \right) + F_1 = 0,$$

$$(20) \qquad (2\mu + \lambda) \frac{1}{g_2} \frac{\partial \Theta}{\partial p_2} - \frac{\mu}{g_3 g_1} \left( \frac{\partial g_1 P_1}{\partial p_3} - \frac{\partial g_3 P_3}{\partial p_1} \right) + F_2 = 0,$$

$$(2\mu + \lambda) \frac{1}{g_3} \frac{\partial \Theta}{\partial p_3} - \frac{\mu}{g_1 g_2} \left( \frac{\partial g_2 P_2}{\partial p_1} - \frac{\partial g_1 P_1}{\partial p_2} \right) + F_3 = 0.$$

The quantities $\Theta$ and $\mathbf{P}$ that occur here have according to (2.25) and (2.26) the meaning

$$(20a) \qquad \Theta = \frac{1}{g_1 g_2 g_3} \left\{ \frac{\partial}{\partial p_1} (g_2 g_3 \, \delta q_1) + \cdots \right\},$$

$$P_1 = \frac{1}{g_2 g_3} \left( \frac{\partial}{\partial p_2} (g_3 \, \delta q_3) - \frac{\partial}{\partial p_3} (g_2 \, \delta q_2) \right), \cdots, \cdots.$$

The boundary conditions to be added to Eqs. (18) or (20) may refer either to displacements or to stresses, or, as in the problem of 44, partly to the one and partly to the other (mixed boundary value problem). When the *displacements* are prescribed along the entire surface of the elastic body by

$$(21) \qquad\qquad \xi = \xi_0, \qquad \eta = \eta_0, \qquad \zeta = \zeta_0,$$

where $\xi_0$, $\eta_0$, $\zeta_0$ are point functions given along the surface, the boundary value problem is similar to that of potential theory (p. 25) but considerably more complicated. In order to prescribe the *stresses* one assumes a system of external forces distributed over the surface of the body in the form $\mathbf{F} = \mathfrak{f}do$ so that the "surface density" $\mathfrak{f}$ of the external forces is a finite point vector given along the surface. The surface stresses developed by the body must be balanced by the given vector $\mathfrak{f}$ everywhere on the surface. Hence we have for each surface element with normal direction $n$

$$(21a) \qquad \sigma_{nx} + f_x = 0, \qquad \sigma_{ny} + f_y = 0, \qquad \sigma_{nz} + f_z = 0.$$

In the particular case that *no external forces* act on the surface, the boundary conditions become

$$(21b) \qquad\qquad \sigma_{nx} = \sigma_{ny} = \sigma_{nz} = 0.$$

Such is the case for the sides of the bent beam in 41 or for the bar in torsion in 42. When the components of $\sigma$ are expressed by the $\xi$, $\eta$, $\zeta$ in (21a, b), three boundary conditions result for the first derivatives of the $\xi$, $\eta$, $\zeta$. Thus the integration problem is similar to that of the second boundary value problem of potential theory (cf. problem I, 5), but again considerably more difficult.

The uniqueness of solution of the problems mentioned, in particular for a simply connected body, can be proved as in potential theory essentially by an application of Green's theorem (cf. p. 25). In such a proof, homogeneity of the material, hence uniform temperature, and strictly elastic behavior are assumed. In reality, however, internal stresses are present even without external loads: they are caused by imperfections of the casting, inhomogeneous cooling etc. and are a constant source of trouble for the engineer; in applied elasticity, they constitute an unknown element which makes the strict uniqueness proofs of the mathematical theory somewhat unrealistic.

We turn now to the energetic aspect of the stress-strain relation and determine first the *strain-energy* function. The calculation proceeds again in Cartesian coordinates $x$, $y$, $z$ and displacement components $\xi$, $\eta$, $\zeta$.

Consider the two $x$-surfaces of the volume element $\Delta\tau = \Delta x \cdot \Delta y \cdot \Delta z$ while an infinitesimal increment of the displacement

$$\xi, \eta, \zeta \to \xi + d\xi, \qquad \eta + d\eta, \qquad \zeta + d\zeta$$

is imposed on the strain field. The work done by the stresses across the negative and positive $x$-surface is, in the same order,

$$dW_x = -(\sigma_{xx}d\xi + \sigma_{xy}d\eta + \sigma_{xz}d\zeta)\Delta y\Delta z$$

and

$$dW_{x+\Delta x} = +\,(\sigma_{xx}\,d\xi + \sigma_{xy}\,d\eta + \sigma_{xz}\,d\zeta)\Delta y\Delta z$$

$$+ \frac{\partial}{\partial x}\,(\sigma_{xx}\,d\xi + \sigma_{xy}\,d\eta + \sigma_{xz}\,d\zeta)\Delta x\Delta y\Delta z.$$

The algebraic sum of the two expressions is clearly the second line of the second equation. On adding the work done by the external force $\mathbf{F}$ in $x$-direction, one obtains

$$(22) \qquad \left\{\frac{\partial}{\partial x}\,(\sigma_{xx}\,d\xi + \sigma_{xy}\,d\eta + \sigma_{xz}\,d\zeta) + F_x\,d\xi\right\}\Delta x\Delta y\Delta z.$$

Two analogous expressions are obtained for the work done in the displacement of the $y$- and $z$-surfaces. On adding all three contributions and carrying out the indicated differentiations, the terms that contain the (not differentiated) factors $d\xi$, $d\eta$, $d\zeta$ cancel the $F$-terms, if we make use of the equilibrium condition (8.12). The work $dW$ per unit of volume is thus found as

$$(23) \qquad dW = \sigma_{xx}\,d\frac{\partial\xi}{\partial x} + \sigma_{yy}\,d\frac{\partial\eta}{\partial y} + \sigma_{zz}\,d\frac{\partial\zeta}{\partial z} + \sigma_{xy}\,d\!\left(\frac{\partial\xi}{\partial y} + \frac{\partial\eta}{\partial x}\right)$$

$$+ \sigma_{yz}\,d\!\left(\frac{\partial\eta}{\partial z} + \frac{\partial\zeta}{\partial y}\right) + \sigma_{zx}\,d\!\left(\frac{\partial\zeta}{\partial x} + \frac{\partial\xi}{\partial z}\right).$$

In this calculation we have employed the symmetry of the stress tensor and the fact that $\partial\,d\xi$ (the gradient of the incremental displacement) is commutative with $d\,\partial\xi$ (the increment of the gradient of the displacement) (cf. 4, Fig. 5; $d$ here corresponds to $\delta$ there). Eq. (23) can be written in the form

$$(24) \qquad dW = \sigma_{xx}\,d\epsilon_{xx} + \sigma_{yy}\,d\epsilon_{yy} + \sigma_{zz}\,d\epsilon_{zz} + 2\sigma_{xy}\,d\epsilon_{xy}$$

$$+ 2\sigma_{yz}\,d\epsilon_{yz} + 2\sigma_{zx}\,d\epsilon_{zz} = \sum_i \sum_k \sigma_{ik}\,d\epsilon_{ik}\,,$$

when the strain tensor is introduced.

It is possible to express relation (24) in terms of the strain tensor alone, by means of the stress-strain relations. Substitution according to (7) yields the form

$$(25) \qquad dW = 2\mu \sum_i \sum_k \epsilon_{ik}\,d\epsilon_{ik} + \lambda\Theta\,d\Theta.$$

*This expression is a total differential* as long as the "constants" $\lambda$ and $\mu$ remain truly constant in the process of loading. We may then speak of a strain-energy function, without referring to the particular way of loading that starts from the (stress-free) initial state and leads to the final (stressed) state. In other words, the strain-energy is a *function of state*. It reads in terms of the strain components:

$$(26) \qquad W = \mu \sum \sum \epsilon_{ik}^2 + \frac{\lambda}{2} \theta^2.$$

As a function of state, $W$ must be independent of the particular choice of coordinates $x$, $y$, $z$. This can be seen immediately through the following rearrangement:

$$\sum \sum \epsilon_{ik}^2 = (\epsilon_{11} + \epsilon_{22} + \epsilon_{33})^2$$
$$+ 2(\epsilon_{12}^2 + \epsilon_{23}^2 + \epsilon_{31}^2 - \epsilon_{11}\epsilon_{22} - \epsilon_{22}\epsilon_{33} - \epsilon_{33}\epsilon_{11})$$
$$= \theta^2 - 2\Delta,$$

where $\Delta$ is the second scalar of the strain tensor, as explained in Eq. (4.9). Thus, instead of (26), we have the invariant representation

$$(26a) \qquad W = \frac{2\mu + \lambda}{2} \theta^2 - 2\mu\Delta.$$

A more symmetrical expression for $W$ which also makes the invariance evident is

$$(26b) \qquad W = \frac{1}{2} \sum \sum \sigma_{ik}\epsilon_{ik}.$$

The right member in this equation is the simplest *simultaneous invariant of the two tensors* $\sigma$ *and* $\epsilon$. On expressing $\sigma$ by $\epsilon$ or vice versa [Eqs. (7) or (7a)] one obtains the quadratic forms

$$(27) \qquad W(\epsilon) \qquad \text{or} \qquad W(\sigma).$$

The first is, of course, identical with (26), the second can be obtained from (26) by formally replacing $\epsilon$, $\theta$ by $\sigma$, $\Sigma$ and $\mu$, $\lambda$ by $\mu'$, $\lambda'$ from (7b).

The factor $\frac{1}{2}$ that occurs, e.g., in (26b) is in obvious connection with Hooke's law, that is, with the assumed *linear* relation between stress and strain.

From the quadratic forms (27) the stress and strain components can be reobtained; they are the partial derivatives of the energy $W$ with respect to the corresponding strain and stress components:

$$\sigma_{ik} = \frac{\partial W(\epsilon)}{\partial \epsilon_{ik}}, \qquad \epsilon_{ik} = \frac{\partial W(\sigma)}{\partial \sigma_{ik}}.$$

For this reason, the negative strain-energy $W$ is also called the *elastic potential* per unit of volume.

We have already observed that $\mu$ and $\lambda$ are temperature-dependent parameters. If the temperature is kept constant while the load is applied (isothermal transition), then $\mu$ and $\lambda$ are true constants, and the preceding determination of $W$ is correct. If on the other hand the transition is carried out without heat transfer to or from the individual volume elements of the body (adiabatic transition), then the strain energy can again be shown to be a function of state, but the "adiabatic" values of $\mu$ and $\lambda$ are different from the "isothermal" ones. This difference is, as one would expect, especially noticeable in the gaseous state and will be discussed in 13 when the velocity of sound is calculated. In the more general case, as in a body with non-uniform temperature distribution, the strain energy is *not a function of state*, but depends on the path of the transition; the elastic potential does not exist.

## 10. Viscous Pressures and Dissipation, Particularly in Incompressible Fluids

The following analysis is based on an assumption about the nature of the viscous forces which, in the special form to be discussed first, is due to Newton. Let the velocity of the fluid $u(y)$ be everywhere parallel to the $x$-axis, increasing in some way with increasing $y$. The reactions due to viscous friction that are transferred through a surface normal to the $y$-axis are, according to Newton, proportional to the velocity gradient, the factor of proportionality $\mu$ depending on the nature of the fluid. We refer the reactive forces to the unit of area and designate them as frictional or viscous pressures. Newton's assumption then is

$$(1) \qquad p_{yx} = -\mu \frac{\partial u}{\partial y}.$$

The minus sign as well as the notation $p_{yx}$ requires some explanation. In Fig. 13 a surface $SS$ has been indicated which, relative to the lower part of the fluid, is a positive $y$-surface, since the outward normal points in the direction of the positive $y$-axis. The fluid above $SS$, flowing faster, tries to take with it the fluid below $SS$; the upper small arrow in Fig. 13 represents therefore the retarding action exerted on the upper fluid body by the lower one. Either arrow should be designated by $\sigma_{yx}$ if we were to use stresses in our analysis, the upper one representing the $+x$-traction across a $+y$-surface, the lower one the $-x$-traction across a $-y$-surface. We use pressures, however, and this explains the minus sign in Eq. (1);

$p_{yx}$ is negative for the positive $y$-surface (since $\partial u/\partial y$ is positive) and positive for the negative $y$-surface.[19]

Our aim is to generalize Newton's formula for the case of an arbitrary fluid flow. To do this we must go back to the fundamental theorem of kinematics of 1, which deals with the resolution of an infinitesimal displacement into translation, rotation, and deformation. The first two, as we know, correspond to rigid body motions and cannot involve internal friction. The friction is entirely due to the third part that can be de-

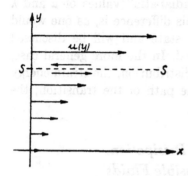

FIG. 13. Transfer of friction pressures in laminar flow. Newton's hypothesis.

scribed in terms of the strain tensor $\epsilon$. Now, in a fluid the time rate of displacement rather than the displacement itself is the physically important quantity (see 5, p. 39), accordingly, we shall not consider $\epsilon$, but $\dot{\epsilon}$ as responsible for the friction. In anticipation of the result of the subsequent discussion we assume now:

$$(2) \qquad p_{ik} = -2\mu\dot{\epsilon}_{ik}, \qquad \left.\begin{matrix} i \\ k \end{matrix}\right\} = x, y, z.$$

The factor of proportionality $\mu$ which has already occurred in (1) characterizes the internal friction and is called *the coefficient of viscosity*. The value of $\mu$ depends on the nature of the fluid and changes very strongly with the temperature (in liquids, increasing temperature makes $\mu$ decrease).

We show first that the set (2) is equivalent to (1) for the special flow of Fig. 13 where

$$u = u(y), \qquad v = 0, \qquad w = 0.$$

Applying (1.11) and the array (1.12) to the deformation velocities $u, v, w$ rather than to the deformations $\xi, \eta, \zeta$, we obtain

[19]Cf. the footnote on p. 39.

$$\begin{pmatrix} 0 & \dfrac{1}{2}\dfrac{\partial u}{\partial y} & 0 \\[2ex] \dfrac{1}{2}\dfrac{\partial u}{\partial y} & 0 & 0 \\[2ex] 0 & 0 & 0 \end{pmatrix}$$

and the content of Eq. (2) reduces to the following equations which are identical with (1):

$$p_{xy} = p_{yx} = -\mu \frac{\partial u}{\partial y} \, ;$$

the other $p_{ik}$ vanish.

Assumption (2) is in agreement with the tensor-analytic principles considered in the preceding article. This is at least true for *incompressible fluids*, and we wish to limit the validity of (2) to that case. Then $\Theta = 0$ by (5.1), and the stress-strain relations (9.7) become identical with (2) as soon as the stresses are replaced by the pressures (change of sign) and the strains by the rates of strain. Hence one would conclude from the uniqueness of the stress-strain relations that a tensor-invariant relation between $p$ and $\dot{\epsilon}$ other than (2) cannot exist. (The phys.cal meaning of the viscosity coefficient $\mu$ has, of course, nothing to do with Lamé's modulus $\mu$ in the last article.)

First, however, the following point must be clarified: in the theory of the elastic solid we were able to show by an equilibrium consideration that $\sigma$ is a *symmetric* tensor. The corresponding argument is questionable in the case of the viscous tensor $p$, since the necessary transition to a sufficiently small volume element is no longer legitimate in the present case, as we shall see below. Suppose then one would admit the possibility that

$$p_{ik} \neq p_{ki} \, .$$

Now, an unsymmetric pressure tensor could always be resolved into a symmetric part $\bar{p}$ and an antisymmetric part $\pi$ such that

$$\bar{p}_{ik} = \bar{p}_{ki} \qquad \text{and} \qquad \pi_{ik} = -\pi_{ki} \, .$$

The latter may be represented by an (axial) vector [cf. Eq. (1.12a)] and can be invariantly connected only with a quantity of the same character. The vortex velocity $\omega$ is such a quantity, and it is the only one that can be used. One would then have to postulate a static-kinematic relation of the form

(3)                                  $\pi_{ik} = -\kappa\omega_{ik} \, .$

Hence to admit the possibility of the asymmetry of the viscous tensor $p$ amounts to the acceptance of a "vortex friction" as in (3), but this concept is inadmissible since it leads to a nonsensical result even in the simplest case.

Suppose a drum filled with some viscous liquid is made to revolve about its axis. The liquid starts to rotate at the circumference of the drum and the motion is gradually communicated to the bulk of the fluid. While this goes on a moment is required to overcome the internal friction of the liquid in addition to the moment required for the acceleration of the moving mass. In the *steady state* which is finally reached, the angular velocity $\omega$ is everywhere the same; at least, one may safely consider this as the final state since no experience points to the contrary. The liquid rotates as a rigid body and surely *no moment* is needed to sustain the motion if bearing friction etc. is disregarded. In contrast to that, Eq. (3) contends the presence of friction also in the steady state. Originating in the fluid, the friction would be conveyed to the drum and would have to be balanced by an external moment.

This whole discussion is of a *phenomenological nature* in accordance with the general point of view taken in this volume, and the same is true for the preceding tensor-geometric considerations. The deformable medium is being considered as a *continuum* and no attention given to the *molecular* origin of the stresses or pressures. The molecular theory of *compressible* fluids will be dealt with in the kinetic theory of gases. It furnishes a *direct explanation of the frictional pressures* in terms of the transport of momentum by thermal motion or, as it is sometimes put, in terms of the diffusion of momentum. Such a theory is physically altogether more profound; for compressible fluids it leads *of necessity* to the relation (21) between pressure and rate of deformation,[20] which will be found at the end of this article. That more general relation comprises (2) as a special case if the fluid is incompressible ($\Theta = \dot{\Theta} = 0$). On the other hand, the gas-kinetic concepts do not directly apply to *incompressible* fluids, that is, not unless arbitrary additional assumptions are made. There is, however, no doubt that also in the incompressible case the single volume element is exchanging momentum with more distant volume elements, and that this constitutes the physical cause of fluid friction. But it is just this idea of interaction by exchange of molecules that is in contradiction with the transition to an *isolated* small fluid element as in 8.

In addition to the frictional pressures and superimposed on them, there is a uniform normal pressure of the same nature as the pressure in the

---

[20] Cf. in particular the critical remarks of E. Fues, Z. f. Physik 118, 409 (1941) and 121, 58 (1943) which start from ideas of Maxwell and Boltzmann, like most investigations in that field.

fluid at rest when the external force $\mathbf{F}$ is present. We denote it by $p$ as before (no subscripts, since it is a scalar), although it is not identical with the hydrostatic $p$ (see below). The total pressure tensor is then

(4)
$$P = \begin{pmatrix} p + p_{xx}, & p_{xy}, & p_{xz} \\ p_{yx}, & p + p_{yy}, & p_{yz} \\ p_{zx}, & p_{zy}, & p + p_{zz} \end{pmatrix}$$

We are now ready to state the relation that is the analogue of the equilibrium condition (8.12). Although it is true that we are not concerned with an elastic body at *rest*, but with a *flow* of fluid particles, we may first assume that (in the frame of reference under consideration) the motion proceeds without change of magnitude and direction of the velocity. Each particle is then in a state of mechanical equilibrium, and we can formulate an "equilibrium condition" also in the present case. It is modelled after (8.12) and reads (with the proper change of sign)

(5)
$$\mathrm{Div}\, P = \mathbf{F}.$$

Here Div stands for the "vector divergence", explained in (8.13) in an invariant way. Written in Cartesian coordinates, (5) takes the form

(6)
$$\frac{\partial p}{\partial x} + \frac{\partial p_{xx}}{\partial x} + \frac{\partial p_{yx}}{\partial y} + \frac{\partial p_{zx}}{\partial z} = F_x,$$
$$\frac{\partial p}{\partial y} + \frac{\partial p_{xy}}{\partial x} + \frac{\partial p_{yy}}{\partial y} + \frac{\partial p_{zy}}{\partial z} = F_y,$$
$$\frac{\partial p}{\partial z} + \frac{\partial p_{xz}}{\partial x} + \frac{\partial p_{yz}}{\partial y} + \frac{\partial p_{zz}}{\partial z} = F_z.$$

These equations determine a steady fluid motion as soon as we express the six unknowns $p_{ik}$ through the three velocity components $u, v, w$ by means of (2). We obtain first

(7)
$$\frac{\partial p}{\partial x} - 2\mu \frac{\partial^2 u}{\partial x^2} - \mu\left(\frac{\partial^2 v}{\partial x\, \partial y} + \frac{\partial^2 u}{\partial y^2}\right) - \mu\left(\frac{\partial^2 w}{\partial x\, \partial z} + \frac{\partial^2 u}{\partial z^2}\right) = F_x,$$

which reduces on account of the condition of incompressibility

(8)
$$\frac{\partial u}{\partial x} + \frac{\partial v}{\partial y} + \frac{\partial w}{\partial z} = 0$$

to

(8a)
$$\frac{\partial p}{\partial x} - \mu \nabla^2 u = F_x.$$

The corresponding equations for the $y$- and $z$-directions are

$$\frac{\partial p}{\partial y} - \mu \nabla^2 v = F_y,$$

(8b)

$$\frac{\partial p}{\partial z} - \mu \nabla^2 w = F_z.$$

Eqs. (8), (8a), (8b) form a system of *four* equations for the four unknowns $u$, $v$, $w$, $p$. The pressure $p$ is, of course, not equal to the hydrostatic pressure value, but is to be found from the quadruplet of equations simultaneously with the unknowns $u$, $v$, $w$.

The same Eqs. (8)-(8b) can also be considered as approximate equations for a flow where the particle accelerations do not exactly vanish, but are so small that their omission appears to be justifiable. This type of fluid motion—"creeping" one might call it—will be studied in 35, 36.

In both cases (either exactly or approximately free of acceleration), boundary conditions must be added to the differential equations to specify a problem, but there is a fundamental difference between the boundary conditions to be imposed on a perfect and on a viscous fluid. While in the first case only the boundary condition $v_n = 0$ is to be fulfilled and arbitrary tangential velocities $v_t$ may be admitted along the rigid walls, we must stipulate in the second case that the fluid adheres to the surface of the rigid boundaries. This implies, of course, that $v_t$ approaches zero *continuously* in the neighborhood of the boundary, since a jump of $v_t$ would mean an infinitely large gradient of $v_t$ and, consequently, *infinitely large values of the viscous pressures* $p_{nt}$. Experiments on the flow in narrow tubes (see below) have justified this form of boundary conditions which we state once more:

(9)                    $v_n = 0, \qquad v_t = 0.$

The fundamental difference between perfect and viscous fluids makes it understandable that the transition from small friction to the limit of vanishing friction is not without analytical difficulties. We shall come back to this point in 33 (Prandtl's boundary layer).

We now discuss the simplest application of the equilibrium conditions (8) and boundary conditions (9), viz., the classical Hagen-Poiseuille flow[21] in a capillary tube. Let the horizontal tube have circular cross section with a radius $a$, so small as to make the flow proceed in straight stream

---

[21] G. Hagen, Poggendorffs Ann. **46** (1839). J. Poiseuille, Comptes Rendus 11 (1840) **12** (1841). Hagen was superintendent of works in Berlin, Poiseuille a physician in Paris. Both found independently the law (15), essentially by experimental investigations.

lines parallel to the axis of the tube (for more details about laminar flow and the conditions of its stability, cf. 16). When the axis of the tube is taken as $x$-axis and the distance $r = \sqrt{y^2 + z^2}$ from the $x$-axis is introduced, the velocity distribution is

(10) $$v = w = 0, \qquad u = u(r).$$

Eq. (8) shows then that $u$ is independent of $x$; because of the cylindrical symmetry, $u$ is a function of $r$ alone. Gravity may be neglected in our case, hence the force $\mathbf{F}$ is zero. Eq. (8b) gives now

(11) $$\frac{\partial p}{\partial y} = \frac{\partial p}{\partial z} = 0, \qquad \text{therefore} \qquad p = p(x).$$

Since the first term of (8a) depends only on $x$, the second only on $r$, and their difference is zero, either term must be a constant, say $-A$. Thus we have

(11a) $$\frac{dp}{dx} = -A, \qquad \nabla^2 u = -\frac{A}{\mu}.$$

According to this, the constant $A$ is the pressure gradient along the tube, or, if multiplied with the length of the tube $l$, the pressure difference $\Pi$ between beginning and end of the tube. We substitute for $\nabla^2 u$ its value in cylinder coordinates (problem I, 3) and note that there is no dependence on $\varphi$ and on the axial coordinate:

$$\nabla^2 u = \frac{1}{r}\frac{d}{dr} r \frac{du}{dr}.$$

The conclusions following from (11a) are then

$$\frac{d}{dr} r \frac{du}{dr} = -\frac{A}{\mu} r, \qquad r\frac{du}{dr} = -\frac{A}{\mu}\frac{r^2}{2} + C_1.$$

Here $C_1$ must vanish since otherwise $du/dr$ would become infinite for $r = 0$. A second integration yields

(12) $$u = -\frac{A}{4\mu} r^2 + C_2.$$

The constant $C_2$ is to be determined from the boundary condition (9) which now reads

(13) $$u = 0 \qquad \text{for} \qquad r = a.$$

*The liquid sticks to the wall of the tube.* There is no doubt that (13) is correct: it follows unequivocally from the observations and holds for water as well as for non-wetting liquids such as mercury on glass. According to (12),

$$C_2 = \frac{A}{4\mu} a^2,$$

so that (12) takes the form

(14)
$$u = \frac{A}{4\mu} (a^2 - r^2).$$

*The velocity profile is parabolic* (cf. Fig. 14). The particles that are at the time $t = 0$ on a plane cross-section, lie at $t > 0$ on a paraboloid of rotation

FIG. 14. Laminar flow in a capillary tube; the liquid adheres to the wall, the velocity profile is parabolic.

the vertex of which travels in the direction of the flow while the points of contact with the wall remain fixed.

The volume discharge $Q$ per unit of time is

(15)
$$Q = 2\pi \int_0^a ur \, dr = \frac{\pi}{8} \frac{A}{\mu} a^4.$$

This formula has served as the basis for most determinations of viscosity constants in the last hundred years. The velocity of flow averaged over the cross section

(16)
$$u_m = \frac{Q}{\pi a^2} = \frac{A}{8\mu} a^2$$

is half the maximum velocity $u_{\max}$, [Eq. (14) with $r = 0$]; the pressure loss follows from (15) and (16) as

(17)
$$\Pi = 8\mu l \frac{u_m}{a^2}.$$

*The pressure loss due to friction is proportional to the mean flow velocity and inversely proportional to the square of the radius.* This law which, in modified form, is also correct for other laminar flows (cf. problems II, 3 and 4 and Figs. 19 and 19a on p. 121) is in characteristic contrast to the law for *turbulent* flow (see 16) where *the pressure loss is approximately proportional to the square of the average velocity and inversely proportional to the radius of the tube.*

We turn to the discussion of the *energetic* aspect of viscous flow. To calculate the *work* done by the *friction forces* we can use Eq. (9.22) as a starting point. Replacing the stresses $\sigma$ by the pressures $p$ (change of sign)

and the *virtual* displacements $d\xi$, $d\eta$, $d\zeta$, by the quantities $u\,dt$, $v\,dt$, $w\,dt$, which now are *actual motions occurring in the time element dt*, we obtain in place of (9.22)

$$\left\{ -\frac{\partial}{\partial x}\left[ (p + p_{xx})u + p_{xy}v + p_{xz}w \right] + F_x u \right\} dt\, \Delta x \Delta y \Delta z.$$

This is the work done by the uniform pressure $p$ and the friction pressures $p_{ik}$ on the $x$-faces augmented by the work of the $x$-component of the external force. On adding the corresponding expressions in $y$ and $z$ and carrying out the differentiations, the resulting expression can be reduced by applying the conditions of equilibrium and incompressibility (6) and (8), and becomes after division by $dt\Delta x \Delta y \Delta z$

(18)
$$- p_{xx}\frac{\partial u}{\partial x} - p_{yy}\frac{\partial v}{\partial y} - p_{zz}\frac{\partial w}{\partial z}$$

$$- p_{xy}\left(\frac{\partial u}{\partial y} + \frac{\partial v}{\partial x}\right) - p_{yz}\left(\frac{\partial v}{\partial z} + \frac{\partial w}{\partial y}\right) - p_{zx}\left(\frac{\partial w}{\partial x} + \frac{\partial u}{\partial z}\right).$$

This energy is, as it were, abundant, it has no mechanical equivalent since no energy is needed for the maintenance of a steady fluid motion; it must therefore appear in the form of heat. *Expression* (18) *thus represents the heat produced by friction per unit of time and of volume (dissipation).*

Introducing a notation similar to the thermodynamic usage, we designate by $dq$ the heat transferred to the unit of volume during the time $dt$. Using the definition of the strain components $\epsilon_{ik}$ in (1.11), we can transform (18) into

(19)
$$\frac{dq}{dt} = -\sum_i \sum_k p_{ik}\dot{\epsilon}_{ik}.$$

On applying the relations (2) between the $p_{ik}$ and the $\dot{\epsilon}_{ik}$, we obtain for the dissipation

(20)
$$\frac{dq}{dt} = 2\mu \sum_i \sum_k \dot{\epsilon}_{ik}^2.$$

The quantity $dq$ defined by (20) is, of course, *not* a total differential like the elastic $dW$ of the corresponding equation (9.25), whence there is no thermal variable of state $\int dq$ that would correspond to the elastic strain energy (9.26).

In conclusion we wish to point out very briefly the changes which the foregoing considerations require in the case of *compressible* fluids. In the first place, comparison with the stress-strain relations in the form of (9.7) shows that the assumption (2) should be replaced by

(21)
$$p_{ik} = -2\mu\dot{\epsilon}_{ik} - \lambda\delta_{ik}\dot{\Theta}$$

because of $\dot{\Theta} = 0$. Thus we have a second coefficient of viscosity,[22] $\lambda$, which is associated with the rate of dilatation $\dot{\Theta}$.

The equilibrium conditions (5) and (6) remain valid, if only $\mathbf{F}$ is replaced by $\rho\mathbf{P}$ as in (7.3), but the form (8a) of these conditions must now be changed to

$$(22) \qquad \frac{\partial p}{\partial x} - \mu\nabla^2 u - (\mu + \lambda)\frac{\partial}{\partial x}\dot{\Theta} = \rho P_x .$$

This equation together with the two others that are obtained by rotating the letters, and the equation of continuity (5.4) determine the four unknowns $u$, $v$, $w$, and $p$ provided that a $p,\rho$-relation is prescribed. That the present set of equations is more involved is seen even in the case of the Hagen-Poiseuille problem for gases; one does not obtain an equation for $u$ as above, but two simultaneous equations for $u$ and $\rho$. As for the dissipation (20), it changes, as one would expect, to

$$(23) \qquad \frac{dq}{dt} = 2\mu \sum \sum \dot{\epsilon}_{ik}^2 + \lambda\dot{\Theta}^2 .$$

---

[22]No further determination of this quantity is intended here. In the literature one finds usually $\lambda = -2\mu/3$ on the basis of a gas-kinetic argument valid for monatomic gases (Enskog, Uppsala thesis, 1917). It is not difficult to draw the general conclusion from (21) that in the case of uniform compression an isotropic frictional pressure of the magnitude $(2\mu/3 + \lambda)\cdot\dot{\theta}$ arises. The number $2\mu/3 + \lambda$ can therefore be designated as volume viscosity, in distinction from $\mu$ which could be called laminar viscosity. According to Enskog the volume viscosity of monatomic gases is zero, but the generalization of this result to gases whose molecules contain more than one atom is not justified.

## CHAPTER III

# DYNAMICS OF DEFORMABLE BODIES

## *11. Euler's Equations for a Perfect Incompressible Fluid*

The transition from statics to dynamics is accomplished by adding the inertial resistances to the external forces in accordance with d'Alembert's principle. The inertial resistance of a particle with mass $\Delta m$ is

$$- \Delta m \frac{d\mathbf{v}}{dt} = -\rho \,\Delta\tau \frac{d\mathbf{v}}{dt} .$$

Hence the inertial resistance per unit of volume becomes

(1) $$-\rho \frac{d\mathbf{v}}{dt} .$$

This quantity must be added to the external force per unit of volume $\mathbf{F}$ in the equilibrium conditions (6.4) if one wishes to obtain the equations of motion. Transposing the inertia term (1) to the other side of the equation, we have

(2) $$\rho \frac{d\mathbf{v}}{dt} + \operatorname{grad} p = \mathbf{F}.$$

Here one must carefully distinguish between the *total* or *material* acceleration $d\mathbf{v}/dt$ and the *local* acceleration $\partial\mathbf{v}/\partial t$. Take for instance the $x$-component of the velocity $u(x, y, z, t)$. While the particle moves through $d\mathbf{s} = \mathbf{i}dx + \mathbf{j}dy + \mathbf{k}dz$ in the time $dt$, the change of $u$ is

$$du = \frac{\partial u}{\partial t} dt + \frac{\partial u}{\partial x} dx + \frac{\partial u}{\partial y} dy + \frac{\partial u}{\partial z} dz,$$

where the particle coordinates $x, y, z$ are to be considered as functions of $t$. Accordingly, the material acceleration in $x$-direction along the path $dx = udt, dy = vdt, dz = wdt$ becomes

(3) $$\frac{du}{dt} = \frac{\partial u}{\partial t} + u \frac{\partial u}{\partial x} + v \frac{\partial u}{\partial y} + w \frac{\partial u}{\partial z} ,$$

and the difference between material and local acceleration in $x$-direction is

(3a) $$\frac{du}{dt} - \frac{\partial u}{\partial t} = u \frac{\partial u}{\partial x} + v \frac{\partial u}{\partial y} + w \frac{\partial u}{\partial z} ,$$

the meaning of the local acceleration being the rate of change of the velocity at a *specified point* of the flow. To illustrate the difference by an example, take the steady flow through a pipe of varying cross-section. Let the axis of the pipe coincide with the $x$-axis; then we are mainly concerned with the velocity component $u$. By hypothesis the flow is steady: hence $\partial u/\partial t = 0$ everywhere, but by no means $du/dt$. On the contrary, the velocity of flow increases where the pipe becomes narrower and decreases where it widens. The difference between the two accelerations is given by the right member of (3a), in particular, by the first term because of the assumed preponderance of $u$. The terms on the right side of (3a) are often called the *convective terms of the acceleration*.

Writing (2) in Cartesian components by means of (3) and the corresponding expressions obtained by rotation of letters, one finds

$$\rho\left(\frac{\partial u}{\partial t} + u\frac{\partial u}{\partial x} + v\frac{\partial u}{\partial y} + w\frac{\partial u}{\partial z}\right) + \frac{\partial p}{\partial x} = F_x ,$$

(4)
$$\rho\left(\frac{\partial v}{\partial t} + u\frac{\partial v}{\partial x} + v\frac{\partial v}{\partial y} + w\frac{\partial v}{\partial z}\right) + \frac{\partial p}{\partial y} = F_y ,$$

$$\rho\left(\frac{\partial w}{\partial t} + u\frac{\partial w}{\partial x} + v\frac{\partial w}{\partial y} + w\frac{\partial w}{\partial z}\right) + \frac{\partial p}{\partial z} = F_z .$$

The fourth differential equation is the condition of incompressibility

(4a)
$$\frac{\partial u}{\partial x} + \frac{\partial v}{\partial y} + \frac{\partial w}{\partial z} = 0.$$

The quadruplet of equations (4), (4a) constitute *Euler's equations*[1] *of perfect fluid motion*. The pressure $p$ is not to be confused with the hydrostatic pressure; it appears as the fourth unknown beside $u$, $v$, $w$.

It suggests itself to abbreviate Euler's equations by writing them in the symbolic form

(5)
$$\rho\left(\frac{\partial}{\partial t} + (\mathbf{v}\,\mathrm{grad})\right)\mathbf{v} + \mathrm{grad}\,p = \mathbf{F},$$

where the symbol $(\mathbf{v}\,\mathrm{grad})$ could be replaced by $(\mathbf{v}\nabla)$ as on p. 23. But this form is misleading as soon as one tries to use it, as it stands, for non-Cartesian coordinates (see problem III, 1). The operation grad applies only to scalars and must not operate on vectorial quantities. We may,

[1]Leonhard Euler (1707-1783). His first two papers on the equilibrium and motion of fluids appeared 1755 in Vol. 11 of the Berlin Academy, a later treatment 1770 in Vol. 14 of the Petrograd Academy.

however, define the *pseudo-vectorial symbol* $(\mathbf{v}\,\text{grad})\,\mathbf{v}$ by explaining it through a *legitimate vector formula*. The definition we have in mind reads

$$(6) \qquad (\mathbf{v}\,\text{grad})\mathbf{v} = \text{grad}\,\frac{\mathbf{v}^2}{2} - \mathbf{v} \times \text{curl}\,\mathbf{v}.$$

In fact, the $x$-component of the first member of (6) can be rearranged in the form

$$u\,\frac{\partial u}{\partial x} + v\,\frac{\partial u}{\partial y} + w\,\frac{\partial u}{\partial z}$$

$$= \frac{1}{2}\frac{\partial}{\partial x}\,(u^2 + v^2 + w^2) + v\!\left(\frac{\partial u}{\partial y} - \frac{\partial v}{\partial x}\right) + w\!\left(\frac{\partial u}{\partial z} - \frac{\partial w}{\partial x}\right)$$

which establishes right away the identity with the $x$-component of the second member of (6). The corresponding relations for the $y$- and $z$-components follow by rotating the letters.

Now we may replace Eq. (5) by

$$(7) \qquad \rho\!\left(\frac{\partial \mathbf{v}}{\partial t} - \mathbf{v} \times \text{curl}\,\mathbf{v}\right) + \text{grad}\left(\rho\,\frac{\mathbf{v}^2}{2} + p\right) = \mathbf{F}.$$

Adding the condition of incompressibility

$$(7a) \qquad\qquad \text{div}\,\mathbf{v} = 0,$$

we have obtained Euler's equations in invariant form that can be specialized for any sort of curvilinear coordinates (e.g. polar coordinates) by the use of the expressions for grad, div, and curl in (2.24)-(2.26), or of the ready formulas in the tabulated solutions of problem I, 3.

Considering the mathematical character of Euler's equations we notice immediately their *non-linearity* which distinguishes them from the many linear equations of mathematical physics, as e.g. in potential theory, heat conduction, electrodynamics, etc. The non-linearity which is caused by the presence of the convective terms

$$u\,\frac{\partial u}{\partial x} + v\,\frac{\partial u}{\partial y} + w\,\frac{\partial u}{\partial z} \qquad \text{etc., or} \qquad \mathbf{v} \times \text{curl}\,\mathbf{v} + \text{grad}\,\frac{\mathbf{v}^2}{2}$$

makes the integration incomparably more difficult, for we can no longer use the *principle of superposition* of solutions by which more general integrals are found in the form of a combination of particular integrals. The integration of Euler's hydrodynamic equations is thus a considerably more difficult mathematical problem than, for example, that of the seemingly more complicated equations of Maxwell in electromagnetic theory.

Only in the case of *irrotational* flow is it possible to give immediately a first integral of Euler's equations. The condition is

(8) $$\boldsymbol{\omega} = \frac{1}{2} \operatorname{curl} \mathbf{v} = 0;$$

it is sufficient that Eq. (8) be fulfilled at a certain time instant only, since in an inviscid fluid it then remains permanently fulfilled as will be shown in Chap. IV.

We shall at first consider a still more restricted case, viz. steady flow, so that

(8a) $$\frac{\partial \mathbf{v}}{\partial t} = 0.$$

It is further assumed as in (6.5) that the external force $\mathbf{F}$ has a potential

(8b) $$\mathbf{F} = - \operatorname{grad} U.$$

Then (7) becomes

$$\operatorname{grad} \left( \rho \frac{v^2}{2} + p + U \right) = 0$$

and can be integrated to

(9) $$\rho \frac{v^2}{2} + p + U = \text{const}$$

which is Bernoulli's famous equation.[2] It was found *before* the discovery of Euler's equations by an argument which could be considered as an anticipation of the energy principle. For Bernoulli, no other potential energy than that of gravity would count; with $U = \rho g z$ ($z$ pos. upwards), Eq. (9) would read

(9a) $$\frac{v^2}{2} + \frac{p}{\rho} + gz = \text{const}.$$

Bernoulli's equation is the most important theorem in elementary fluid dynamics and finds application in the solution of numerous technical problems of turbine design, aerodynamics, etc. The first term of Eq. (9) is the kinetic energy per unit of volume, the third term the potential energy of the external force, again per unit of volume. The pressure $p$ in (9) appears, as it were, as the potential energy of internal forces that are active in the unit volume cell and account for the dynamic interaction

---

[2]Daniel Bernoulli, Hydrodynamica, Strasbourg 1738. This work contains also the first attempt at a molecular theory of gases and the first treatment of the bending of a beam (cf. 41).

between neighboring incompressible fluid elements. A deeper mathematical interpretation of $p$ will be the object of the next article.

Let us now illustrate the physical content of (9a) by some simple examples. Imagine a tank filled with liquid having an orifice at the level of the bottom. Let the free surface be $z = 0$ and the bottom $z = -h$. As long as the orifice is closed, $v = 0$ everywhere in the tank and $p = 0$ at the surface (where $p$ denotes the overpressure relative to the atmosphere). The constant in (9a) is then equal to zero, the pressure $p = \rho g h = \gamma h$ at the bottom and is identical with the hydrostatic pressure in 6. When the orifice is opened we may assume that, at the surface, $v$ remains approximately zero so that const $= 0$ as before. But since $p$ is now zero also at the orifice, Eq. (9a) gives for $z = -h$

$$(10) \qquad v = \sqrt{2gh}$$

as in the case of a freely falling body. This is the well known content of *Torricelli's* theorem.[3]

Another example is the steady flow through a horizontal tube of variable cross-section. Since the volume flux through each cross section is the same for an incompressible fluid, $v$ increases with decreasing cross section and vice versa, as already observed on p. 84. (The present $v$ is essentially the same as the axial component $u$ there). According to (9a) the pressure must exhibit the converse change; this is easily demonstrated if the tube is furnished with a series of vertical open manometers in which the liquid is allowed to rise. The levels in the manometer tubes indicate the variations of the pressure along the tube, showing smaller pressure when the cross-section is smaller. The comparison with the behavior of a crowd of people trying to force their way through a narrow passage is inevitable: the way the fluid does it is more rational.

The following little experiment which can be improvised at any time may also serve as an illustration of Bernoulli's equation. Put a piece of paper (say $2 \times 2$ inches) on your left hand and hold the second and third finger of your right hand closely above it. Blow now vigorously through the narrow slit formed by the two fingers against the center of the paper. Contrary to expectation, the paper is not pressed against the left hand but lifted toward the fingers of the right hand where it remains floating for some time. Explanation: the air that emerges between the fingers and the paper flows in a channel of increasing cross section (cf. Fig. 15). Its velocity decreases, hence the pressure increases. At the end of the channel there is atmospheric pressure, $p_0$ , hence the pressure in

---

[3] Torricelli was a pupil of Galilei, he lived before Bernoulli's time and could not have used Bernoulli's equation in obtaining this result.

the channel $p < p_0$. Since underneath the paper there is atmospheric pressure $p_0$, an overpressure $p_0 - p$ acts from below, or as one might also put it, there is suction from above. When this experiment is carried out on a larger scale with a jet of compressed air directed against a plate that is constrained to move normal to the jet, then the plate can be heavily loaded, e.g., with one's own weight.

In this experiment and in the preceding one we have applied Bernoulli's equation to air, although the formulation given in (9) is only correct for incompressible fluids. Yet the error caused in this way is not very great as long as the velocity of flow is not too large. In general, the

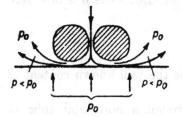

FIG. 15. Suction produced by blowing against a piece of paper through the slit formed by two fingers. The paper is not blown away but lifted toward the fingers.

corrections due to compressibility are small when the velocity of flow is small compared to the velocity of sound (cf. 13, and Appendix).

We have still to set up the generalized form of Bernoulli's equation in the case of a non-steady flow. Accordingly, we drop assumption (8a), but maintain (8) (irrotationality) and (8b) (**F** has a potential). Now (8) is the necessary and sufficient condition for the expression $udx + vdy + wdz$ to be a total differential, hence we may put

(11) $$udx + vdy + wdz = -d\Phi$$

and call $\Phi$ the *velocity potential*. Then

$$u = -\frac{\partial\Phi}{\partial x}, \qquad v = -\frac{\partial\Phi}{\partial y}, \qquad w = -\frac{\partial\Phi}{\partial z},$$

or, more briefly,

(12) $$\mathbf{v} = -\operatorname{grad}\Phi \qquad \text{and therefore also} \qquad \frac{\partial\mathbf{v}}{\partial t} = -\operatorname{grad}\frac{\partial\Phi}{\partial t}.$$

It will be noticed that the negative sign in (11) and (12) is unessential;[4] it is put in only to maintain the analogy with the potential of a force as in (8b).

Because of div $\mathbf{v} = 0$, $\Phi$ satisfies the potential equation (3.17)

(13) $$\nabla^2\Phi = 0.$$

---

[4] Its use has, in fact, been discontinued by many authors.

On account of (8), (8b), and (12), Euler's equations (7) take now the form

$$(14) \qquad \text{grad}\left(-\rho\,\frac{\partial\Phi}{\partial t} + \rho\,\frac{v^2}{2} + p + U\right) = 0,$$

which is integrated to

$$(15) \qquad -\frac{\partial\Phi}{\partial t} + \frac{1}{2}\,D\Phi + \frac{1}{\rho}\,(p + U) = \text{const.}$$

Here we have used the notation (3.9c) for the first differential parameter.

Eq. (15) is the *generalization of Bernoulli's equation for non-steady motions*. Like (9), it is a first integral of Euler's equations. Since (15) is found by spatial integration of (14), the constant in (15) is certainly independent of $x$, $y$, $z$, but it may in general still depend on the time. In other words, *const* is a function of $t$ that has a uniform value for all points of the fluid. It must be found from the boundary conditions which may very well change with time. It will be noticed that this time dependence can be included in the definition of the velocity potential $\Phi$ since the potential equation (13) determines only the spatial, but not the temporal behavior of $\Phi$.

So far we have only considered the case of *irrotational motion*; the question remains what can be done in the more general case where curl **v** *is not zero*.

A glance at equation (7) shows that also in this case integration becomes possible if one integrates in the *direction of a stream line*, or, in the non-steady case along the field line of the vector field **v**. When this is done, the integral over the second term of the left member of (7) vanishes since **v** × curl **v** is perpendicular to **v** and does not contribute anything to the integration. If we, moreover, maintain conditions (8a) and (8b) (steady flow in a field of force that has a potential) Eq. (9) is still correct, but in a different sense: the constant on the right side has no longer a uniform value for the entire space filled by the fluid, but changes from *one streamline to the next*. We shall call the equation (9), if understood in this sense, the *modified Bernoulli equation*. Its generalization for non-steady motion will not be discussed here.

## 12. Derivation of Euler's Equations from Hamilton's Principle The Pressure, a Lagrange Multiplier

There is no doubt that the concept of pressure in an *incompressible* fluid presents certain difficulties to the physical understanding. The pressure is considered a variable of state, but is denied any influence upon the way in which the fluid occupies the space, a difficulty, which is not present in a compressible fluid where the pressure determines the density. We shall therefore try another approach toward understanding the concept of pressure in the case of an incompressible fluid, using the ideas of Hamiltonian mechanics.

We subject the mass particles of our fluid in motion to a *virtual* displacement

$$\delta \mathbf{s} = \delta \xi, \; \delta \eta, \; \delta \zeta,$$

which must not violate, however, the condition of incompressibility, $\theta = 0$ (1.24). The expression

(1)      $$\delta \Theta = \frac{\partial \delta \xi}{\partial x} + \frac{\partial \delta \eta}{\partial y} + \frac{\partial \delta \zeta}{\partial z} = \text{div } \delta \mathbf{s}$$

must therefore vanish. This can be taken care of by providing (1) with a Lagrange multiplier $\lambda$ and adding it to the integrand of Hamilton's principle. This we write in the form of Eq. (33.10) of Vol. I and obtain[5]

(2)      $$\int_{t_0}^{t_1} dt \int d\tau (\delta T + \delta W + \lambda \delta \Theta) = 0.$$

The integration with respect to $d\tau$ refers to the entire volume of the fluid. The variation of the kinetic energy $\delta T$ and the virtual work of the external forces $\delta W$ refer to the unit of volume in the same way as $\delta \Theta$ in (1). Hence we have

(3)      $$T = \frac{\rho}{2} \mathbf{v}^2, \qquad \delta T = \rho \mathbf{v} \cdot \delta \mathbf{v}$$

and obtain for the virtual work

(4)      $$\delta W = \mathbf{F} \cdot \delta \mathbf{s}.$$

In (3) we substitute for the actual velocity

(4a)      $$\mathbf{v} = \frac{d\mathbf{s}}{dt}$$

and, accordingly, for its virtual variation

(4b)      $$\delta \mathbf{v} = \delta \frac{d\mathbf{s}}{dt} = \frac{d}{dt} \delta \mathbf{s}.$$

We now transform the term $\delta T$ by partial integration with respect to $t$, observing that $\delta \mathbf{s}$ vanishes at the limits $t_0$ and $t_1$ according to the operational rules of Hamilton's principle. Thus we obtain

_____

[5] See Jeffreys and Jeffreys, *op. cit.*, 10.06.

(5)     $$\int_{t_0}^{t_1} \rho \mathbf{v} \cdot \delta \mathbf{v} \, dt = \int_{t_0}^{t_1} \rho \mathbf{v} \cdot \frac{d}{dt} \delta \mathbf{s} \, dt = -\int_{t_0}^{t_1} \rho \frac{d\mathbf{v}}{dt} \cdot \delta \mathbf{s} \, dt.$$

The last term in the integrand of (2) is transformed according to the relation

$$\lambda \, \mathrm{div} \, (\delta \mathbf{s}) = \mathrm{div} \, (\lambda \delta \mathbf{s}) - \mathrm{grad} \, \lambda \cdot \delta \mathbf{s}.$$

On integrating and using Gauss's theorem we obtain

(6)     $$\int \lambda \, \mathrm{div} \, (\delta \mathbf{s}) \, d\tau = \int \lambda \, \delta s_n \, d\sigma - \int \mathrm{grad} \, \lambda \cdot \delta \mathbf{s} \, d\tau.$$

The surface integral on the right side will be taken up eventually; right now we are only concerned with

(7)     $$\int \lambda \, \mathrm{div} \, (\delta \mathbf{s}) \, d\tau = \cdots - \int \mathrm{grad} \, \lambda \cdot \delta \mathbf{s} \, d\tau.$$

Introducing (1), (4), (5) and (7) in Hamilton's integral (2), we have

(8)     $$\int_{t_0}^{t_1} dt \int d\tau \left[ \left( -\rho \frac{d\mathbf{v}}{dt} + \mathbf{F} - \mathrm{grad} \, \lambda \right) \cdot \delta \mathbf{s} \right] + \cdots = 0.$$

Thanks to the introduction of the multiplier $\lambda$ the variation $\delta \mathbf{s}$ may be chosen arbitrarily within the volume element $\tau$ and the time interval $t_0$ to $t_1$ ; but, with arbitrary displacements, Eq. (8) can only be fulfilled, if the volume and the surface integral vanish separately. Thus the factor of $\delta \mathbf{s}$ in the scalar product must be zero everywhere inside the domain of integration:

(9)     $$\rho \frac{d\mathbf{v}}{dt} + \mathrm{grad} \, \lambda = \mathbf{F}.$$

*This is Euler's equation written in the form* (11.2) *provided* $\lambda$ *is identified with the pressure* $p$; hence from the point of view of general mechanics, the hydrodynamic pressure represents the *reaction against the condition of incompressibility*, since $\lambda$ was introduced to eliminate that condition. This corresponds to the way in which $\lambda$ was interpreted in the case of the spherical pendulum in Vol. I. Eq. (18.7) where the constraint was a rigid spherical surface on which the mass point had to remain. The multiplier $\lambda$ turned out to be a measure of the force that keeps the mass point on the constraint.[6] In the present investigation, the condition of incompressibility has also been shown to have the character of a rigid constraint without energetic consequence; this may help to remove the difficulty associated with the concept of pressure in an incompressible fluid.

---

[6] See e.g., Joos, G., Theoretical Physics, Hafner, New York, 1934, p. 110.

We should, however, be able to base our argument on the usual form of the incompressibility condition and its variation

(10)          $\text{div } \mathbf{v} = 0. \qquad \delta \text{ div } \mathbf{v} = \text{div } \delta\mathbf{v} = 0$

rather than on condition (1). Retaining the expressions (3) and (4) for $\delta T$ and $\delta W$, we rewrite Hamilton's principle (2) in the form

(11)          $\int_{t_0}^{t_1} dt \int d\tau \{\rho\mathbf{v}\cdot\delta\mathbf{v} + \mathbf{F}\cdot\delta\mathbf{s} + \lambda' \text{ div } \delta\mathbf{v}\} = 0.$

We have denoted the multiplier by $\lambda'$ to indicate that it is not identical with our previous $\lambda$ (if for no other, then for dimensional reasons). We must now transform the middle term in (11) so as to have the same independent variation $\delta\mathbf{v}$ everywhere. We put

(12)          $$\overline{\mathbf{F}} = \int \mathbf{F} \, dt$$

and obtain

$$\int_{t_0}^{t_1} dt \, \mathbf{F}\cdot\delta\mathbf{s} = - \int_{t_0}^{t_1} dt \, \overline{\mathbf{F}}\cdot\delta\mathbf{v};$$

in the last transformation Eq. (4b) was used and the fact applied that the terms, which occur in the partial integration and refer to the limits $t_0$ and $t_1$ , vanish as before in (5). When we also transform the term with the factor $\lambda'$ as in (7), the volume integral in (11) becomes

$$\int_{t_0}^{t_1} dt \int d\tau(\rho\mathbf{v} - \overline{\mathbf{F}} - \text{grad } \lambda')\cdot\delta\mathbf{v}.$$

The same conclusion that leads from (8) to (9) leads now to

(13)          $\rho\mathbf{v} - \overline{\mathbf{F}} - \text{grad } \lambda' = 0.$

It is only necessary to differentiate this equation with respect to $t$ and to observe the meaning of $\overline{\mathbf{F}}$ in (12), to reobtain Euler's equations ($\rho$ is, of course, independent of $t$ because of the incompressibility). The result is

(14)          $\rho \dfrac{d\mathbf{v}}{dt} + \text{grad } p = \mathbf{F} \qquad \text{with} \qquad p = -\dfrac{d\lambda'}{dt}.$

In this argument the pressure is again given by the *Lagrange multiplier* $\lambda'$ although in a form different from the one we had before. This should not be surprising in view of the remark following (11).

We still have to consider the surface integral occurring in Eq. (6) which should give us the *surface conditions* required for the complete determination of the pressure. On replacing $\lambda$ by $p$ and $\delta s_n$ by $\delta v$, the integral reads

$$(15) \qquad \int p \, \delta v \, d\sigma.$$

Before continuing, we wish to broaden the physical basis of this argument by considering another expression of the same structure as (15) which takes care of the cohesion of the fluid. In doing so we must for once resort to ideas taken from molecular physics. The fluid particles act upon each other not only with the pressure $p$ (a consequence of their permanent volume as we may put it now), they also interact with their close neighbors through short range attractive forces which have their origin in the electrical structure of the molecules. For an inner volume element these forces cancel each other because they are in average directionally uniform. Not so for an element adjacent to the surface, where these forces add up to a resultant $N$ in the direction of the inward surface normal. If $N$ is referred to the unit of surface layer, the virtual work done by $N$ in the displacement $\delta v$ equals

$$\delta W = -N \, d\sigma \, \delta v.$$

Thus there appears in Hamilton's principle in addition to the *volume* integral over $\delta W$, in which the cohesive forces cancel, the *surface* integral

$$(16) \qquad \int \delta W \, d\sigma = - \int N \, \delta v \, d\sigma.$$

This and the surface integral (15) (which resulted from the condition of incompressibility) can be written as one integral, viz.

$$(17) \qquad \int (p - N) \, \delta v \, d\sigma.$$

According to Hamilton's principle the time integral of (17) must vanish together with the first term in (8), whence

$$(18) \qquad \int_{t_0}^{t_1} dt \int (p - N) \, \delta v \, d\sigma = 0.$$

The conclusions to be drawn from (18) can be arranged according to the following three types of possible boundaries:
a) The fluid is bounded by a rigid wall.
b) The fluid has a free surface; it is bounded by vacuum or air.
c) The fluid is bounded by another fluid, e.g., oil above water, but the boundary surface may have arbitrary shape.

a) A rigid wall prevents the fluid from moving perpendicularly to it; if this be true for the real motion, it must also be valid for the virtual displacement. Hence $\delta\nu = 0$ and condition (18) is automatically fulfilled. *The rigid wall does not provide a surface condition for the pressure.*

b) At a free surface $\delta\nu$ is arbitrary. Eq. (18) requires then

$$(19) \qquad p = N$$

for each element $d\sigma$ of the free surface at any time $t$.

c) At the surface of separation between fluid 1 and 2 with hydrodynamic pressures $p_1$, $p_2$ and cohesive pressures $N_1$, $N_2$, we first require according to (18) that for any $d\sigma$ and any $t$

$$(20) \qquad (p_1 - N_1)\delta\nu_1 + (p_2 - N_2)\delta\nu_2 = 0.$$

However, the contact of the fluids must not be disturbed by the virtual displacements. In other words, there must be

$$\delta\nu_1 = -\delta\nu_2 ,$$

the negative sign originating in our convention about the direction of $n$. Thereupon (20) gives

$$(21) \qquad p_1 - p_2 = N_{12} , \qquad N_{12} = N_1 - N_2 .$$

*Along the surface of separation of two fluids there is a pressure difference caused by the cohesive forces.* It depends on the *nature of the fluids* and on the shape of the *surface of separation,* but is independent of the motion of the fluids.

Thus not only the differential equations of the problem (in the present case Euler's equations) follow from Hamilton's principle, but also such surface conditions for the pressure as may be imposed, can be found in the same way; this will be taken up again in 17 (surface tension).

## 13. Euler's Equations for the Perfect Compressible Fluid and Their Application to Acoustics

The transition from statics to dynamics is made as before in 11 by adding the inertial resistance to the external force. We have found it an advantage in the compressible case, to refer the external force to the unit of mass rather than to the unit of volume, hence we should do the same with the inertial resistance which is then simply $-d\mathbf{v}/dt$. Replacing $\mathbf{P}$ in the equilibrium condition (7.5) by $\mathbf{P} - d\mathbf{v}/dt$, we obtain

$$(1) \qquad \frac{d\mathbf{v}}{dt} + \frac{1}{\rho}\,\mathrm{grad}\,p = \mathbf{P}.$$

These equations supplemented by the equation of continuity (5.4)

$$(2) \qquad \frac{\partial \rho}{\partial t} + \text{div} \, (\rho \mathbf{v}) = 0$$

constitute *Euler's equations for the compressible fluid.* The $p,\rho$-relation may be represented by the sufficiently general polytropic relation (7.1). The pressure $p$ as well as the density $\rho$ do now depend on $x$, $y$, $z$, and $t$.

In acoustics we are usually concerned only with *small oscillations* of the air. The difficulties arising from the non-linear terms (11.3a) in Euler's equations can then be avoided: we replace

$$\frac{d\mathbf{v}}{dt} \qquad \text{by} \qquad \frac{\partial \mathbf{v}}{\partial t} \,.$$

Quadratic terms must be kept, however, when we investigate, as in 37, shock-like phenomena such as occur in the firing of a gun. In the present case it is useful to take the normal atmospheric pressure $p_0$ as reference pressure for $p$; $p_0$ will then not occur in (1), since it is constant in space and drops out if the gradient is formed, while $p$ has the character of a small disturbance of $p_0$. As we neglect terms of second order, we may write $\rho_0$ instead of $\rho$ in (1) and in the second term of (2), $\rho_0$ being the normal density. In other words, we consider $p$, the derivatives of $\rho$, and the components of $\mathbf{v}$ as small quantities. Finally, it is permitted in acoustics to neglect body forces altogether, i.e., to set $\mathbf{P} = 0$.

Thus we obtain the simple system of equations

$$(3) \qquad \rho_0 \frac{\partial \mathbf{v}}{\partial t} + \text{grad} \, p = 0,$$

$$(4) \qquad \frac{\partial \rho}{\partial t} + \rho_0 \, \text{div} \, \mathbf{v} = 0.$$

Instead of grad $p$ we may write

$$(5) \qquad \text{grad} \, p = \left(\frac{dp}{d\rho}\right)_0 \text{grad} \, \rho = c^2 \, \text{grad} \, \rho, \qquad c^2 = \left(\frac{dp}{d\rho}\right)_0,$$

where the constant $c$ is related to the normal state of the atmosphere. On substituting in (3) we obtain

$$(6) \qquad \rho_0 \frac{\partial \mathbf{v}}{\partial t} + c^2 \, \text{grad} \, \rho = 0.$$

We are now ready to eliminate $\mathbf{v}$ from (4) and (6). Taking the partial derivative of (4) with respect to $t$ and substituting for $\rho_0 \partial v / \partial t$ according to (6), we obtain

(7)
$$\frac{\partial^2 \rho}{\partial t^2} = c^2 \operatorname{div} \operatorname{grad} \rho = c^2 \nabla^2 \rho.$$

The same equation is valid for $p$, since grad $p$, and therefore $\nabla^2 p$ and $\partial^2 p/\partial t^2$ are equal to the corresponding expressions in $\rho$ if the latter quantities are multiplied with $c^2$ [cf. (5)]. Thus

(8)
$$\frac{\partial^2 p}{\partial t^2} = c^2 \nabla^2 p.$$

The same differential equation also governs the velocity field. It is, of course, not the velocity $\mathbf{v}$ that satisfies equation (8), but the velocity potential $\Phi$, provided it exists. For, on putting $\mathbf{v} = -\operatorname{grad} \Phi$ according to (11,12), (4) and (6) become[7]

$$\frac{\partial \rho}{\partial t} = \rho_0 \nabla^2 \Phi \qquad \text{and} \qquad \rho_0 \frac{\partial \Phi}{\partial t} = c^2 \rho.$$

Here $\rho$ can be eliminated by differentiation as before, and the result is

(8)
$$\frac{\partial^2 \Phi}{\partial t^2} = c^2 \nabla^2 \Phi.$$

Equations of the type (7), (8), (9) are called *wave equations*. They play a very important part in the fundamentals of mathematical physics in the form of the *equation of the vibrating string* (one-dimensional, $\nabla^2 = \partial^2/\partial x^2$) or *vibrating membrane* (two-dimensional, $\nabla^2 = \partial^2/\partial x^2 + \partial^2/\partial y^2$).

Our present problem reduces to the equation of the vibrating string if we consider a process that depends on one coordinate only. Eq. (8), for instance, takes the form

(10)
$$\frac{\partial^2 p}{\partial t^2} = c^2 \frac{\partial^2 p}{\partial x^2}.$$

The integral of this equation can be given in the form of *d'Alembert's solution*

(11)
$$p = F_1(x + ct) + F_2(x - ct),$$

where $F_1$ and $F_2$ are real functions, entirely arbitrary except for certain continuity requirements (existence of first and second derivatives). In fact, any function of $x \pm ct$ satisfies the differential equation (10); the

---

[7]In the second of the two following equations a "constant," independent of $x$, $y$, $z$ but dependent on $t$ should be added; it can be absorbed in the definition of $\Phi$, however. (cf. p. 89).

same is true for a linear combination of such functions thanks to the fact that we neglected the quadratic terms in Euler's equations.

Expression (11) is the *general* solution of the one-dimensional wave equation, since it can be adapted to an arbitrarily given initial state. Let it be required that for $t = 0$

$$p = f_1(x), \qquad \frac{\partial p}{\partial t} = f_2(x).$$

Then we have only to make

$$F_1(x) + F_2(x) = f_1(x), \qquad F_1'(x) - F_2'(x) = \frac{1}{c} f_2(x).$$

This gives directly

$$(12) \qquad F_{1,2}(x) = \frac{1}{2} \left( f_1(x) \pm \frac{1}{c} \int_{x_0}^{x} f_2(\xi)\, d\xi \right).$$

The result is illustrated by Fig. 16 in which $f_2 = 0$ is assumed. Half of the initial pressure disturbance $f_1(x)$ moves to the right and half of it to the left, both travel without change of shape (undistorted, as one says) with the velocity $c$. The physical meaning of the constant $c$ becomes now apparent; it is the *velocity of sound*. In Fig. 16 the initial disturbance is

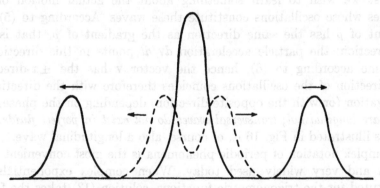

FIG. 16. Illustrating d'Alembert's solution. The initial pressure distribution (solid curve at the center) travels with constant velocity to both sides so that the ordinates of the two pressure hills are half the original ordinates, the width being the same. At first, the two hills partly overlap, (dotted curve) later they leave an undisturbed space between each other (solid curves left and right).

represented by the "hill" in the middle. A later state is illustrated by the half size pressure hills left and right. In terms of acoustics, the figure illustrates the propagation of a *noise*; the initial disturbance may have any shape as a function of $x$, the pressure being the same at all points of a plane normal to $x$. Or, in terms of the vibrating string, the string is plucked (rather, distorted into an arbitrary plane curve) at the time $t = 0$ and then left alone. In the latter case we should have to complete the picture by adding the reflections that the disturbances suffer at the ends of the (finite) string.

In acoustics of speech and music one is not so much concerned with a single but with a periodically sustained excitation of the air. Denoting the pitch (= frequency) by $\nu = 1/\tau$ and introducing $\omega = 2\pi\nu$, we assume $F_1$ and $F_2$ in the form of trigonometric functions

(13)
$$F_2(x - ct) = a \cos (kx - \omega t + \alpha),$$

$$F_1(x + ct) = b \cos (kx + \omega t + \beta).$$

($a$ and $b$ are amplitudes, $\alpha$ and $\beta$ phase constants, $k = \omega/c = 2\pi/\lambda$ is the *wave number*, that is, the number of waves on a segment of length $2\pi$). The plane wave $F_2$ travels in the positive and the wave $F_1$ in the negative $x$-direction. A superposition of the two waves results in a standing wave if the amplitudes $a$, $b$ are equal. Note the formula

(14)
$$c = \frac{\omega}{k} = \frac{\lambda}{\tau} = \lambda\nu.$$

First we wish to learn something about the actual motion of the particles whose oscillations constitute these waves. According to (5), the gradient of $\rho$ has the same direction as the gradient of $p$, that is, the $\pm x$-direction; the particle acceleration $d\mathbf{v}/dt$ points in this direction at any time according to (6), hence the vector $\mathbf{v}$ has the $\pm x$-direction. The direction of the oscillations coincides therefore with the direction of propagation (or with the opposite direction, depending on the phase): *our waves are longitudinal; transversal waves do not exist in perfect fluids.* The process illustrated in Fig. 16 is, of course, also a longitudinal wave.

Complex notation of periodic phenomena is the most convenient symbolism and very widely used today. When complex exponentials are substituted for the trigonometric functions, solution (13) takes the form

(15)
$$Ae^{i(kx-\omega t)}, \qquad A = ae^{i\alpha},$$

$$Be^{-i(kx+\omega t)}, \qquad B = be^{-i\beta}.$$

Physically meaningful is, of course, only the real part of either expression; it agrees with the corresponding right member in (13). This leaves the sign of $i$ still undetermined; it has been chosen in (15) so as to make the time dependent term in both expressions equal to $e^{-i\omega t}$. One may even suppress the time factor altogether, and obtains then the following symbols

$Ae^{ikx}$, a plane wave traveling in positive $x$-direction

(15a)

$Be^{-ikx}$, a plane wave traveling in negative $x$-direction

This representation is valid for pressure, density and velocity potential with the correspondingly changed meaning of $A$ and $B$. It will contribute greatly to shorten our formulas in Chap. V.

We still have to say something about the *numerical value of the sound velocity*. Under the assumption of an isothermal change of state the polytropic exponent $n$ equals 1. Then from (7.1)

$$\frac{dp}{d\rho} = \frac{p_0}{\rho_0},$$

and by (5)

$$(16) \qquad c = \sqrt{\frac{p_0}{\rho_0}}.$$

On the basis of standard pressure and temperature we have (p. 44)

$$p_0 = 1033 \times 981 \frac{\text{gr-wt}}{\text{cm sec}^2}, \qquad \rho_0 = 1.293 \times 10^{-3} \frac{\text{gr}}{\text{cm}^3},$$

and consequently

$$c = \sqrt{\frac{1012}{1,293}} \times 10^3 \frac{\text{cm}}{\text{sec}} = 280 \frac{\text{m}}{\text{sec}}.$$

Newton had already obtained this value which is much too small. The theoretical formula upon which he based his calculation is

$$(16a) \qquad c = \sqrt{\frac{\text{Elasticity}}{\text{Density}}},$$

which requires some clarification. Actually, the term elasticity means in the present case the same as modulus of compression. For an elastic solid, this is given by (9.13); the corresponding definition for a gas would be

$$K = \frac{dp}{|d\Theta|} \qquad \text{with} \qquad d\Theta = -\frac{d\rho}{\rho},$$

or,

$$(16b) \qquad K = \rho \frac{dp}{d\rho}.$$

Thus (16a) becomes

$$c = \sqrt{\frac{dp}{d\rho}} = \sqrt{\left(\frac{dp}{d\rho}\right)_0} \, ,$$

in accordance with definition (5).

The question arises, however, whether the assumption of an *isothermal* change of state can be justified. It was Laplace who first noticed that heat exchange between the fluid particles should be practically impossible because of the rapidity of the oscillations; hence the change of state should be considered *adiabatic*. The polytropic exponent is then no longer 1 but 1.4 (p. 50) and we have from (7.1)

$$(17) \qquad \left(\frac{dp}{d\rho}\right) = n \, \frac{p_0}{\rho_0} \, ,$$

The value of the velocity of sound takes up the factor $\sqrt{1.4}$ and turns out to be

$$(17a) \qquad c = \sqrt{1.4} \times 280 = 332 \, \frac{\text{m}}{\text{sec}} \, ,$$

in satisfactory agreement with the experiment.[8]

In conclusion we wish to generalize d'Alembert's method of integration for a *spherical wave*. Let the origin of the wave coincide with the origin $O$ of a system of polar coordinates $r$, $\vartheta$, $\varphi$, and assume a spherically symmetric distribution of pressure (no dependence on $\vartheta$ and $\varphi$). The expression for $\nabla^2 p$ (problem I, 3) is then

$$\nabla^2 p = \frac{1}{r} \frac{\partial^2 (rp)}{\partial r^2}$$

and equation (8) takes the form

$$\frac{\partial^2 (rp)}{\partial t^2} = c^2 \, \frac{\partial^2 (rp)}{\partial r^2} \, .$$

Exactly as in (11) the integral is

$$rp = F_1(r + ct) + F_2(r - ct).$$

Any distribution of $p$ and $\partial p/\partial t$ given for $t = 0$ and $r > 0$ can be represented according to this formula and continued for $t > 0$; the values of $F_2$ for negative argument must be adjusted so as to keep $p$ finite for $r = 0$.

---

[8]This difference in the two values of $c$, which is due to a difference in $K$, illustrates well the fact that the parameters of elasticity depend on the character of the change of state (see end of 9).

Here again, we are mainly concerned with the periodic case which we now write according to (15), but without the time factor $e^{-i\omega t}$

$$(18) \quad p = \frac{A}{r} e^{ikr} \quad \text{and} \quad (18a) \quad p = \frac{B}{r} e^{-ikr} \quad \text{respectively.}$$

The same representation is valid for $\rho$ and $\Phi$. Since the gradient of $\Phi$ coincides with the $r$-direction the oscillation is *longitudinal* also in this case.

Eq. (18) represents an outgoing spherical wave: the spherical surfaces of equal phase $kr - \omega t = $ const travel away from $O$ with a phase velocity

$$\frac{dr}{dt} = \frac{\omega}{k} = c.$$

Such a radiation process can be considered as produced by a pulsating sphere at $O$.

On the other hand, (18a) means an incoming spherical wave. The surfaces of equal phase $-kr - \omega t = $ const travel with the velocity

$$\frac{dr}{dt} = -\frac{\omega}{k} = -c$$

toward the origin, but this radiation pattern can hardly be realized physically.

# Appendix

## Comparison of Compressible and Incompressible Flows.[9]

Since the analysis of flows of compressible fluids presents much greater difficulties because of the variable density, the question arises: *when is it permissible to consider a compressible medium as incompressible within given limits of accuracy?* This means that a certain prescribed error is admitted for the quantity of interest which may be either $\rho$ or $p$ or $v$.

We restrict the following discussion to steady flow and start from the equation of continuity. Cross-sections of stream filaments in the compressible case do not change in the same way as in the incompressible case, which is a consequence of the different form of the continuity equation, viz., div $(\rho \mathbf{v}) = 0$ according to (5.4). On integrating the divergence of the mass flow over a stream tube between the small cross-sections $F$ and $F_0$ assumed to be orthogonal to the streamlines, one obtains by Gauss's theorem

[9]Cf. also Prandtl-Tietjens, Fundamentals of Hydro- and Aeromechanics, McGraw-Hill, New York, p. 224.

(19)                 $$\rho F v = \rho_0 F_0 v_0 = \text{const} = M$$

where $v$, $v_0$ and $\rho$, $\rho_0$ are the absolute velocity and density values at $F$ and $F_0$.

In the incompressible case, (19) reduces to

(19a)                $$F v = F_0 v_0 = \text{const} = V.$$

The quantities $M$ and $V$ in (19) and (19a) are the mass flux and the volume flux through the stream tube.

We now compute the pressure in the steady flow of a compressible fluid and compare it with the pressure in an incompressible flow. Bernoulli's equation (with no other body force than gravity) is according to (11.9a) and (7.6b)

(20)         $$\frac{1}{2}(v^2 - v_0^2) + g(h - h_0) + \int_{p_0}^{p} \frac{dp}{\rho} = 0,$$

but in an incompressible medium one would have instead of (20)

(20a)        $$\frac{1}{2}(v^2 - v_0^2) + g(h - h_0) + \frac{p - p_0}{\rho} = 0.$$

The quantities without subscripts refer to an arbitrary cross-section of a stream tube while the subscript 0 denotes the reference cross-section of the same stream tube. The evaluation of the integral in (20) according to the polytropic assumption $p = C\rho^n$ yields

(21)
$$\int_{p_0}^{p} \frac{dp}{\rho} = \frac{n}{n-1}\left(\frac{p}{\rho} - \frac{p_0}{\rho_0}\right) = \frac{n}{n-1}\frac{p_0}{\rho_0}\left[\left(\frac{\rho}{\rho_0}\right)^{n-1} - 1\right]$$

$$= \frac{n}{n-1}\frac{p_0}{\rho_0}\left[\left(\frac{p}{p_0}\right)^{(n-1)/n} - 1\right].$$

Substituting this in (20) and solving for $p$, we obtain first

(21a)    $$p = p_0\left[1 - \frac{n-1}{n}\frac{\rho_0}{p_0}\left\{\frac{1}{2}(v^2 - v_0^2) + g(h - h_0)\right\}\right]^{n/(n-1)}.$$

If we assume

(21b)         $$\frac{1}{2}(v^2 - v_0^2) + g(h - h_0) < \frac{n}{n-1}\frac{p_0}{\rho_0},$$

we may expand the right member of (21a) according to the binomial theorem and have

(21c)    $$p = p_0\left[1 - \frac{\rho_0}{p_0}\left\{\frac{1}{2}(v^2 - v_0^2) + g(h - h_0)\right\} + \frac{1}{2n}\frac{\rho_0^2}{p_0^2}\left\{\quad\right\}^2 \pm \cdots\right].$$

If all but the linear term of the series are neglected, one reobtains the result (20a) as one would expect.

Let us now estimate the error in the important case of the flow past a rigid body. The reference point $p_0$, $v_0$ is assumed to be at a great distance from the body, and differences in altitude are from now on neglected so that $h = h_0$. With a view to the later Figs. 46 and 48 we compute the pressure $p$ at the stagnation point or simply the "stagnation pressure" $p - p_0$. Since $v^\bullet = 0$ [cf. also the definition of the stagnation point in (29.7)], we obtain from (21c)

$$p - p_0 = \frac{1}{2}\,\rho_0 v_0^2 + \frac{1}{2n}\frac{\rho_0}{p_0}\frac{\rho_0 v_0^4}{4} \mp \cdots$$

Here we introduce the *sound velocity associated with the polytrope under consideration*. At the reference point which we identify with what was called the normal state in (13.17), the sound velocity is

$$(21\text{d}) \qquad c_0 = \sqrt{\left(\frac{dp}{d\rho}\right)_0} = \sqrt{n\,\frac{p_0}{\rho_0}}\,.$$

Thus the previous equation takes the form

$$(22) \qquad p - p_0 = \frac{1}{2}\,\rho_0 v_0^2\left(1 + \frac{v_0^2}{4c_0^2} \mp \cdots\right),$$

and can be compared with (20a), which we now write

$$(22\text{a}) \qquad p - p_0 = \frac{1}{2}\,\rho_0 v_0^2\,.$$

The error incurred in computing the stagnation pressure from (22a) instead of from (22) amounts to 1% at the most if $v_0 \leqq 0.2c_0$. In the case of standard air ($n = 1.4$, $c_0 = 332\text{m/sec}$) this means $v_0 \leqq 66.4$ m/sec; if, on the other hand, an error of 10% is admitted, $v_0$ should be kept smaller than $0.63c_0 = 209$ m/sec. These estimates have been carried out on the basis of the quadratic term in (22).

The change of density associated with the pressure change is obtained from (21a) by the polytropic relation. The analogous expansion yields

$$(23) \qquad \rho = \rho_0\left(1 - \frac{v^2 - v_0^2}{2c_0^2} + \cdots\right).$$

We now determine the change of $v$ along the stream tube. In the incompressible case (20a) gives with $h = h_0$

$$(24) \qquad v = \sqrt{v_0^2 + \frac{2\Delta p}{\rho_0}} \qquad \text{with} \qquad \Delta p = p_0 - p \qquad \text{and} \qquad \rho = \rho_0\,.$$

On the other hand, (20) gives, with the last value of the integral (21),

$$(24a) \qquad v = \sqrt{v_0^2 + \frac{2n}{n-1} \frac{p_0}{\rho_0} \left[ 1 - \left(\frac{p}{p_0}\right)^{(n-1)/n} \right]}$$

in the compressible case. On putting again $p = p_0 - \Delta p$ and expanding as before, one has

$$(24b) \qquad v = \sqrt{v_0^2 + \frac{2\Delta p}{\rho_0} \left( 1 + \frac{1}{2n} \frac{\Delta p}{p_0} + \cdots \right)},$$

which agrees with (24) except for the quadratic term in $\Delta p$.

Returning once more to the stream tube cross-section in Eqs. (19) and (19a), we obtain by logarithmic differentiation of the equation of continuity (19)

$$(25) \qquad \frac{d\rho}{\rho} + \frac{dF}{F} + \frac{dv}{v} = 0,$$

Bernoulli's equation (20) with $h = h_0$ can also be written in differential form, viz.

$$(26) \qquad v\, dv + \frac{dp}{\rho} = 0,$$

and so can the polytropic equation:

$$(27) \qquad \frac{dp}{p} = n \frac{d\rho}{\rho}.$$

On substituting for the differentials $dp$ and $d\rho$ in (25) from (27) and (26), one has

$$(28) \qquad \frac{dF}{F} = \frac{dp}{\rho v^2}\left(1 - \frac{\rho v^2}{np}\right) = \frac{dp}{\rho v^2}\left(1 - \frac{v^2}{c^2}\right).$$

Note however that $c$ is now the sound velocity at the cross-section of interest, $F$. The corresponding relation for the incompressible case is obtained by putting $d\rho = 0$ in (27), or $c \to \infty$ in (28):

$$(28a) \qquad \frac{dF}{F} = \frac{dp}{\rho v^2}.$$

Comparison of the two results shows first that the laws govering the change of cross-section are practically identical as long as $v \ll c$, similarly to the laws for the pressure changes given in (22) and (22a). But Eqs. (28) and (28a) allow another important conclusion to be drawn. *For an incompressible fluid, dF and dp have always the same sign, i.e., the cross-*

*section enlarges with increasing pressure and becomes smaller with decreasing pressure.* For a *compressible fluid* this is only true if $v < c$, i.e., in *subsonic flow. In supersonic flow* ($v > c$) *the cross-section enlarges with decreasing pressure and becomes smaller with increasing pressure.* It is this fact that is responsible for the fundamental difference between supersonic and subsonic flows of compressible fluids, a difference which does not exist for incompressible fluids for which the sound velocity is infinitely large according to the definition $c = \sqrt{dp/d\rho}$. We also infer from (28) that the cross-section of a stream tube has an extremum when the sound velocity is reached ($v = c$), and we can easily see that it must be a *minimum*.

Formulas (19) and (19a) play an important role in hydraulics. They serve to determine such flow cross-sections as occur in piping systems, valves, turbine bladings, etc. Eqs. (24) and (24a) serve to calculate the velocity of discharge from the nozzle of a pressure tank where $p_0$ is the (constant) inside pressure and $p$ the (constant) external pressure; $v_0$ may be assumed $\sim 0$ in good approximation.—Formula (24a) is named after de Saint-Venant and Wantzel (1839).

The Laval[10] nozzle is of particular interest in this connection. It serves to change energy of high pressure and low velocity into energy of low pressure and high velocity as any nozzle does, but the peculiar feature of the Laval nozzle is that the velocity of discharge can exceed the sound velocity. In accordance with our results, its design must be such that the cross-sections decrease until sonic speed is reached and then increase again.

## 14. Dynamics of the Elastic Body

The dynamic equations of the elastic body follow again from the static equilibrium condition by adding the inertia force to the external force **F**. Displacements in a solid can always be considered as small (only at the yield point, that is, beyond the limit of purely elastic behavior considerable displacements occur, cf. 39). The material derivative may, therefore, be identified with the local derivative and the inertia force per unit of volume becomes $-\rho \partial^2 \mathbf{s}/\partial t^2$, where $\mathbf{s} = \mathbf{i}\xi + \mathbf{j}\eta + \mathbf{k}\zeta$.

Writing the static equilibrium condition in the form (8.12), we obtain as the differential equation of elasticity

$$(1) \qquad \rho \frac{\partial^2 \mathbf{s}}{\partial t^2} = \text{Div } \sigma + \mathbf{F},$$

---

[10]Devised by the Swedish engineer de Laval, who also invented a good separator and the first usable steam turbine.

or, spelled out in Cartesian coordinates as in (8.11),

(1a)
$$\rho \frac{\partial^2 \xi}{\partial t^2} = \frac{\partial \sigma_{xx}}{\partial x} + \frac{\partial \sigma_{yx}}{\partial y} + \frac{\partial \sigma_{zx}}{\partial z} + F_x ,$$

$$\rho \frac{\partial^2 \eta}{\partial t^2} = \frac{\partial \sigma_{xy}}{\partial x} + \frac{\partial \sigma_{yy}}{\partial y} + \frac{\partial \sigma_{zy}}{\partial z} + F_y ,$$

$$\rho \frac{\partial^2 \zeta}{\partial t^2} = \frac{\partial \sigma_{xz}}{\partial x} + \frac{\partial \sigma_{yz}}{\partial y} + \frac{\partial \sigma_{zz}}{\partial z} + F_z .$$

Using, however, form (9.18) of our fundamental equation which is preferable for what follows, we obtain the elastic differential equation in the form

(1b) $\quad \rho \dfrac{\partial^2 \mathbf{s}}{\partial t^2} = (\mu + \lambda)\, \text{grad } \Theta + \mu \nabla^2 \mathbf{s} + \mathbf{F}, \qquad \Theta = \text{div } \mathbf{s}.$

This equation refers to Cartesian coordinates; a more general equation is obtained when one starts from Eq. (9.19) instead of (9.18).

We are here mainly interested in free oscillations, therefore put $\mathbf{F} = 0$. From (1b) it follows directly that there will be waves of dilatation as well as of distortion, the first having *longitudinal*, the latter *transversal character*.

*Waves of dilatation.* On taking the divergence of every term in (1b), we obtain

$$\rho \frac{\partial^2 \Theta}{\partial t^2} = (\mu + \lambda)\nabla^2 \Theta + \mu \text{ div } \nabla^2 \mathbf{s}$$

and may replace

$$\text{div } \nabla^2 \mathbf{s} \qquad \text{by} \qquad \nabla^2 \text{ div } \mathbf{s} = \nabla^2 \Theta.$$

In this way we obtain the simple scalar equation

(2) $\qquad\qquad\qquad \rho \dfrac{\partial^2 \Theta}{\partial t^2} = (2\mu + \lambda)\nabla^2 \Theta.$

By comparison, e.g. with (13.8), the *velocity of propagation of the waves of dilatation*, which we denote by $a$, is seen to be

(3) $\qquad\qquad\qquad a^2 = \dfrac{2\mu + \lambda}{\rho} .$

For a one dimensional or spherically symmetric state the periodic solutions obtained by d'Alembert's method (13.11) for the infinite elastic body are of the form

(4) $\qquad\qquad e^{ikz} \quad \text{or} \quad \dfrac{1}{r} e^{ikr}, \quad k = \dfrac{\omega}{a} .$

They are plane or spherical waves with circular frequency $\omega$ that travel with the constant velocity $a$.

*Waves of distortion.* On taking the curl of Eq. (1b) term by term and putting as in (2.19)

$$\text{curl } \mathbf{s} = 2\mathbf{\phi},$$

Eq. (1b) yields because of curl grad $= 0$

(5) $$\rho \frac{\partial^2 \mathbf{\phi}}{\partial t^2} = \mu \nabla^2 \mathbf{\phi}.$$

Here $\mathbf{\phi}$ represents in magnitude and axis the angle of torsion of the volume element under consideration. Comparison with (13.8) shows that the velocity of propagation of the waves of distortion, which we denote by $b$, is given by

(6) $$b^2 = \frac{\mu}{\rho}.$$

This is the reason why Lamé's modulus $\mu$ is sometimes called the torsion modulus, as already mentioned on p. 68. Again d'Alembert's method gives plane and spherical waves that are represented by (4) in the same way as the compression waves, provided that the wave number $k$ is now $\omega/b$. The comparison of (3) and (6) shows immediately that

$$a > b, \quad \text{since} \quad \frac{a^2}{b^2} = 2 + \frac{\lambda}{\mu} = \frac{2 - 2\nu}{1 - 2\nu};$$

as for the last term, cf. (9.11). For iron with $\nu$ about 0.3 (see p. 65), we obtain $a/b = \sqrt{7/2} = 1.87$.

The meaning of this is obvious: the elastic resistance of solids against changes of volume is considerably larger than that against changes of the relative orientation. The absolute value of $b$ in a solid is, of course, still much larger than the sound velocity $c$ in air. With the data of pp. 65, 68, and with $\rho = 7.8\text{gr/cm}^3$ for iron, the value of $b$ is found to be 3100m/sec. Both formulas (3) and (6) for $a$ and $b$ can be subsumed under Newton's rule (13.16a), provided the term elasticity is suitably specified in each case.

The compressional waves and the waves of distortion considered here are *three-dimensional waves* progressing in an elastic body without boundaries. In addition, there exist *surface waves* that are bound to the free surface of the body and have a special importance for seismic phenomena; they are dealt with in 45. The surface waves travel more slowly than the three-dimensional waves, that is to say, the earth puts up more elastic resistance against the latter than the former. They constitute the principal parts of the observed earthquakes, while the three-dimensional waves that arrive earlier are observed as preliminaries.

## 15. The Quasi-Elastic Body as Model of the Ether

In 19th century physics, a material carrier was assumed for the optical phenomena, equipped as far as possible with the properties of ordinary elastic bodies. This construction, however, led to difficulties even in the most elementary problem of reflexion and refraction, about which more in 45. As early as 1839 MacCullagh tried to drop the connection with the ordinary theory of elasticity with the aim to develop a representation of optics that would be free of the difficulties mentioned. It turned out later that his theory agreed formally with Maxwell's electro-magnetic optics (1864), in particular as far as the optics of transparent bodies is concerned. The following remarks should be considered as an interpretation of MacCullagh's equations.

Let us go back to the beginning of 1. There the general locomotion of a continuous medium was decomposed into the three parts of translation, rotation, and deformation. The elastic body responds to a deformation with a stress tensor which is determined by the deformation tensor; it is not sensitive to rotation (and, of course, not to translation). We now try to imagine a "quasi-elastic" body, supposedly insensitive to deformations but *responsive to rotations relative to absolute space*! Since the rotation has the character of an antisymmetric tensor (1.12a) we shall assume that the stress acting on the volume element as a result of the rotation is also an antisymmetric tensor, as indicated in the following array:

$$(1) \qquad \begin{pmatrix} 0 & \sigma_{xy} & \sigma_{xz} \\ \sigma_{yx} & 0 & \sigma_{yz} \\ \sigma_{zx} & \sigma_{zy} & 0 \end{pmatrix}, \qquad \sigma_{ik} = -\sigma_{ki} .$$

The *stress-rotation relation* here assumed is illustrated in Fig. 17. Let the volume element $\Delta\tau$ be twisted through the angle $\varphi_z$ (indicated by the right hand screw arrow about the positive z-axis). To produce or to maintain this rotation we need—this is our hypothesis—a moment about the z-axis

$$(2) \qquad M_z = k\varphi_z \Delta\tau.$$

The quantity $k$ could be called the "twist modulus" of the quasi-elastic body. The moment is associated with two shear stresses $\sigma_{xy}$ and $-\sigma_{yx}$ across the positive x- and y-surfaces and with corresponding anti-parallel shear stresses across the negative x- and y-surfaces, all indicated in the figure. In order to be in accordance with (2) and (1) we have to put

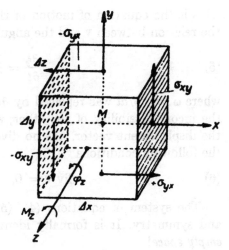

Fig. 17. Strain and rotation in the "quasielastic" body.

$$(3) \qquad \sigma_{zy} = -\sigma_{yz} = \frac{k}{2}\,\varphi_z\,.$$

This assumed, the moment originating at the two $x$-surfaces (see the analogous argument that accompanies Fig. 10) is

$$2\sigma_{zy}\Delta y\Delta z \cdot \frac{\Delta x}{2} = \frac{k}{2}\,\varphi_z\Delta\tau$$

and the moment originating at the two $y$-surfaces

$$-\,2\sigma_{yz}\Delta x\Delta z \cdot \frac{\Delta y}{2} = \frac{k}{2}\,\varphi_z\Delta\tau,$$

so that the total moment is in fact that of equation (2). By rotation of the letters we obtain from (3)

$$(3a) \qquad \sigma_{yz} = -\sigma_{zy} = \frac{k}{2}\,\varphi_x\,, \qquad \sigma_{zx} = -\sigma_{xz} = \frac{k}{2}\,\varphi_y\,.$$

The equation of motion of the quasi-elastic body follows now from (14.1a). In applying this equation we have to assign inertia to the volume element of the ether and consider its displacement as small, since we previously neglected the quadratic convective terms in putting $du/dt = \partial u/\partial t$. We also disregard any external forces ($\mathbf{F} = 0$). In this way we obtain from (14.1a) with the use of (3) and (3a)

$$\rho\,\frac{\partial u}{\partial t} = \frac{\partial\sigma_{yz}}{\partial y} + \frac{\partial\sigma_{zz}}{\partial z} = -\frac{k}{2}\left(\frac{\partial\varphi_z}{\partial y} - \frac{\partial\varphi_y}{\partial z}\right),$$

or, written in vector form,

$$(4) \qquad \rho\,\frac{\partial\mathbf{v}}{\partial t} = -\frac{k}{2}\,\text{curl}\ \phi.$$

This is the equation of motion of the ether; it must be supplemented by the relation between **v** and the angular velocity **ω**, which reads

(5)
$$\frac{\partial \mathbf{\phi}}{\partial t} = \frac{1}{2} \operatorname{curl} \mathbf{v}$$

where $\mathbf{\omega} = d\mathbf{\phi}/dt$ was replaced by $\partial \mathbf{\phi}/\partial t$. As a further assumption we add the incompressibility of the ether; we also note that $\mathbf{\phi}$, being the curl of the displacement vector, has no divergence. Thus **v** and $\mathbf{\phi}$ are subject to the following conditions

(6)
$$\operatorname{div} \mathbf{v} = 0, \qquad \operatorname{div} \mathbf{\phi} = 0.$$

The system of equations (4), (5) and (6) is of impressive simplicity and symmetry. It is formally identical with *Maxwell's equations for the empty space!*

Before elaborating this statement, let us introduce the following notations: **E** = electric field strength, **H** = magnetic field strength; $\alpha, \beta$ = factors of proportionality that depend on the choice of units for $E$ and $H$. Now we put

$$\textit{either} \qquad \text{a)} \qquad \mathbf{v} = \pm \alpha \mathbf{E}, \qquad \mathbf{\phi} = \mp \beta \mathbf{H},$$

$$\textit{or} \qquad \text{b)} \qquad \mathbf{v} = \pm \alpha \mathbf{H}, \qquad \mathbf{\phi} = \pm \beta \mathbf{E}.$$

The signs in these equations depend on the signs chosen for the units of the electric charge and magnetic pole strength. Eqs. (4), (5), and (6) assume then the form

(7)
$$\epsilon_0 \frac{\partial \mathbf{E}}{\partial t} = \operatorname{curl} \mathbf{H}, \qquad \operatorname{div} \mathbf{E} = 0,$$

$$\mu_0 \frac{\partial \mathbf{H}}{\partial t} = - \operatorname{curl} \mathbf{E}, \qquad \operatorname{div} \mathbf{H} = 0$$

for the choice a) as well as for the choice b). The abbreviations $\epsilon_0$, $\mu_0$ are known as the dielectric constant and permeability of the vacuum; in our notation they are given by

(a)
$$\epsilon_0 = \frac{\rho}{k} \frac{2\alpha}{\beta}, \qquad \mu_0 = \frac{2\beta}{\alpha},$$

(b)
$$\mu_0 = \frac{\rho}{k} \frac{2\alpha}{\beta}, \qquad \epsilon_0 = \frac{2\beta}{\alpha}.$$

Their product is independent of the choice of the units $\alpha$, $\beta$, since we have in either case

$$(8) \qquad \epsilon_0 \mu_0 = \frac{4\rho}{k} = \frac{1}{c^2} .$$

The quantity $c$ defined in (8) has, physically, the meaning of the *velocity of light in vacuum*. It can be included in Newton's rule (13.16a) in analogy to the sound velocities in 13 and 14, when the word elasticity is given the meaning of one fourth of the "twist modulus" $k$.

It is by no means our intention to assign any physical reality to this "ether model". Physicists had convinced themselves by the turn of the century that all attempts at a mechanical explanation of Maxwell's equations were doomed to failure. What we mean here is not a mechanical *explanation* but, at best, a mechanical *analogy*. Maxwell's equations are among the fundamentals of the electrical theory of matter, so one should not expect that they can be derived from the macroscopic properties of ponderable bodies. On the contrary, they seem to stem from the same root as general gravitation, that is, from the space-time metric, according to a recent paper by Schrödinger.[11] Our remarks, however, may have some justification inasmuch as they show: if we were to construct an "ether" as a substratum for Maxwell's equation, then we would have to furnish it with qualities that are diametrically opposed to those of ordinary matter, viz., an *absolute directional orientation relative to space* in constrast to the *relative orientation of the volume elements toward each other* possessed by elastic bodies.

The following historical remark may be of interest. In an extension of MacCullagh's ideas Lord Kelvin,[12] in the eighties, developed the concept of the quasi-elastic or, as he sometimes put it, "quasi-rigid" ether. He was not satisfied, only to postulate an ether with reactive responses (2), but attempted to construct a gyroscopic model that actually would react in the required way. As is well known, a fast spinning top can be so arranged as to acquire directional stability and will then respond to fairly strong moments with small angular changes only. But an ether model, based on gyroscopic effects, becomes desperately complicated. Each volume element has to be equipped with several tops that must be oriented relative to each other in such a way that the desired rotational stiffness is achieved not only for one but for all three axes. A construction as complicated as that would be the only way to realize a "gyrostatic" ether.

Lord Kelvin took the point of view marked by a) on p. 110 and correlated the "twist" of his gyrostatic ether to the magnetic vector **H**. He

---

[11]E. Schrödinger, Proc. R. Irish Academy **49**, 43 (1943).

[12]Sir William Thomson (Lord Kelvin), Mathematical and Physical Papers, Vol. 3, Art. 49, 50, 52.

did not commit himself to a definite interpretation of the electric vector
$\mathbf{E}$, thus abstaining from a complete mechanical representation of Maxwell's equations. There is no doubt that standpoint a) is physically more
evident, since $\mathbf{H}$ has the character of an axial vector like $\mathbf{\dot{\xi}}$, while $\mathbf{E}$ possesses
that of a polar vector like $\mathbf{v}$. On the other hand, standpoint b) seemed to
recommend itself because it took care of the pure ether and the insulators,
but also left space for the conductors in the quasi-elastic picture.[13] It was
Boltzmann[14] who pointed out the difficulties involved in standpoint b)
which are caused by the existence of the *true electric charge*. Disregarding
the electrodynamics of conductors, we shall only pursue assumption a)
somewhat further in the following. In 20 we shall come back to the correlation b) in connection with Helmholtz's vortex theory.

For the ponderable insulator the two principal equations (7) are still
valid as in vacuum, only the values $\epsilon_0$ , $\mu_0$ must now be replaced by $\epsilon$, $\mu$.
The divergence conditions, however, undergo an essential change. In-
stead of div $\mathbf{H} = 0$ we have now

$$(9) \qquad \operatorname{div} \mathbf{B} = 0, \qquad \mathbf{B} = \mu \mathbf{H} = \text{magnetic induction}$$

and we would therefore correlate $\mathbf{B}$ rather than $\mathbf{H}$ with the rotation $\mathbf{\dot{\xi}}$,
which would introduce no difficulty. On the other hand the condition
div $\mathbf{E} = 0$ must be altered into

$$(10) \qquad \operatorname{div} \mathbf{D} = \rho_e , \qquad \mathbf{D} = \epsilon \mathbf{E} = \text{dielectric displacement.}$$

Here $\rho_e$ is the spatial density of the true electric charge. When we now
correlate $\mathbf{D}$ instead of $\mathbf{E}$ to the flow velocity $\mathbf{v}$ and set up suitable relations
between the constants $\epsilon$, $\mu$ and $k$, $\rho$, $\alpha$, $\beta$ we are still able to obtain complete
formal analogy with Maxwell's equations for the ponderable insulator. It
is clear, however, that the existence of the true charge causes difficulty
also for the standpoint a), if it is desired to maintain the incompressibility
of the ether. There is, of course, a mathematically permissible but physically rather drastic remedy for this difficulty. One simply prescribes
that the quasi-elastic fluid leaves or enters the field at points that carry
true charges according to the signs. Where the fluid goes or whence it
comes is left in the dark. As an apology for this assumption one could
quote the example of the great Bernhard Riemann[15] who makes the same
hypothesis in an attempt at a theory of gravitation and electrostatics, in
his paper: "Neue mathematische Prinzipien der Naturphilosophie", but

[13]A. Sommerfeld, Ann. d. Phys. 46, 139 (1892).

[14]L. Boltzmann, Ann. d. Phys. 47, 743 (1892), and Vorlesungen über Maxwells
Theorie, Leipzig 1893, Vol. II, 1, p. 6.

[15]Ges. Werke, 2nd ed., Teubner, Leipzig, 1892, p. 528. The paper was written
shortly after Riemann had obtained the doctorate, and published posthumously.

it seems to be more correct to admit that no mechanical or quasi-mechanical picture is a suitable representation of the fundamental fact of the electric charge.

We shall have no reason to come back in Vol. III to the model of the ether discussed here. The electric charge and the structure of the electromagnetic field must be accepted as entities that transcend mechanics.

## 16. Dynamics of Viscous Fluids. Hydrodynamics and Hydraulics. Reynolds' Criterion of Turbulence

In order to set up the general equations of motion of viscous fluids, we have to introduce the inertia force in the equations of uniform motion (10.8a,b). F must be replaced by

$$\mathbf{F} - \rho \frac{d\mathbf{v}}{dt} = \mathbf{F} - \rho \left\{ \frac{\partial \mathbf{v}}{\partial t} + (\mathbf{v}\,\mathrm{grad})\,\mathbf{v} \right\}.$$

In this way we obtain the Navier-Stokes equations[16]

$$(1) \qquad \rho \frac{\partial \mathbf{v}}{\partial t} + \rho(\mathbf{v}\,\mathrm{grad})\,\mathbf{v} - \mu \Delta \mathbf{v} + \mathrm{grad}\,p = \mathbf{F}$$

which are written here in pseudo-vectorial form. In order to put them in invariant shape one has to make use of Eqs. (11.6) and (3.10a), viz.

$$(\mathbf{v}\,\mathrm{grad})\,\mathbf{v} = \mathrm{grad}\,\frac{v^2}{2} - \mathbf{v} \times \mathrm{curl}\,\mathbf{v}$$

and

$$\nabla^2 \mathbf{v} = \mathrm{grad}\,\mathrm{div}\,\mathbf{v} - \mathrm{curl}\,\mathrm{curl}\,\mathbf{v}.$$

Eq. (1) has to be supplemented in the incompressible case by

$$(1a) \qquad \mathrm{div}\,\mathbf{v} = 0.$$

In the case of compressible fluids, (1a) must be replaced by (5.4); at the same time the additional term $-(\mu + \lambda)\,\mathrm{grad}\,\dot\theta$, that is $-(\mu + \lambda)\,\mathrm{grad}\,\mathrm{div}\,\mathbf{v}$, has to be added to the left member of (1) according to (10.22).

Today we consider the Navier-Stokes equations as fundamental for the entire theory of fluid flow. Engineers throughout the 19th century were, on the contrary, convinced that there was a big discrepancy between theoretical hydrodynamics on the one side and hydraulics on the other side. It was the British engineer and physicist Osborne Reynolds

---

[16]Navier, 1822, Stokes, 1845.

who bridged the gap with his profound experimental[17] and theoretical[18] investigations.

He experimented with glass tubes of various diameters at varying pressure gradients which amounts to changing the average velocities of the flow. (Average velocity means here the mean value of the velocity with respect to the cross-section.) Introducing a colored fluid filament into the entrance cross-section he observed the behavior of this filament throughout the tube. At small diameters and not too large average velocities the colored filament remains straight and parallel to the axis of the tube. At a larger diameter or larger velocity there appear irregular deviations that have a tendency to fill the whole tube. The initially straight filament becomes fuzzy and finally indistinguishable. In the first case the motion is called *laminar* (proceeding in regular laminations as in the Hagen-Poiseuille flow), in the last case it is called *turbulent*, as suggested by Reynolds.

Reynolds would not have been able to bring order in his experimental results, if he had not considered them from the point of view of a *"law of similitude."* Considerations of this character already appear in the first beginnings of western physics, as in the work of Galilei.[19] Generally speaking, they consist in the comparison of two experimental arrangements differing only in scale.

In our case, consider two tubes of radii $a_1$ and $a_2$. Putting

$$(2a) \qquad a_2 = \alpha a_1 ,$$

the number $\alpha$ is the change of the scale of length in the transition from experiment 1 to experiment 2. It pertains not only to the radii, but to all dimensions of length, such as the coordinates of two "corresponding points" in the two tubes. Any two such points are thus connected by

$$x_2 = \alpha x_1 , \qquad y_2 = \alpha y_1 , \qquad z_2 = \alpha z_1 .$$

The average velocities in corresponding cross-sections being $v_1$ and $v_2$, we put

$$(2b) \qquad v_2 = \beta v_1 .$$

[17]An Experimental Investigation of the Circumstances which determine whether the Motion of Water shall be direct or sinuous, and the Law of Resistance in Parallel Channels, Phil. Trans. R. Soc. 174 (1883).

[18]On the Dynamical Theory of Incompressible Viscous Fluids and the Determination of the Criterion, Phil. Trans. R. Soc. 186 (1895). In connection with this H. A. Lorentz, Ges. Werke, Vol. I, Leipzig, p. 43.

[19]In his last work "Discorsi e Dimostrazioni Matematiche," 1638.

The dimension of velocity being length/time, the number $\beta$ determines also the change of the time scale in our transition; this change is evidently $\alpha/\beta$, according to (2a) and (2b). In addition we allow the two tubes to carry two different liquids of different density and viscosity and put

$$(2c) \qquad\qquad \rho_2 = \gamma\rho_1 .$$

The factor $\gamma$ defines the change of the mass scale which is $\gamma\alpha^3$ because of (2a) and (2c).

It is more convenient to use instead of the coefficient of viscosity $\mu$ the *kinematic* viscosity $\nu$ defined by

$$\nu = \frac{\mu}{\rho} .$$

Again we put

$$(2d) \qquad\qquad \nu_2 = \delta\nu_1 .$$

We finally compare the pressures in corresponding cross-sections by setting

$$(2e) \qquad\qquad p_2 = \epsilon p_1 .$$

Obviously, $\epsilon$ could be expressed by $\alpha$, $\beta$, $\gamma$, because of the dimension of $p$.

We now write Eq. (1) after division by $\rho$ once more in a schematic way and omit the external force which is here immaterial:

$$(3) \qquad\qquad \text{acceleration} \quad -\nu\nabla^2\mathbf{v} + \frac{1}{\rho}\,\text{grad}\,p = 0.$$

The dimension of the acceleration term may be written either as velocity/time or velocity$^2$/length. In the transition from (1) to (2), this term takes up, according to (2a) and (2b), the factor

$$(4a) \qquad\qquad \frac{\beta^2}{\alpha} ;$$

according to (2b) and (2d) the second term of (3) takes up the factor

$$(4b) \qquad\qquad \delta\,\frac{\beta}{\alpha^2} ,$$

since the repeated differentiation with respect to the coordinates occurring in $\nabla^2\mathbf{v}$ means obviously a repeated division by a length. The change of scale in the last term of (3) is finally [by (3), (2a), (2c), (2e)]

$$(4c) \qquad\qquad \frac{1}{\gamma}\frac{\epsilon}{\alpha} .$$

Suppose now, Eq. (3) is obeyed by the flow in the experimental arrangement 1. If it is to be obeyed also by the flow 2, the ratio of the three factors (4a, b, c) must be equal to one:

(5)
$$\frac{\beta^2}{\alpha} : \delta\frac{\beta}{\alpha^2} : \frac{1}{\gamma}\frac{\epsilon}{\alpha} = 1 : 1 : 1.$$

This amounts to the following two relations:

(5a)          $$\frac{\beta\alpha}{\delta} = 1.$$          (5b)          $$\frac{\epsilon}{\gamma\beta^2} = 1.$$

The meaning of (5a) in terms of the experimental parameters involved is because of (2a, b, d)

(6)
$$\frac{v_1 a_1}{\nu_1} = \frac{v_2 a_2}{\nu_2}.$$

Similarly, (5b) leads by (2b, c, e) to

(7)
$$\frac{p_1}{\rho_1 v_1^2} = \frac{p_2}{\rho_2 v_2^2}.$$

Eqs. (6) and (7) constitute the results of Reynolds theory of similitude. In the literature the term *Reynolds criterion* is usually applied to equation (6) only, although (7) is a part of it as well, as we have just seen. It is the two relations together that furnish the necessary and sufficient conditions of hydrodynamic similitude. In what follows we shall make extensive use also of equation (7).

If these two criteria are satisfied, the flows 1 and 2 are either both laminar or both turbulent, since the experimental arrangement 2 is nothing but a "mechanical image" of 1 on a different scale.

The dimensionless expression defined by (6) is known as Reynolds number and denoted[20] by $R$:

(8)
$$R = \frac{va}{\nu},$$

where the length $a$ may designate any linear dimension characteristic of the experiment (such as a distance of two plates, a depth of a channel, radii of falling spheres, etc.). The number defined by (7) will be denoted by $S$ so that

(9)
$$S = \frac{p}{\rho v^2}.$$

---

[20]Sometimes denoted by *Re*.

The transition from laminar to turbulent flow in particular is a me-
chanically *similar* event for the tubes 1 and 2 which is therefore characterized by the same numerical value of $R$. This is the *critical Reynolds
number*:

$$(10) \qquad R_{\text{crit}} = \left(\frac{va}{\nu}\right)_{\text{crit}}.$$

Its value is rather sensitive to the particular form of the entrance section
of the tube. When a well rounded entrance section in the form of a trumpet
bell is used, a laminar flow is established from the start and stays laminar
up to rather large values of $R$. If the fluid enters the tube from the tank
through a straight, sharp edged connection, the initial flow is disturbed
by lateral components which are not damped out by friction at once;
the transition to full turbulence occurs then at a comparatively low $R$.
The critical Reynolds number is between 1200 (very irregular inlet) and
20,000 (mouth piece with a well-rounded fairing). Thus our statement
regarding the constancy of the critical Reynolds number should actually
be restricted to flows with similar initial conditions.

The question arises: what is the cause of the transition from laminar
to turbulent flow and vice versa? Our treatment of the laminar flow in
capillary tubes on p. 78ff does not point to anything like a limitation of
the flow pattern to small diameters or small velocities; on the contrary, it
appears that the Hagen-Poiseuille flow is a possible form of the motion
under all circumstances, but no longer *stable* if $R > R_{\text{crit}}$. Thus Reynolds's
criterion appears as a *stability criterion* dictated by experience; the following remarks should help to bring it closer to our physical understanding:

Viscosity tends to smooth out lateral components either initially
present or produced by the roughness of the tube, it thus favors the laminar
pattern. The inertia of the fluid requires the conservation of such side
components once they have come into existence, thus it acts in favor of
turbulence. In the static treatment of the capillary flow in 10, there was
no inertia, and side components were entirely out of consideration; thus
it appeared as if the results obtained there could be extended to arbitrarily large tubes and arbitrarily fast flows. Note also the antagonism
between viscosity and inertia that becomes apparent even in the structure
of the expression $\nu = \mu/\rho$ which occurs in the denominator of (8): An
increase of $\mu$ permits one to increase the value of $va$ without leaving the
region of stability; increasing $\rho$ diminishes that value and favors turbulence.

This view is also in accordance with the apparent lack of definition
of the transition point, a fact that could be compared with the well known
thermo-dynamic phenomenon of super-cooled water. Water can be cooled

below 0°C without freezing provided that it is left entirely undisturbed. Similarly, the regime of the laminar flow can be extended to larger $R$ by carefully avoiding the production of lateral components at the inlet or or by the roughness of the walls.

In order to decide whether there exists at all a sharply defined ideal value of the critical Reynolds number one would have to proceed as follows: one studies the flow in a sufficiently long tube with sufficiently smooth walls at a sufficiently large distance from the inlet. Initial flow irregularities in the region of the inlet will have enough time to attenuate in the stable case without the interference of side components newly created at the wall. If now at certain $R$-values laminar and at other $R$-values turbulent flow is observed and if the upper limit of the *former* $R$-values agrees with the lower limit of the *latter* $R$-values, the common limit would appear as the ideal value of $R_{crit}$ .[21] Such an experiment cannot be carried out practically, therefore the large variation of the values of $R_{crit}$ given in the literature.[22]

The transition to turbulent flow becomes apparent not only in the change of the flow pattern, but also in the change of the *pressure-law*, to which we now turn. Here the second law of similitude (9) comes into the foreground. We replace $p$ in (9) by the pressure difference $\Pi$ between the beginning and the end of the tube observing that the dimensions of $\Pi$ and $p$ are, of course, the same. In doing so the length $l$ of the tube must now also be considered (the tube in Eq. (9) could have had any arbitrary length). Now $l$ cannot be related to the dimensionless number in (9), but we may use the ratio $l/a$ to rewrite the definition of $S$ as

$$(11) \qquad \Pi = \rho v^2 \frac{l}{a} S,$$

which is valid for the similar motions 1 and 2 at the same $S$-value but for different subscripts of $\Pi$, $\rho$, $v$, $l$, $a$.

We now compare (11) with the formula used in hydraulics to compute the loss of pressure head in pipe or channel systems:

$$(11a) \qquad h = \lambda \frac{v^2}{2g} l \frac{U}{F} .$$

[21]It seems not impossible that with improved experimental conditions (better regulation at the inlet and smoother walls) $R_{crit}$ may be raised to arbitrarily high values. If this should be true, an ideal value would not exist and Reynolds's stability limit would only be of practical value, depending on experimental conditions. On the other hand, there seems to exist a definite Reynolds number below which turbulent flow cannot be maintained.

[22]They have been systematically investigated according to inlet conditions by L. Schiller.

Here $\lambda$ is a pure number, the "hydraulic friction coefficient", $F$ is the cross-sectional area of the flow, $U$ the wetted circumference of the conduit. For a circular cross-section filled with fluid, $U/F$ becomes $2/a$ so that

(11b) $$h = \lambda \frac{v^2}{g} \frac{l}{a}.$$

On identifying $S$ with $\lambda$ and expressing $\Pi$ by the hydrostatic level difference $h$ ($\Pi = \gamma h = \rho g h$), (11b) becomes identical with (11).

If, on the other hand, $S$ is identified with the pure number $8/R$, so that

$$S = \frac{8\nu}{va} = \frac{8\mu}{\rho va},$$

the Hagen-Poiseuille formula

(12) $$\Pi = 8\mu l \frac{v}{a^2},$$

is obtained with the only difference that, previously, we had the more precisely defined average velocity $u_m$ where we now have $v$.

In (11b) and (12) the different forms of the pressure dependence in laminar and turbulent flow that was already mentioned on p. 80, become again apparent: in (11b) the dependence on $v$ and $a$ is in the form $v^2/a$, in (12) in the form $v/a^2$. We have been able to overcome this contrasting behavior by choosing the number $S$ in two different ways. (Note that the definition of $S$ in (9) does not, in itself, teach anything about the interdependence of the flow parameters; only the identification of $S$ with another non-dimensional combination of flow parameters has the character of a physical law.)

It will be noticed that the hydraulic formula (11a) has the same structure as a law of air resistance that had already been used by Newton. According to this law, a body having the velocity $v$ relative to air experiences a resistance proportional to $v^2$ (the same is true for the wind pressure against a surface at rest). Also in ballistics the resistance law is, at subsonic velocities, nearly a $v^2$-law.

A more accurate representation of the pressure losses observed in turbulent flow through smooth pipes is possible within the framework of Eq. (11) if a *weak* dependence of $S$ on $R$ is assumed, (that is, $S$ is supposed neither independent of $R$ as in (11b) nor $\sim R^{-1}$ as in (12)). The following assumptions have been tested for larger $R$-values

(13) $$S = \lambda R^{-\kappa} \quad \text{and} \quad S = \lambda_0 + \lambda_1 R^{-\kappa}$$

with $\kappa = 1/4$ to $1/5$ in the first formula (13) and with somewhat larger values of $\kappa$ in the second one.

The hydraulic $v^2$-law (11a) is modified by the first formula (13) (Blasius formula, $\kappa = 1/4$) to $v^{2-\kappa}$ which agrees almost exactly with old observations of Hagen and Reynolds (cf. also problem III, 2). For more details about the pressure in the transition region and the influence of roughness see L. Prandtl's "Strömungslehre", Braunschweig 1949, which can be considered as a source book in this field.[23]

The change of the pressure law at the critical limit is schematically represented in Fig. 18: At small velocities $v < v_{crit}$ the linear law of

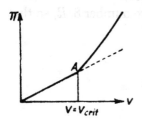

FIG. 18. The pressure gradient as a function of the mean velocity $v$. At the critical point $A$ the laminar flow becomes turbulent.

Hagen-Poiseuille, at velocities $v > v_{crit}$ the increase according to $v^{2-\kappa}$. The broken line that continues the straight branch beyond $A$ represents the possible increase of the stability limit $A$.

The change of the pressure law at $A$ is accompanied ("caused" would perhaps be the better word) by a noticeable change of the velocity distribution over the cross-section. In contrast to the parabolic distribution of the laminar case, the time average of the velocity in turbulent flow is practically constant over the interior of the tube; only in the close vicinity of the wall it drops quickly to zero, satisfying the boundary condition $v = 0$.

The contrast laminar-turbulent occurs not only in circular tubes or pipes with other cross-sections, but can also be observed in the flow between two plates at rest. This flow pattern, a two-dimensional analogy to the Hagen-Poiseuille flow, was treated in problem II, 3a. The parabolic velocity profile which is also characteristic for this case is shown in Fig. 19.

If the parabolic cylinder of Fig. 19 is cut along the symmetry plane $y = h$, that is, if only the lower half of the flow pattern is considered, one obtains the velocity distribution in a stream in laminar motion (strictly speaking, the stream is infinitely wide). Let the bottom of the stream form the angle $\alpha$ with the horizontal (Fig. 19a), so that the component of gravity $\rho g \sin \alpha$ replaces the pressure gradient of the previous figure. L. Hopf[24] (Munich thesis, 1909) made a careful investigation of

---

[23]For pipe flow compare J. C. Hunsaker and B. G. Rightmire, Engineering application of Fluid Dynamics, McGraw-Hill, New York, 1947, Chap. VIII.

[24]Partly published in Ann. d. Phys. **32**, 777 (1910).

the experimental realization of such a laminar stream stabilized by viscous forces, and of the limits of its stability. The critical Reynolds number in this investigation was found to be $R_{crit} = vh/\nu \cong 330$.

An even simpler case of laminar motion develops from Fig. 19 when one of the plates is at rest while the other moves with constant velocity

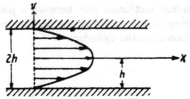

FIG. 19. Laminar flow between two plates at rest.

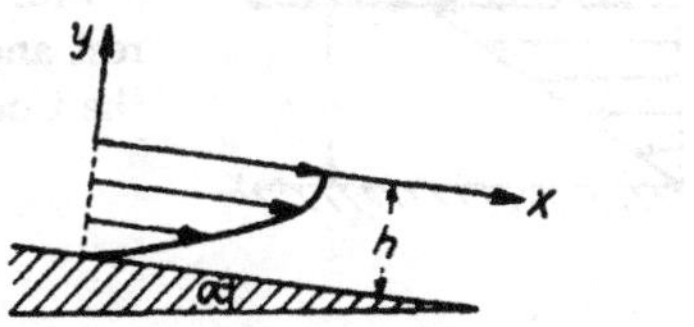

FIG. 19a. The "river in laminar flow" under the influence of gravity; free surface at $y = 0$.

$U$. The velocity profile is here linear for laminar flow (cf. Fig. 19b). On denoting the flow velocity by $u$, we have

$$(14) \qquad u = \frac{y}{h} U \qquad \text{and, of course,} \qquad v = w = 0.$$

The flow proceeds parallel to the $x$-direction; it should be noticed, however, that this flow is not irrotational although there seems to be no rotation involved in it:

$$2\omega_z = \frac{\partial v}{\partial x} - \frac{\partial u}{\partial y} = -\frac{U}{h}.$$

Because of its particular analytic simplicity, (14) is the preferred flow for theoretical investigations of stability (cf. 38).

Following Couette,[25] an elegant realization of this flow can be made by means of two coaxial cylinders. When the inner cylinder is at rest and the outer cylinder moves with a circumferential velocity $U$, transition to turbulent flow occurs according to Couette at a Reynolds number $R_{crit} = Uh/\nu = 1900$ where $h$ is the thickness of the fluid layer. The converse case where the inner cylinder moves has been thoroughly investigated by Taylor.[26]

In this discussion we have only scratched the surface of the turbulence problem. Central questions such as: "What is the reason for the instability?" "Is it possible to give a mathematical description of the turbulent fluctuations of the velocity?" "Is the turbulent state contained among

[25]M. Couette, Ann. Chim. Phys. 21 (1890).
[26]Sir G. I. Taylor, Phil. Trans. 223, 289 (1923).

the solutions of the Navier-Stokes equations?" have not been touched. No complete answer to these questions to which we shall return in 38 seems possible today.

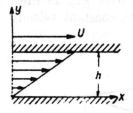

FIG. 19b. The laminar rectilinear flow between a plate at rest and in motion may be considered as the limiting case of the Couette flow between coaxial cylinders.

## 17. Some Remarks on Capillarity[27]

The molecular forces of cohesion that are responsible for the phenomena of capillarity have already been encountered at the end of 12 in the form of the normal pressure $N$ acting on the fluid surface (Laplace's theory). For the purpose of a short survey, however, it is more convenient not to adopt Laplace's[28] starting point but to follow Gauss[29] who proceeds from a minimum principle for the surface energy; the normal pressure $N$ of Eq. (12.16) will be derived from this minimum principle.

We stipulate that there acts within the surface $F$ of the fluid an everywhere equal *surface tension T* in tangential direction: $F$ is, as it were, covered by a membrane which does not by itself possess elasticity of form. Its resistance against deformation (comparable with that of a thin rubber skin) is furnished solely by a tangential tension $T$ that acts along the border of the membrane in the direction that is normal to the border and lies in the tangential plane.

We start with a simplified case, restricting ourselves to a *cylindrical* fluid surface or, which amounts to the same, to a membrane of cylindrical form. The tension equilibrium in such a membrane is the same as that of a *vibrating string* the instantaneous shape of which coincides with the profile of the cylinder.

Let $u$ be the displacement of the string between two neighboring points $P$ and $P'$ and $\Delta s$ their distance. We separate this segment from the rest of the string by cuts at $P$ and $P'$ (cf. Fig. 20a). To make up for the

---

[27]Since we wish to treat the phenomena of capillarity only to such extent as is indispensable in a textbook on general hydrodynamics, we refer for a more complete representation to H. Minkowski's carefully written article in Enzyklopädie d. Math. Wiss. Vol. V, 1 p. 558: there the reader will also find a molecular-theoretical explanation of capillarity which is more complete and cuts deeper than our purely phenomenological remarks.

[28]Mécanique céleste, Supplément de l'action capillaire, 1806.

[29]Principia generalia theoriae fluidorum in statu aequilibrii, (1830).

effect of the two remaining pieces, we introduce the tensions **T** and **T'** acting in the tangential directions of the string.[30] They are both equal in magnitude to the tension to which the string is subjected (say, by the peg at the end) but are not parallel to each other. Let $\Delta\epsilon$ be the change

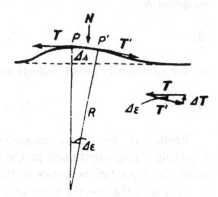

FIG. 20a. The tension **T** of a vibrating string and the corresponding restoring force **N**. At the same time this is a cross-sectional diagram of a cylindrical liquid surface.

of the slope corresponding to $\Delta s$, and $\Delta$**T** the closing side of the force triangle with sides **T** and **T'** (insert of Fig. 20a). The magnitude of the vector $\Delta$**T** is then

$$\Delta T = T \mid \Delta\epsilon \mid$$

and its direction is perpendicular to the arc element $\Delta s$. The resultant force acting on $\Delta s$ equals $-\Delta$**T**. Instead of this, we introduce the resultant force per unit of length **N** so that **N**$\Delta s = -\Delta$**T** and consequently

(1) $$N\Delta s = T \mid \Delta\epsilon \mid.$$

On the other hand, the curvature $1/R$ is given by

(2) $$\pm \frac{1}{R} = \lim_{\Delta s \to 0} \frac{\Delta\epsilon}{\Delta s} = \frac{\frac{\partial^2 u}{\partial x^2}}{\left(1 + \left(\frac{\partial u}{\partial x}\right)^2\right)^{3/2}}$$

where $u$ is the deflection of the string and $x$ the abscissa taken along the undeflected string; in the denominator of the last member we must take the positive value of the square root if $s$ increases with $x$, which we assume.

Eqs. (1) and (2) determine the magnitude of **N**:

$$N = \frac{1}{\mid R \mid} T.$$

The direction of **N** points always to the concave side of the string.

---

[30]Note that the dimension of $T$ here is that of a force while in the following two-dimensional case $T$ has the dimension force/length.

Let us now slightly extend the meaning of $N$ by providing it with a $+$ or $-$ sign according to whether $\Delta\epsilon$ is negative or positive. (In Fig. 20a, $N$ is thus positive.) If, in addition, we adopt the lower sign in (2) which makes $R \gtrless 0$ if $\Delta\epsilon \lessgtr 0$, we may rewrite the last equation for the modified $N$:

$$(3) \qquad\qquad N = \frac{1}{R}\,T,$$

or, for sufficiently small deflection,

$$(4) \qquad\qquad N = -\frac{\partial^2 u}{\partial x^2}\,T.$$

From (4) we derive immediately the differential equation of the vibrating string mentioned in the context of Eq. (1°.9). We have only to state the equilibrium between the external force in $+u$-direction (that is $-N$) and the inertia force of the unit length of the string $-\rho\,\partial^2 u/\partial t^2$, and obtain

$$(5) \qquad\qquad \rho\,\frac{\partial^2 u}{\partial t^2} = T\,\frac{\partial^2 u}{\partial x^2}$$

where $\rho$ is the linear density of the string.

If we now pass from a cylindrical to a surface of double curvature, (4) has to be replaced by

$$(6) \qquad\qquad N = \left(\frac{1}{R_1} + \frac{1}{R_2}\right)T,$$

where $R_1$ and $R_2$ are the two principal radii of curvature at the surface point under consideration. While in general the surface normals at two infinitesimally distant surface points do not intersect, but are skew to

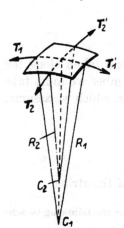

FIG. 20b. A vibrating membrane of double curvature; $R_1$ and $R_2$ are the radii of curvature of the two principal sections. At the same time this represents a liquid surface of the same shape.

each other, surface normals along the principal directions do intersect, and it is exactly this circumstance that defines the principal directions. The radii $R_1$, $R_2$ are then explained as the reciprocals of the curvatures of the two principal (normal) sections. Denoting their centers of curvature by $C_1$ and $C_2$ respectively (see Fig. 20b), we now investigate the equilibrium of the surface tensions **T** acting on the infinitesimal rectangle $\Delta s_1 \Delta s_2$ of Fig. 20b oriented parallel to the principal directions 1 and 2. If one constructs the force triangles for either principal section as in Fig. 20a one obtains two concluding sides $\Delta \mathbf{T}_1$ and $\Delta \mathbf{T}_2$ and from them two normal pressures $N_1$ and $N_2$ as in equation (4). The normal pressure $N$ in Eq. (6) is evidently obtained by adding $N_1$ and $N_2$.

From (6), *the differential equation of the vibrating membrane* follows directly in the form mentioned in the context of Eq. (13.9), if one puts according to (3) and (4)

$$(7) \qquad \frac{1}{R_1} \sim -\frac{\partial^2 u}{\partial x^2}, \qquad \frac{1}{R_2} \sim -\frac{\partial^2 u}{\partial y^2}$$

and substitutes for $N$ the inertia force

$$N = -\rho \frac{\partial^2 u}{\partial t^2}, \qquad (\rho = \text{surface density}).$$

The equation thus obtained from (6)

$$(8) \qquad \frac{\partial^2 u}{\partial t^2} = c^2 \left( \frac{\partial^2 u}{\partial x^2} + \frac{\partial^2 u}{\partial y^2} \right), \qquad c^2 = \frac{T}{\rho}$$

is not only correct for coordinates taken in the principal directions, but because of the invariance of the $\nabla^2$-operator for arbitrary Cartesian coordinates $x$, $y$, too.

The quantity

$$(9) \qquad M = \frac{1}{R_1} + \frac{1}{R_2}$$

that occurs in (6) is the *mean curvature* of the surface, while the measure of the curvature[31] introduced by Gauss

$$(9a) \qquad K = \frac{1}{R_1 R_2}$$

is called *Gauss's or total curvature*.

---

[31]Minkowski *loc. cit.* defines the mean curvature $M = 1/2 (1/R_1 + 1/R_2)$ which corresponds better to the idea of a "mean" curvature. We retain the usual definition (9), which is, of course, irrelevant for the following results.

Our remarks about the vibrating string and membrane seem to be a digression from the subject of capillarity, but actually Eqs. (4) or (6) are applicable, as they stand, for the capillary surface skin in a cylindrical or more general form. They can be introduced in Eqs. (12.19) or (12.21), which were derived from Hamilton's principle, and determine therefore the surface condition for the hydrodynamic pressure $p$.

Let us first take a *free surface* such as water bounded by air; then $p = N$ according to (12.19). For a cylindrical surface (realized by a capillary wave traveling, e.g., in $x$-direction), $N$ is given by (3) or (4), and we have

$$(10) \qquad p = \frac{T}{R} \quad \text{or} \quad p = -T\frac{\partial^2 u}{\partial x^2},$$

where in the second equation the displacement $u$ relative to the rest position is supposed to be sufficiently small. We shall come back to this equation in 25. Likewise we have for a surface of general shape according to (6)

$$(11) \qquad p = T\left(\frac{1}{R_1} + \frac{1}{R_2}\right).$$

Here as well as in (10) $p$ is the (hydrodynamic) pressure in the fluid over which the capillary lamina is spread out.[32]

A soap film subtended by a rigid boundary curve (a wire loop bent in some way) can be considered as an isolated capillary lamina devoid of the fluid body by which it is otherwise supported. Since the air pressure on both sides cancels, $p = 0$, and according to (11) the mean curvature $M$ is then also zero. The same condition would, of course, be true for an actual membrane in equilibrium if subjected to the same boundary conditions; the equilibrium condition $\nabla^2 u = 0$ as it follows from (8) agrees with (11) when $p = 0$, provided the membrane is approximately plane.

The vanishing of $M$ means that the two principal radii of curvature are numerically equal but opposite. A surface with this property is called a *minimal surface*. Its area $F$ is smaller than that of any neighboring surface with the same boundary curve. This is easily understood from the capillary standpoint: The tensional energy $U$ stored in the capillary film is

$$(12) \qquad U = T \cdot F.$$

---

[32]As a supplement to Eqs. (10) and (11) we should like to emphasize that there is also a pressure load on the plane, undisturbed surface termed internal pressure by van der Waals. As indicated on p. 93, it is a consequence of the molecular attraction. The surface tension $T$ is present on a plane surface too, but it becomes active only on a wavy surface where it changes direction from point to point.

A virtual displacement at constant $T$ with the boundary kept fixed produces a change of energy

(13) $$\delta U = T\delta F.$$

This, on the other hand, is the virtual work which is supposed to vanish in equilibrium. Therefore $\delta F = 0$ is the equilibrium condition, indicating that *the surface F occurring in the energy expression* (12) *must be a surface of smallest area.*

This is shown in the following well known experiment:[33] on a plane soap film with an otherwise arbitrary boundary a loop of thread is placed. The film inside the loop is destroyed, e.g., by piercing it. Upon that, the loop is immediately pulled out into circular form. Since the circle is the curve of maximum area for a given circumference the remaining part of the membrane has the smallest area for the given (rigid) boundary and length of thread.

In the interior of a *closed soap bubble* there is overpressure relative to the external pressure. The shape of the bubble is spherical if gravity is neglected; since for reasons of symmetry $R_1 = R_2$, the radius of the bubble is by (11)

$$R = \frac{2T}{p}.$$

In a structure consisting of many adjacent bubbles (*foam*), the pressures in the single bubbles or cells will be slightly different. The pressure difference along the walls of the structure is therefore constant for any region that bounds not more than two cells. According to (11) the entire structure is then composed of surface pieces of *constant mean curvature.*

Condition (13) applies also to phenomena involving the *angle of contact.* Consider e.g. an arrangement as in Fig. 21 and suppose that, of the three

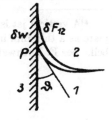

FIG. 21. Equilibrium between a liquid (1), air (2), and a rigid wall (3); $\vartheta$ is the angle of contact.

media abutting on each other at $P$, 1 is a liquid, 2 is air and 3 a solid. $P$ is a point of the contact line which we imagine perpendicular to the plane of the drawing; if the point $P$ is given a virtual displacement up-

---

[33]A great many other experiments with accurate description of the soap bubble technique in C. V. Boys, Soap Bubbles, London (1931).

ward, $\delta w$, the surface of separation $F_{12}$ is increased per unit of length of the contact line by

$$\delta F_{12} = \cos \vartheta \cdot \delta w \cdot 1,$$

where $\vartheta$ is the angle of contact, and the surface of contact $F_{13}$ is increased by

$$\delta F_{13} = \delta w \cdot 1.$$

As in (13) we stipulate

$$T_{12}\delta F_{12} + T_{13}\delta F_{13} = (T_{12} \cos \vartheta + T_{13})\delta w = 0,$$

or, since $\delta w$ is an arbitrary variation,

$$(14) \qquad \cos \vartheta = - \frac{T_{13}}{T_{12}}.$$

In order to obtain an acute angle for $\vartheta$ (water and glass), $T_{13}$ must be negative and $|T_{13}| < T_{12}$. For non wetting liquids (mercury and glass) $\vartheta$ is obtuse.

Capillary elevation and depression are caused by the combined action of gravity and surface tension, cf. problem III, 5. As familiar a phenomenon, however, as a liquid droplet under the action of gravity leads to a complicated differential equation that admits only approximate solutions. A less complicated problem is the theory of the beautiful phenomena known as "water bells".[34]

A complete theory of capillarity is not possible without the use of thermodynamic considerations (Gibbs).

We should like to add that in the original theory of Laplace there is, beside the surface contribution to the capillary energy [our equation (12)], a second term which is proportional to the *volume* of the liquid. This term is essential in van der Waals's theory of the liquid state; it corresponds to the constant $a$ that occurs in van der Waals's equation of state of a real gas.

---

[34]A vertical jet is directed centrally against a circular plate. The liquid leaves the plate as a thin film in the form of a surface of revolution, sometimes in the shape of a bell. See E. Buchwald and H. König, Ann. d. Phys. **23**, 557 (1935) and **26**, 659 (1936).

## CHAPTER IV

# VORTEX THEORY

## 18. Helmholtz's Vortex Theorems

In the paper of 1858 in which Helmholtz gives the kinematic analysis of vortex motion (cf. the footnote on p. 1) he also completes the dynamic theory of vortices in its essentials. Simplifications in method were found in the following decades, but new results were not discovered.

The main content of Helmholtz's theory are the conservation laws:[1] *It is impossible to produce or destroy vortices,* or, expressed in more general terms, *the vortex strength is constant in time.* This theorem is correct under the following conditions: the fluid is inviscid and incompressible; the external forces possess a single-valued potential within the space filled by the fluid. Apart from the conservation of the vortex strength *in time* we shall see that there is also a *spatial* conservation: the vortex strength is constant *along each vortex line or vortex tube,* which must be either closed or end at the boundary of the fluid.

## 1. The Differential Form of the Conservation Theorem

Following Helmholtz we start from Euler's equations in the form (11.2). Since the external force **F** has a potential $U$ by hypothesis, Euler's equations can be written

$$(1) \qquad \frac{d\mathbf{v}}{dt} = -\operatorname{grad}\left(\frac{p+U}{\rho}\right).$$

The differential quotient on the left side is again the material acceleration, that is, the total rate of change of the velocity of the material particle under consideration. This leads to the form (11.7) of Euler's equations which can now be written in the following way:

$$(2) \qquad \frac{\partial \mathbf{v}}{\partial t} - \mathbf{v} \times \operatorname{curl} \mathbf{v} = -\operatorname{grad}\left(\frac{\mathbf{v}^2}{2} + \frac{p+U}{\rho}\right).$$

---

[1]Corresponding theorems have been found in posthumous papers of Lejeune Dirichlet, +1859, and were published by Dedekind who edited Dirichlet's collected works.

We take the curl of this equation: the right member vanishes because of curl grad $= 0$. On setting $\omega = \frac{1}{2}$ curl $\mathbf{v}$ we obtain

$$(3) \qquad \frac{\partial \omega}{\partial t} - \text{curl } (\mathbf{v} \times \omega) = 0.$$

We make use of the following well known formula to transform the vector product of any two vectors $\omega$ and $\mathbf{v}$:

$$(4) \qquad \text{curl } (\omega \times \mathbf{v}) = (\mathbf{v} \text{ grad}) \, \omega - (\omega \text{ grad}) \, \mathbf{v} + \omega \text{ div } \mathbf{v} - \mathbf{v} \text{ div } \omega.$$

(This formula is proved, just as the related formula (11.6), by verification in Cartesian coordinates.)

In our case, (4) simplifies on account of

$$\text{div } \mathbf{v} = 0 \qquad \text{and} \qquad \text{div } \omega = \text{div curl } \mathbf{v} = 0,$$

and one obtains by substitution in (3)

$$(5) \qquad \frac{\partial \omega}{\partial t} + (\mathbf{v} \text{ grad}) \, \omega = (\omega \text{ grad}) \, \mathbf{v}.$$

The left member is the material rate of change of $\omega$ so that (5) can also be written as

$$(6) \qquad \frac{d\omega}{dt} = (\omega \text{ grad}) \, \mathbf{v}.$$

*The material rate of change of $\omega$ is zero wherever $\omega$ is zero.* From this point Helmholtz immediately proceeds to the conclusion, that *fluid particles which were vortex-free at some time remain so for ever.* The reasoning he may have had in mind could be this: Let $\omega = 0$ for $t = t_0$, then by (6) $d\omega/dt = 0$. These two statements seem to imply that, for the mass particle under consideration, $\omega$ is still zero for $t = t_0 + \Delta t$. Then (6) implies $d\omega/dt = 0$ also for $t_0 + \Delta t$, which is equivalent to $\omega = 0$ for $t = t_0 + 2\Delta t$, and so on.

For a rigorous mathematical proof we need more preparation: Let $\mathbf{A}$ be a vector associated with the moving fluid (not a "point function",

FIG. 22. The flux of a vector through a surface element $\sigma$ that moves with the fluid.

but a "particle function") and $d\sigma$ a surface element that likewise moves with the fluid, changing its shape and position in the course of time. Let $A_n$ denote the component of $\mathbf{A}$ normal to $d\sigma$ and consider the flux of $\mathbf{A}$ through $d\sigma$ at successive instances $t_1$ and $t_2$. Corresponding line elements

$ds_1$ and $ds_2$ of the boundaries of $d\sigma_1$ and $d\sigma_2$ are correlated to each other by the displacement $w = vdt$, which is the length of the infinitesimal region in Fig. 22. Its top and bottom surfaces are $d\sigma_2$ and $d\sigma_1$ and its volume $d\tau = w_n\, d\sigma$ . (In view of the limiting process we are going to perform, it does not make any difference whether we identify $d\sigma = d\sigma_1$ and $w_n \perp d\sigma_1$, or $d\sigma = d\sigma_2$ and $w_n \perp d\sigma_2$).

Gauss's theorem or the definition of the divergence (both amount to the same) yields

$$(7)\qquad \int A_n\, d\sigma = \text{div } \mathbf{A}\, d\tau = (\text{div } \mathbf{A})w_n\, d\sigma.$$

The integral to the left refers to the boundary of the volume swept by $d\sigma$ in the time $dt$. Note, however, that the vector field $A$ is for the present kept constant in time. (Think of it as "frozen" at the instant $t_1$ .)

In Fig. 22 one element of the *side surface* of $d\tau$ has been marked. Its contribution to the integral is

$$\mathbf{A}\cdot(d\mathbf{s} \times \mathbf{w}) = d\mathbf{s}\cdot(\mathbf{w} \times \mathbf{A}) = ds(\mathbf{w} \times \mathbf{A})_s,$$

where the subscript $s$ to the last term denotes the component in the direction of $d\mathbf{s}$, which makes a right hand screw with the outward normal of the top surface. The integral over the side wall thus becomes equal to the line integral over the boundary of the surface

$$(7a)\qquad \int ds\,(\mathbf{w} \times \mathbf{A})_s = d\sigma\, \text{curl}_n\,(\mathbf{w} \times \mathbf{A}),$$

a transformation that corresponds to Stokes's theorem.

The contributions of the top and bottom surfaces to the integral in (7) are simply

$$(A_n\, d\sigma)_2 - (A_n\, d\sigma)_1 = \delta(A_n\, d\sigma)$$

(note the opposite orientation of the normals). The right member $\delta(A_n\, d\sigma)$ denotes the change of the A-flux due to the motion and distortion of $d\sigma$. From (7) and (7a) this change is now found as

$$(7b)\qquad \delta(A_n\, d\sigma) = (\mathbf{w}\,\text{div } \mathbf{A} - \text{curl}\,(\mathbf{w} \times \mathbf{A}))_n\, d\sigma.$$

We have still to take account of the change of the field $\mathbf{A}$ itself. The change of flux due to this cause is during the time $dt$

$$\frac{\partial A_n}{\partial t}\, dt\, d\sigma$$

On denoting the total change of flux by

$$\frac{d}{dt}\,(A_n\, d\sigma)\, dt$$

and substituting for the displacement **w** its value **v** $dt$, we obtain finally

(7c)          $\dfrac{d}{dt}(A_n \, d\sigma) = \left(\dfrac{\partial \mathbf{A}}{\partial t} + \mathbf{v} \operatorname{div} \mathbf{A} - \operatorname{curl}(\mathbf{v} \times \mathbf{A})\right)_n d\sigma.$

This formula, too, was found by Helmholtz,[2] but in another connection. It appears in his posthumous paper about the principle of least action in electrodynamics.

Upon identifying the vector **A** with the vortex vector **ω**, Eq. (7c) which is correct for any direction of the normal $n$ of $d\sigma$, becomes

(8)          $\dfrac{d}{dt}(\omega_n \, d\sigma) = \left(\dfrac{\partial \boldsymbol{\omega}}{\partial t} - \operatorname{curl}(\mathbf{v} \times \boldsymbol{\omega})\right)_n d\sigma$

since div **ω** = 0. The right member vanishes because of (3), thus

(9)          $\dfrac{d}{dt}(\omega_n \, d\sigma) = 0.$

This equation expresses the theorem of *conservation of vortices*. The flux of the vortex vector has a certain unalterable value for every surface element that moves with the fluid. *Vortices cannot be created or destroyed*, or, *the vorticity is a convective quantity of the flow; it adheres to the individual fluid particle and moves along with it.*

## 2. The Integral Form of the Conservation Theorem

We start from the concept of *circulation* (p. 17) established by Lord Kelvin[3] with the aim of simplifying Helmholtz's vortex theory. Identifying the general vector **A** of (2.21) with our **v**, we consider the circulation

(10)          $\Gamma = \oint v_s \, ds = \oint (\mathbf{v} \cdot d\mathbf{s}) = \oint (u \, dx + v \, dy + w \, dz)$

around a circuit, i.e., a closed oriented curve in the fluid; we think of it as of a "material curve" that consists, like a string of beads, of *material fluid particles*, and floats along with them. Let $C$ be the circuit at the

[2]Ges. Werke, Vol. III, p. 476, cf. particularly the equations (8a), (8b) that refer to the vector of the virtual displacement $\mathbf{A} = \{\delta\xi, \, \delta\eta, \, \delta\zeta\}$. Helmholtz writes everything in components, which does not add to lucidity; our notation and derivation corresponds to H. A. Lorentz's presentation in Enzykl. d. Math. Wiss. Vol. V, 2, p. 75, Eq. (5); there the equation is shown to be of fundamental importance in the electrodynamics of moving bodies.

[3]W. Thomson (Lord Kelvin), On Vortex Motion, Transactions R. Soc. Edinburgh, Vol. 25 (1869). Reprinted in Mathem. and Phys. Papers, Vol. IV, p. 13. In the same volume on p. 1, there is also the paper on vortex atoms often referred to in the popular literature: the vortex atoms left no impression on the development of atomic physics.

instant $t$ and $C'$ the distorted circuit at $t + \Delta t$ (see Fig. 23). Let $ds$ be an arc element of $C$ and $ds'$ the corresponding element of $C'$. The straight arrows in the figure represent the paths of the fluid particles during $\Delta t$, drawn for two neighboring particles $i$ and $i + 1$; the lengths are

$$\mathbf{v}_i \Delta t \quad \text{and} \quad \mathbf{v}_{i+1} \Delta t.$$

The arrows correspond, of course, to the vector $\mathbf{w}$ in Fig. 22. Fig. 23 differs from Fig. 22 only in that the infinitesimal boundary curves of the elements $d\sigma$ are now pulled out into finite circuits $C$ and $C'$. The following calculation is likewise closely related to the previous one.

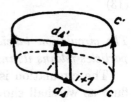

Fig. 23. Illustrating the concept of circulation.

First, we see from the quadrilateral formed by the arrows in Fig. 23 that

$$\mathbf{v}_i \Delta t + ds' = ds + \mathbf{v}_{i+1} \Delta t$$

or

(11)
$$\frac{ds' - ds}{\Delta t} = \mathbf{v}_{i+1} - \mathbf{v}_i .$$

We now calculate

(12)
$$\frac{d\Gamma}{dt} = \lim_{\Delta t \to 0} \frac{\Gamma' - \Gamma}{\Delta t} = \frac{d}{dt} \oint \mathbf{v} \cdot ds = \oint \frac{d\mathbf{v}}{dt} \cdot ds + \oint \mathbf{v} \cdot \frac{d}{dt} ds,$$

where $\Gamma'$ stands for the circulation about $C'$. The expression that appears in the last integrand

$$\frac{d}{dt} ds$$

is nothing else but the limit of the first member of (11) for $\Delta t \to 0$. An approximate value of the integral itself is thus found in the sum

$$\sum_i \mathbf{v}_i \cdot (\mathbf{v}_{i+1} - \mathbf{v}_i).$$

If we now go to the limit $i \to \infty$, the polygon $\cdots i, i+1 \cdots$ approaches the circuit $C$ and our sum approaches the line integral over $C$

$$\oint \mathbf{v} \cdot d\mathbf{v} = \frac{1}{2} \mathbf{v}^2 \Big|_A^B ,$$

where $A$ is the arbitrary starting point (corresponding to $i = 0$) and $B$ the end point of the integration, which coincides with $A$. *Thus the second integral in the last member of* (12) *vanishes.*

But the first integral vanishes likewise. According to Euler's Eqs. (1) it equals the line integral of $-\text{grad}\,[(p + U)/\rho]$, that is

$$\frac{p + U}{\rho}\bigg|_B^A = 0.$$

Thus we have according to (12)

(13) $$\frac{d\Gamma}{dt} = 0, \qquad \Gamma = \text{const.}$$

*The circulation taken about an arbitrary circuit that consists of material particles retains its initial value throughout the flow.*

The circulation is directly connected with the *vorticity* present in the fluid as we shall show now. According to *Stokes's* theorem (3.2) we have

(14) $$\Gamma = \oint \mathbf{v}\cdot d\mathbf{s} = \oint v_s\, ds = \int \text{curl}_n\, \mathbf{v}\, d\sigma = 2\int \omega_n\, d\sigma.$$

If we now consider an arbitrary surface bounded by the circuit $C$ and take at every element $d\sigma$ the normal component of the vortex vector $\boldsymbol{\omega}$, $\Gamma$ equals twice the vorticity that passes the surface $\sigma$. Hence *the flux of the vortex vector through the surface $\sigma$ remains constant throughout the flow.* Helmholtz's vortex theorem appears here in the form of *Kelvin's circulation theorem*; the contents of both are identical.*

It is, of course, not necessary to follow a particular surface $\sigma$ chosen in (14) throughout its motion, since any surface through $C$ gives the same vortex flux $\omega_n\, d\sigma$. This is again a consequence of Stokes's theorem which

---

*It is now easy to see that the assumption $\rho = \text{const}$ is *not necessary* for the validity of the conservation theorem. There is only one point in the foregoing proof that has not a purely kinematic character, and this is the vanishing of $\oint (d\mathbf{v}/dt) \cdot d\mathbf{s}$ in (12); here the compressibility might have an effect. But in the compressible case the integrand may be written in the form $-\text{grad}\,(\mathcal{P} + V)$ when the notation of Eqs. (7.6b) and (7.7) is used, provided that a $p,\rho$-relation is given.

The condition for the validity of Helmholtz's (or Kelvin's) theorem is therefore the existence of a single-valued "acceleration potential"

$$\int_{P_1}^{P} \frac{1}{\rho}\,(\text{grad}\,p + \mathbf{F})\cdot d\mathbf{s},$$

and it does exist when

1. the fluid is inviscid
2. the external force per unit of mass, $\mathbf{F}/\rho$, has a single-valued potential $V$
3. a $p,\rho$-relation exists.

permits us to use an *arbitrary* surface $\sigma$. But it is required that the circuit $C$ consist permanently of the same material particles.

Eqs. (13) and (14) become obviously identical with the previous Eq. (9) when $\sigma$ becomes infinitesimal. Conversely, we could have derived the present result from (9) by an integration over $d\sigma$, but the reasoning used here is simpler and less abstract than the previous one and might be considered as a supplement not entirely superfluous.

While Eqs. (1) and (2) represent the *differential form* of Euler's equations, the circulation theorem (13) could be called their *integral form*, using a terminological distinction taken from electrodynamics. Likewise Eqs. (7) appearing in 15 represent the *differential form of Maxwell's equations*. Besides, there is an integral form of the same equations expressing their physical content in the form of relations between line and surface integrals, more suitable for a number of purposes.

## 3. The Spatial Distribution of the Vorticity

Following an idea of Helmholtz we consider now the field lines of the vector field $\omega$: The vector $\omega$ determines at a given time at every point of the field of flow a certain positive direction of the vorticity. Proceeding from one point to the next along these directions we describe a *vortex line* in exactly the same way as we describe a line of force in a field of force by proceeding along the force directions. Now we take a surface element $\Delta q$ perpendicular to the vortex line which we let pass through a border point of $\Delta q$, and construct the vortex lines that go through the other border points of $\Delta q$. In this way a *vortex tube* is obtained. The circulation about the boundary of $\Delta q$ is according to (14)

$$(15) \qquad \Gamma = 2\mu \qquad \text{with} \qquad \mu = \omega \Delta q.$$

*The flux of the vortex vector through the tube* or the *vortex strength*,[4] as we shall also call it, has been denoted by $\mu$. According to (13), $\mu$ is constant in time regardless of the changes experienced by the vortex tube and the vortex lines of which its boundary consists.

However, the vortex strength remains constant also in space, that is,

---

[4]The vortex carried by an element $\Delta q$ can be characterized by $2\omega\Delta q$ as well as by $\omega\Delta q$. When the first possibility is chosen, the term *vorticity* is frequently used as a name for $2\omega\Delta q$, so that the differential circulation $\Delta\Gamma$ is directly equal to the vorticity. In this book $\omega\Delta q$ has been chosen as the characteristic quantity and the term *vortex strength* introduced to denote it, while the term vorticity is regarded as synonymous with the term vortex vector $\omega = \frac{1}{2}$ curl v.

This definition of $\mu$ also presupposes that the boundary of $\Delta q$ is oriented so as to form a right hand screw with $\omega$; $\mu$ is then essentially positive. In the two-dimensional case it is more convenient to give a sign to $\mu$ [cf. (19.8) and 21].

as long as one proceeds along the same vortex tube (cf. Fig. 24). Since div $\omega = 0$, we obtain, by applying Gauss's theorem to a piece of the vortex tube between the normal cross-sections $\Delta q$ and $\Delta q'$,

$$(16) \qquad 0 = \int \operatorname{div} \omega \, d\tau = \int \omega_n \, d\sigma$$

where the integration to the right includes $\Delta q$ and $\Delta q'$ and the side walls of the tube segment. But $\omega_n$ vanishes on the side wall according to the

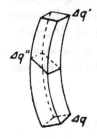

FIG. 24. The flux of the vortex vector through an orthogonal and an oblique cross-section of a vortex tube.

definition of the vortex tube. Furthermore, the component $\omega_n$ has the value $-\omega$ at $\Delta q$ and the value $+\omega'$ at $\Delta q'$, where $\omega$ and $\omega'$ are the amounts of $\omega$ at the respective places. Thus Eq. (16) means

$$(17) \qquad 0 = \omega' \Delta q' - \omega \Delta q, \qquad \text{that is} \qquad \mu' = \mu.$$

This finishes the proof. Note also that the same value $\mu$ is obtained when the tube is cut in oblique direction. With $\Delta q''$ denoting the area of such a cut and $\omega_n''$ the normal component of $\omega$ at $\Delta q''$, Gauss's theorem gives as before

$$\omega_n'' \Delta q'' = \omega \Delta q = \mu.$$

Thus the product of the area of the cut and the normal component of the vortex vector is the same for an arbitrary cut through the vortex tube. This points to an *analogy* between vortex tubes and electric currents which will be discussed in 20.

It is an immediate consequence of the constancy of the vortex strength that a vortex tube can never end within the fluid. *It either reaches the boundary or it must be closed.*

So far we have silently assumed that $\omega$ is continuously distributed in space; the vortex strength $\mu$ is then of the same order of magnitude as the tube cross-section $\Delta q$. But we can consider the limiting case, that $\omega$ is concentrated at single, infinitely narrow, linear regions, in analogy to the so-called linear conductors in electricity. The vortex vector $\omega$ becomes infinitely large within those regions while it may have any finite value outside, in particular it may be zero. In the latter case we speak of a *vortex filament.* Its strength has a finite value that does not go to zero with $\Delta q$ and can be written as

(18) $$\mu = \lim_{\Delta q \to 0} \omega \Delta q.$$

## 19. Two- and Three-dimensional Potential Flow

A flow that is in general irrotational is called a *potential flow.* The term *"in general"* means here "with the possible exception of singular points or lines." As discussed already in 11, p. 88, the condition of irrotationality, $2\omega = \text{curl } \mathbf{v} = 0$, implies the existence of a *velocity potential* $\Phi$; in the preceding article we found in addition that a flow remains irrotational if it starts out as an irrotational flow. In the present article, however, we shall consider only *steady* flow problems so that the question of time dependence of $\Phi$ does not arise. Since we shall further assume an incompressible fluid all through this chapter, the potential equation (11.13)

(1) $$\nabla^2 \Phi = 0;$$

is in general fulfilled in the space filled by the fluid. (For the meaning of "in general" see above.)

We wish to investigate the properties of the flow that result from (1). This is particularly simple in the case of two-dimensional flow where

(2) $$\frac{\partial^2 \Phi}{\partial x^2} + \frac{\partial^2 \Phi}{\partial y^2} = 0.$$

Let us first compare (2) with the wave equation (13.9)

(2a) $$\frac{\partial^2 \Phi}{\partial x^2} - \frac{1}{c^2} \frac{\partial^2 \Phi}{\partial t^2} = 0.$$

By putting $t = y$ and $c^2 = -1$, (2a) is changed into (2). Since the general integral of (2a) is known, viz. d'Alembert's solution (13.11)

(2b) $$\Phi = F_1(x + ct) + F_2(x - ct),$$

we also know the *general solution* of (2); it is:

(3) $$\Phi = \tfrac{1}{2}\{f(x + iy) + f^*(x - iy)\}.$$

The two arbitrary functions of (2b) must be assumed as conjugate complex in (3) (indicated by *) in order to make $\Phi$ real; besides, we have added the factor $\tfrac{1}{2}$ so that $\Phi$ is directly equal to the real part of $f(x + iy)$.

The general solution of (2) is therefore given by the *real part of an arbitrary analytic function f* of the complex variable $z = x + iy$. It is then convenient to write this function in the form

(4) $$f(z) = \Phi(x, y) + i\Psi(x, y).$$

The imaginary part $\Psi$, that is, the function conjugate to $\Phi$, is sometimes called the conjugate potential. In hydrodynamics $\psi$ is known as the *stream function*.

To find interrelations between $\Phi$ and $\Psi$ we differentiate (4) once with respect to $x$ and once with respect to $y$, indicating the differentiation with respect to the complex argument $z$ by $f'(z)$ as usual. This gives

(4a)
$$\frac{\partial}{\partial x} f(z) = f'(z) \frac{\partial z}{\partial x} = f'(z) = \frac{\partial \Phi}{\partial x} + i \frac{\partial \Psi}{\partial x},$$

$$\frac{\partial}{\partial y} f(z) = f'(z) \frac{\partial z}{\partial y} = if'(z) = \frac{\partial \Phi}{\partial y} + i \frac{\partial \Psi}{\partial y}.$$

By elimination of $f'(z)$ results

$$\frac{\partial \Phi}{\partial x} - \frac{\partial \Psi}{\partial y} + i\left(\frac{\partial \Psi}{\partial x} + \frac{\partial \Phi}{\partial y}\right) = 0.$$

Since $\Phi$ and $\Psi$ are real, this relation is equivalent to the two relations

(5)
$$\frac{\partial \Phi}{\partial x} = \frac{\partial \Psi}{\partial y}, \qquad \frac{\partial \Phi}{\partial y} = -\frac{\partial \Psi}{\partial x},$$

the *Cauchy-Riemann equations of theory of complex functions*. They can be expressed in the comprehensive form of one symbolic equation

(5a)
$$\frac{\partial \Phi}{\partial s} = \frac{\partial \Psi}{\partial n},$$

valid for an arbitrary pair of orthogonal line elements of the $x$, $y$-plane that follow each other in the same sense as the positive real and the positive imaginary axes. When $ds$ coincides with the $x$-direction (and $dn$ consequently with the $y$-direction), (5a) becomes identical with the first

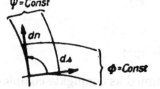
$\Psi = Const$
$dn$
$ds$
$\phi = Const$

Fig. 25. Equipotential lines $\Phi$ = const and stream-lines $\Psi$ = const in two-dimensional irrotational flow.

equation (5); when $ds$ coincides with the $y$-direction and $dn$ with the negative $x$-direction, (5a) becomes the second Eq. (5). That the relation between the directional derivatives (5a) is correct for any other pair $ds$, $dn$ can be easily derived from (5). Thus (5a) is the appropriate vectorial expression of the Cauchy-Riemann equations in invariant writing. When the direction $ds$ is made to coincide with a contour line of the velocity potential as indicated in Fig. 25, the derivative $\partial\Phi/\partial s = 0$; hence, by (5a), the contour line of the stream function coincides with the $n$-direction since in this case

$$(6) \qquad\qquad \frac{\partial\Psi}{\partial n} = 0.$$

Thus it is seen that the *two families of curves* $\Phi = const$ *and* $\Psi = const$ *are orthogonal to each other*. This is usually shown by cross multiplication of Eqs. (5), resulting in

$$(6a) \qquad\qquad \operatorname{grad} \Phi \cdot \operatorname{grad} \Psi = 0,$$

but (5a) is a direct expression of that fact as we have just seen.

To justify the term *stream function*, we turn again to Fig. 25. If $ds$ coincides with the contour line of the velocity potential as above, the vector $\mathbf{v} = -\operatorname{grad} \Phi$ points in the direction of $dn$, but this is the direction of the curve $\Psi = const$ by (6); hence the lines $\Psi = const$ are the *stream lines* of the flow.

The families $\Phi = const$ and $\Psi = const$ are not only *orthogonal* but also *isometric*, that is, the rate at which the functions $\Phi$ and $\Psi$ increase in the directions $ds$ and $dn$ respectively are equal; in other words, if the increments of both functions are equal, say $\delta = d\Phi = d\Psi$, then $ds = dn$ according to (5a). This means geometrically that the curves $\Phi = c_1$, $\Psi = c_2$, $\Phi = c_1 + \delta$, $\Psi = c_2 + \delta$ form an infinitesimal *square*. As a consequence, the image of an infinitesimal figure drawn in the $x + iy$ -plane appears geometrically similar in the $\Phi + i\Psi$-plane: The mapping of the $x + iy$-plane on the $\Phi + i\Psi$-plane is conformal, that is, geometrically similar in the small. The scale of the mapping is given by $|\,f'(x + iy)\,|$; it changes from point to point, but, at a given point, is independent of the direction.[5]

---

[5]It is no exaggerated claim that the theory of analytic functions of a complex variable is identical with two-dimensional potential theory or, in terms of hydrodynamics, with two-dimensional theory of potential flow. In particular the methods introduced by Riemann in his dissertation (1851), where analytic functions are characterized by their intrinsic properties such as singularities, connectivities at branch points, etc. rather than by their analytic expressions, have been prompted by his deep interest in mathematical physics. The following story reveals in a characteristic way the attitudes of two great minds of the epoche toward Riemann's discovery. On a vacation trip in Switzerland, sometime in the seventies, Helmholtz meets Weierstrass, the great mathematician. The latter, recently engaged in the study of Riemann's dissertation, complains that R.'s methods were so hard to understand. Upon this Helmholtz borrows the paper and when he meets Weierstrass next time, he declares that Riemann's ideas appear to *him* perfectly natural and self-evident.

The simplest example in which we apply the methods of the theory of functions to hydrodynamics is the *line vortex* discussed on p. 35. We put

(7)          $$f(z) = iA \log z \quad (A \text{ real}).$$

Setting $z = re^{i\varphi}$ we obtain

$$f(z) = \Phi + i\Psi = -A\varphi + iA \log r,$$

and consequently

(7a)          $$\Phi = -A\varphi, \qquad \Psi = A \log r.$$

The stream lines are given by $\Psi = \text{const}$ or $r = \text{const}$. The flow velocity follows from $\mathbf{v} = -\text{grad } \Phi$. Since in our case $\Phi$ depends only on $\varphi$, the velocity

(7b)          $$\pm v = v_\varphi = -\text{grad}_\varphi \, \Phi = -\frac{1}{r}\frac{\partial \Phi}{\partial \varphi} = \frac{A}{r}, \qquad v_r = 0$$

in agreement with (4.29).

The origin is a singular point of the flow since $v$ becomes infinite there. This singularity appears already in Eq. (7) since $f(z)$ goes logarithmically to infinity for $z = 0$.

The flow is everywhere irrotational except for $r = 0$. This must be so by virtue of its derivation from a complex function $f$, and can be verified by computing the vortex vector $\boldsymbol{\omega}$, which is here normal to the $x$, $y$-plane, in application of the results of problem I, 3:

$$2\omega = \frac{1}{r}\frac{\partial(rv_\varphi)}{\partial r} - \frac{1}{r}\frac{\partial v_r}{\partial \varphi} = 0.$$

The corresponding calculation in Cartesian coordinates is unnecessarily involved; it is carried out here for once:

$$v_x = \mp v \sin \varphi = \mp v \frac{y}{r} = -A \frac{y}{r^2},$$

$$v_y = \pm v \cos \varphi = \pm v \frac{x}{r} = +A \frac{x}{r^2}.$$

Thus we have at every point with the exception of the origin

$$2\omega = \frac{\partial v_y}{\partial x} - \frac{\partial v_x}{\partial y} = \frac{A}{r^2} - \frac{2Ax^2}{r^4} + \frac{A}{r^2} - \frac{2Ay^2}{r^4} = 0.$$

According to the invariant definition in (2.21), $\omega$ can always be computed from the circulation around an infinitely small circuit enclosing the point under consideration. This implies that *the circulation around any*

*point with the exception of r = 0 vanishes provided the circuit is sufficiently
close to the point.* If on the other hand the circulation is taken around any
finite (arbitrarily small) circle enclosing the origin one obtains from (18.10)

$$(8) \qquad \Gamma = \oint \mathbf{v}\cdot d\mathbf{s} = \int_0^{2\pi} vr\, d\varphi = 2\pi A.$$

$\Gamma$ is independent of $r$ and therefore the same for any two circles $r_1$ and
$r_2$ ; this is necessarily so since the circular ring between the two is vortex-
free. *The entire vorticity is thus concentrated at the singular origin.* The
vortex strength* at this point is according to our definition of $\mu$ in (18.15)

$$(8a) \qquad \mu = \pi A.$$

Viewed in three dimensions, the singular point corresponds to a *vortex
filament* normal to the $x,y$-plane. A second vortex filament of opposite
strength $\mu = -\pi A$ is at infinity, when infinity is considered as a point
of the complex plane as one usually does in theory of functions. Note
the formula following from (7b) and (8a):

$$(9) \qquad v_\varphi = \frac{\mu}{\pi r}.$$

We can have both vortex filaments located in a finite domain, e.g.,
at the points $z = \pm c$, if we subject $z$ to a bilinear transformation. Consider instead of (7) the function

$$(10) \qquad f(z) = iA \log \frac{z-c}{z+c}.$$

and put

$$(10a) \qquad z - c = r_1 e^{i\varphi_1}, \qquad z + c = r_2 e^{i\varphi_2},$$

Defining the *bipolar* coordinates $\rho$ and $\varphi$ by

$$(10b) \qquad \rho = \log \frac{r_1}{r_2}, \qquad \varphi = \varphi_1 - \varphi_2,$$

we obtain according to (4)

$$(11) \qquad \Phi = -A\varphi, \qquad \Psi = A\rho = A \log \frac{r_1}{r_2}.$$

Since all angles $\varphi$ subtended by the segment $(-c, +c)$ have their
vertices on a circular arc, the equipotential lines $\Phi = $ const are circles
through the singular points $z = \pm c$; but the streamlines $\Psi = $ const are

*In the two-dimensional case, we take all circulation integrals in the same sense
("enclosed" area to the left hand); the vortex strength $\mu$ can then be positive or negative.

also circles since the locus of constant ratio of distances from two given points is a circle. Note that the intersection points of any circle $\Psi = $ const with the real axis separate the points $-c$, $+c$ harmonically. Fig. 26 shows the orthogonality of the two families of circles, it also gives an idea of the network of infinitesimal quadratic meshes that is obtained if one allows the parameters of either family of circles to change in equal increments $\delta$.

Problem IV, 1 shows how to handle bipolar coordinates. They are infrequently used in mathematical physics, one of the reasons being that

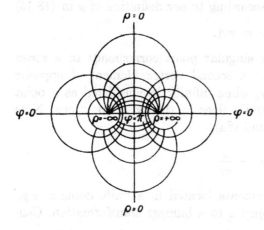

FIG. 26. Superposition of two vortices of equal strength and opposite orientation. The diagram also shows bipolar coordinate lines.

the two-dimensional wave equation is not "separable" in these coordinates; this is discussed in the same problem.

Our present function $f(z)$ can be obtained by the superposition of two simple vortices of the type (7), one having the center $z = +c$ and the constant $+A$, the other the center $z = -c$ and the constant $-A$:

$$(12) \qquad f(z) = iA \log (z - c) - iA \log (z + c).$$

A circuit of arbitrary shape enclosing $+c$, but not $-c$, gives according to (8)

$$(12a) \qquad \Gamma_1 = 2\pi A$$

A circuit of arbitrary shape enclosing $-c$, but not $+c$, gives according to (8)

$$(12b) \qquad \Gamma_2 = -2\pi A$$

A circuit that encloses both or none of the two singularities gives $\Gamma = 0$. As in (8a), the strength of the vortex filament in $+c$ or $-c$ is

$$(13) \qquad \mu_1 = \pi A \qquad \text{or} \qquad \mu_2 = -\pi A.$$

The field belonging to our present function $f(z)$ can also be obtained by the superposition of two fields of the type (9) if $r$ is replaced by $|z - c|$ and $|z + c|$, respectively. One obtains by superposition

(14)
$$\mathbf{v} = \frac{\mu \mathbf{\phi}_1}{\pi \, |z - c|} - \frac{\mu \mathbf{\phi}_2}{\pi \, |z + c|} = \frac{\mu \mathbf{\phi}_1}{\pi r_1} - \frac{\mu \mathbf{\phi}_2}{\pi r_2}$$

Here $\mathbf{\phi}_1$ and $\mathbf{\phi}_2$ are unit vectors in the direction of increasing angles $\varphi_1$ and $\varphi_2$. This result will be compared in problem IV, 2 with the representation of the same flow in bipolar coordinates according to (11)

(14a)
$$\mathbf{v} = A \operatorname{grad} \varphi = \frac{\mu}{\pi} \operatorname{grad} \varphi.$$

We turn now to confocal elliptic coordinates, which are of importance for the flow patterns to be treated in Chapter VI. A simple approach to these orthogonal coordinates takes on the form of a geometrical analysis of the mapping of the plane $z = x + iy$ on the plane $\zeta = \xi + i\eta$ given by the following relation

(15)
$$z = c \cosh \zeta$$

which may also be written as

(15a)
$$x + iy = \frac{c}{2} (e^{\xi + i\eta} + e^{-\xi - i\eta}).$$

The substitution

$$e^{\pm i\eta} = \cos \eta \pm i \sin \eta,$$

leads immediately to

$$x + iy = c (\cosh \xi \cos \eta + i \sinh \xi \sin \eta).$$

Separation of real and imaginary parts on both sides gives now

(16)    $x = c \cosh \xi \cos \eta,$     $y = c \sinh \xi \sin \eta.$

From (16), $\eta$ can be eliminated by the use of $\cos^2 \eta + \sin^2 \eta = 1$, resulting in

(17a)
$$\frac{x^2}{c^2 \cosh^2 \xi} + \frac{y^2}{c^2 \sinh^2 \xi} = 1.$$

The elimination of $\xi$ from (16) by the use of $\cosh^2 \xi - \sinh^2 \xi = 1$ gives

(17b)
$$\frac{x^2}{c^2 \cos^2 \eta} - \frac{y^2}{c^2 \sin^2 \eta} = 1.$$

For constant $\xi$, (17a) represents an ellipse with the semi-axes

$$a = c \cosh \xi, \qquad b = c \sinh \xi.$$

Its eccentricity is given by

(18a) $$\sqrt{a^2 - b^2} = c,$$

and thus independent of $\xi$. Thus, if $\xi$ varies, one obtains a system of confocal ellipses with foci $x = \pm c$, $y = 0$. On the other hand, for constant $\eta$ (17b) represents a hyperbola with the semi-axes

$$a = c \cos \eta, \qquad b = c \sin \eta;$$

its eccentricity is

(18b) $$\sqrt{a^2 + b^2} = c$$

therefore independent of $\eta$ and equal to that of the ellipses given in (18a). With variable parameter $\eta$, (17b) represents a system of hyperbolas, confocal among themselves and with the system of ellipses, cf. Fig. 27.

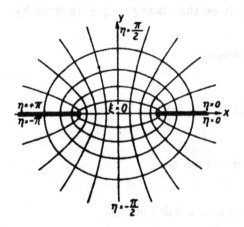

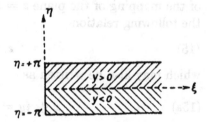

Fig. 27 (*left*). The system of confocal ellipses and hyperbolas $\xi = $ const and $\eta = $ const.

Fig. 27a (*above*). The mapping of the $(x + iy)$-plane on the $(\xi + i\eta)$-plane.

It is of interest to consider the limiting parameter values $\xi = 0$ and $\xi = \infty$ in the family of ellipses: the value $\xi = 0$ gives by (17a) the focal line

$$y = 0, \; - c < x < + c$$

and the value $\xi = \infty$ gives an infinitely extended ellipse.

As regards the parameter $\eta$, note the particular values

$$\eta = 0, \qquad \eta = \pm \frac{\pi}{2}, \qquad \eta = \pm \pi.$$

The value $\eta = 0$ characterizes, by (17b), the part of the real axis to the right of $c$: $y = 0$, $x > c$. $\eta = \pm \pi/2$ corresponds to $x = 0$, the hyperbola coincides with the $y$-axis. $\eta = \pm \pi$ gives the part of the $x$-axis to the left of $c$: $y = 0$, $x < -c$.

In order to obtain a unique relation between the coordinates $\xi$, $\eta$ and the points of the $x,y$-plane, $\xi$ and $\eta$ must be restricted to a certain domain. One way of doing this is to assign the values $0 < \eta < \pi$ to the upper halves of the hyperbolas, and the values $-\pi < \eta < 0$ to the lower halves. By this rule the entire $x,y$-plane is mapped on the strip of the $\xi,\eta$-plane

(19) $$0 < \xi < +\infty, \qquad -\pi < \eta < +\pi$$

indicated in Fig. 27a. Note that the one-to-one correspondence between the $x,y$-plane and the $\xi,\eta$-strip does not include the boundary of the strip; either the upper or the lower half of the borderline must be omitted. Or one might also consider boundary points that are mapped on the same point in the $x,y$-plane as identical (viz. $0, \pm\eta; \xi \neq 0, \pm\pi$).

The advantage of elliptic over bipolar coordinates with regard to the separability of the wave equation will be discussed in connection with problem IV,3. In Chapter VI problems arising from the flow past a plate will be treated by means of elliptic coordinates. We shall use a complex potential of the form

(20) $$\Phi + i\Psi = f(\zeta), \qquad \text{where} \qquad f(\zeta) = \text{const} \cdot \sinh \zeta.$$

Here, $\zeta$ is thought to be related to $z$ according to Eq. (15) so that $f(\zeta)$ becomes a complex function of $z$.

Let us return once more to the simple potential (7), but without the factor $i$ on the right hand side. Then

(21) $$f(z) = A \log z, \qquad \Phi = A \log r, \qquad \Psi = A\varphi.$$

The rays $\varphi = $ const originating at $O$ are now the streamlines. The contour lines of the potential are the circles $r = $ const about $O$. Depending on the sign of $A$, $O$ is either a source or a sink. Note in this connection that the general central symmetric solution of the potential equation is the logarithmic potential $\Phi = \log r$, save for a multiplicative and an additive constant.

A few remarks about the potential flow in space follow here as a supplement.

The central symmetric solution of the potential equation in space is the so-called Newtonian potential

(22) $$\Phi = \frac{A}{r}.$$

No historical implication is intended, and Newton never spoke of this potential; the term should imply that (22) is associated with the field of the Newtonian attractive force

(22a) $$\mathbf{F} = -\operatorname{grad} \Phi = -A \operatorname{grad} \frac{1}{r}.$$

The hydrodynamical meaning of $\Phi$ is again the velocity potential of a source or sink at $O$.

In general, no stream function $\Psi$ can be associated with the velocity potential in the three-dimensional case. Only in the case of an axial symmetric flow is it possible to construct a stream function.[6] For this purpose cylindrical coordinates $\rho$, $\varphi$, $z$ are introduced, $z$ being the axis of symmetry. If the problem is axial-symmetric, $\Phi$ will not depend on $\varphi$ but only on $\rho$ and $z$. The potential equation (1), the mathematical equivalent of incompressibility, reads then (see problem I, 3)

$$(23) \qquad \frac{\partial^2 \Phi}{\partial \rho^2} + \frac{1}{\rho}\frac{\partial \Phi}{\partial \rho} + \frac{\partial^2 \Phi}{\partial z^2} = 0.$$

It may also be written in the form

$$(23a) \qquad \frac{\partial}{\partial \rho}\, \rho\, \frac{\partial \Phi}{\partial \rho} + \frac{\partial}{\partial z}\, \rho\, \frac{\partial \Phi}{\partial z} = 0.$$

From this form we can conclude that there exists a function $\Psi = \Psi(\rho, z)$ such that

$$\frac{\partial^2 \Psi}{\partial \rho\, \partial z} = \frac{\partial}{\partial \rho}\, \rho\, \frac{\partial \Phi}{\partial \rho} = -\frac{\partial}{\partial z}\, \rho\, \frac{\partial \Phi}{\partial z}\, .$$

If we now put in particular

$$(24) \qquad \frac{\partial \Psi}{\partial z} = \rho\, \frac{\partial \Phi}{\partial \rho}\, , \qquad \frac{\partial \Psi}{\partial \rho} = -\rho\, \frac{\partial \Phi}{\partial z}\, ,$$

we obtain by cross multiplication

$$\rho\left( \frac{\partial \Phi}{\partial \rho}\frac{\partial \Psi}{\partial \rho} + \frac{\partial \Phi}{\partial z}\frac{\partial \Psi}{\partial z} \right) = 0.$$

This implies that in any meridian plane $\varphi = \text{const}$ the curves $\Phi = \text{const}$ and $\Psi = \text{const}$ are mutually orthogonal. In other words, in such a plane the lines $\Psi = \text{const}$ coincide with the direction grad $\Phi$ and are *stream lines*. The function $\Psi$ may therefore be considered as a stream function. From (24) and $\mathbf{v} = -\text{grad}\ \Phi$, one obtains the velocity components in the form

$$(25) \qquad v_\rho = -\frac{1}{\rho}\frac{\partial \Psi}{\partial z}\, , \qquad v_z = \frac{1}{\rho}\frac{\partial \Psi}{\partial \rho}\, .$$

It is impossible, however, to combine $\Phi$ and $\Psi$ to form a complex function $\Phi + i\Psi$ of the variable $\rho + iz$ as in the two-dimensional case. If this could be done, then the relations (24) should coincide with the Cauchy-Rie-

---

[6]Stokes, Trans. Cambridge Phil. Soc. Vol. 7, 1842, p. 439.

mann Eqs. (5), which is not the case because of the factor $\rho$ in (24). Also, $\Phi$ and $\Psi$ would have to satisfy the equation $\partial^2 u/\partial\rho^2 + \partial^2 u/\partial z^2 = 0$, but actually $\Phi$ satisfies Eq. (23) and $\Psi$ the equation obtained from curl $\mathbf{v} = 0$ by (25):

$$(26) \qquad \frac{\partial^2\Psi}{\partial\rho^2} - \frac{1}{\rho}\frac{\partial\Psi}{\partial\rho} + \frac{\partial^2\Psi}{\partial z^2} = 0.$$

The powerful tool of the theory of complex functions cannot be used in three-dimensional potential theory.[7]

## 20. A Fundamental Theorem of Vector Analysis

The theorem which we wish to prove here is this: *A continuous vector field* $\mathbf{V}$, *defined everywhere in space and vanishing at infinity together with its first derivatives, can be represented as the sum of an irrotational field* $\mathbf{V}_1$ *and a solenoidal field* $\mathbf{V}_2$ :

$$(1) \qquad \mathbf{V} = \mathbf{V}_1 + \mathbf{V}_2 ,$$

where

$$(1a) \qquad \text{curl } \mathbf{V}_1 = 0, \text{ div } \mathbf{V}_2 = 0.$$

Our decomposition charges all sources and sinks of the given field $\mathbf{V}$ to the component field $\mathbf{V}_1$ and all vortices to the component field $\mathbf{V}_2$ so that:

$$(1b) \qquad \text{div } \mathbf{V}_1 = \text{div } \mathbf{V}, \qquad \text{curl } \mathbf{V}_2 = \text{curl } \mathbf{V}.$$

The representation (1) is *unique except for a vectorial constant.*

This fundamental theorem was proved in its essentials by Stokes[8] in

---

[7] The great mathematician David Hilbert (1862-1943) sharply characterized the futility of all attempts in this direction by the following remark: time is one-dimensional, space is three-dimensional, however the number, that is, the perfect complex number, has two dimensions.

[8] In his big paper: On the dynamical theory of diffraction, Trans. Cambridge Phil. Soc. Vol. 9, p. 1, reprinted in Math. and Phys. Papers, Vol. II, Cambridge, 1883, p. 243 and particularly p. 254 ff. Stokes does not yet give an explicit definition of the vector potential $\mathbf{A}$. Instead, $\mathbf{V}_2$ is calculated by the following formula obtained through a combination of our equations (5) and (9)

$$4\pi\mathbf{V}_2 = \int \frac{\text{curl curl } \mathbf{V}}{r}\, d\tau.$$

For a rigorous proof see: O. Blumenthal, Ueber die Zerlegung unendlicher Vektorfelder, (Mathem. Annalen 61, 235, 1905). His only restriction is that $\mathbf{V}$ and its first derivatives vanish at infinity while no additional assumption is made about how quickly they vanish. It turns out that the component fields $\mathbf{V}_1$ and $\mathbf{V}_2$ need not vanish themselves, they may even become in a restricted way infinite. In the following we shall make the somewhat vague assumption that $\mathbf{V}$ vanishes "sufficiently strongly" at infinity.

148          MECHANICS OF DEFORMABLE BODIES          [IV.20]

1849. In a more complete form, it is the basis of Helmholtz's paper on
vortex motion of 1858, where it is also proved (implicitly however, and,
of course, without the use of vector notation). This theorem penetrates
deeper into the integration methods of mathematical physics than anything we have done so far; it is of great importance in hydrodynamics,
but even more so in electro-magnetic theory. We shall prove it in three
steps.

**1.** *Calculation of* $\mathbf{V}_1$. The irrotational vector $\mathbf{V}_1$ can be derived from a
*scalar potential* $\Phi$ in the form

(2) $$\mathbf{V}_1 = - \text{ grad } \Phi + \text{const.}$$

The potential $\Phi$ satisfies the differential equation

$$\text{div grad } \Phi = \nabla^2\Phi = - \text{ div } \mathbf{V}_1 ,$$

which because of our condition (1b) may also be written as

(3) $$\nabla^2\Phi = - \text{ div } \mathbf{V}.$$

The constant added in (2) for the sake of generality means nothing for
the determination of $\Phi$. From the *inhomogeneous potential equation* (3) the
right member of which is a known function, $\Phi$ is determined by Green's
theorem in the form

(4) $$4\pi\Phi = \int \frac{\text{div } \mathbf{V}}{r} d\tau.$$

(see **1a** below). Here $r = r_{PQ}$ is the distance of the field point $P$, at which
$\Phi$ is to be determined, from the "source" point $Q$ which designates the
place of the volume element $d\tau$, that is to say, the coordinates of $Q$ are
the integration variables. If $\Phi$ is known, $\mathbf{V}_1$ is also known from (2) except
for a constant.

**2.** *Calculation of* $\mathbf{V}_2$. One introduces a *vector potential* $\mathbf{A}$ by putting

(5) $$\mathbf{V}_2 = \text{curl } \mathbf{A} + \text{const}$$

and adds as a further condition

(6) $$\text{div } \mathbf{A} = 0.$$

Note: if $\mathbf{A}$ fulfills (5), $\mathbf{A} + \text{grad } \lambda$, with $\lambda$ being an arbitrary point function,
does the same; hence an additional condition like (6) may be stipulated
  Because of (5)

$$\text{curl curl } \mathbf{A} = \text{curl } \mathbf{V}_2 ,$$

which, according to (1b), may be written

(7) $$\text{curl curl } \mathbf{A} = \text{curl } \mathbf{V}.$$

Also here the additive constant in (5) is immaterial. To (7) we now apply the vector rule (3.10a) valid for Cartesian components:

$$\nabla^2 \mathbf{A} = \text{grad div } \mathbf{A} - \text{curl curl } \mathbf{A}.$$

Because of (6) this reduces to $\nabla^2 \mathbf{A} = - \text{curl curl } \mathbf{A}$, hence (7) is equivalent to

$$(8) \qquad\qquad \nabla^2 \mathbf{A} = - \text{curl } \mathbf{V}.$$

The components of $\mathbf{A}$ then satisfy, like $\Phi$, inhomogeneous potential equations that can be integrated by the following formula, analogous to (4):

$$(9) \qquad\qquad 4\pi\mathbf{A} = \int \frac{\text{curl } \mathbf{V}}{r}\, d\tau.$$

We shall prove in **2a** that this representation of $\mathbf{A}$ satisfies by itself the condition (6). When $\mathbf{A}$ is known, $\mathbf{V}_2$ is obtained by (5) except for a constant.

**3.** *Uniqueness of the resolution* $\mathbf{V} = \mathbf{V}_1 + \mathbf{V}_2$. Suppose there were another resolution $\mathbf{V}_1'$, $\mathbf{V}_2'$ beside $\mathbf{V}_1$, $\mathbf{V}_2$. The irrotational vector $\mathbf{V}_1' - \mathbf{V}_1$ would then at the same time be solenoidal, since we should have according to (1b)

$$\text{div } (\mathbf{V}_1' - \mathbf{V}_1) = 0.$$

In the same way we conclude that the solenoidal vector $\mathbf{V}_2' - \mathbf{V}_2$ should at the same time be irrotational since, again by (1b),

$$\text{curl } (\mathbf{V}_2' - \mathbf{V}_2) = 0.$$

The calculation of the potentials $\Phi$ and $\mathbf{A}$ for the vectors $\mathbf{V}_1' - \mathbf{V}_1$ and $\mathbf{V}_2' - \mathbf{V}_2$ carried out as in (4) and (9) would yield $\Phi = 0$, $\mathbf{A} = 0$. Hence, by (2) and (5), we have

$$\mathbf{V}_1' - \mathbf{V}_1 = \text{const}, \qquad \mathbf{V}_2' - \mathbf{V}_2 \doteq \text{const}.$$

Thus the resolution is indeed *unique* except for a vectorial constant. For more details see Blumenthal, loc. cit.

**1a.** *The calculation of* $\Phi$ *and* $\mathbf{A}$ *by Green's theorem.* Let us put in Green's theorem (3.15)

$$U = \Phi, \qquad V = \frac{1}{r}$$

and apply it to a large sphere with radius $R$ and center $O$. We know from (19.22) that $\nabla^2(1/r) = 0$ except for the singular point $r = 0$ which has to be excluded from the integration by a small spherical surface with radius $\rho$. First we calculate the surface integral over the $\rho$-sphere which appears on the right side of (3.15). Let $\Omega$ denote the solid angle, that is, $d\Omega$ be the surface element of the unit sphere. The surface element of the $\rho$-sphere is $d\sigma = \rho^2 d\Omega$. Since $dn$ is the outward normal seen from the region of integration, we have

$$dn = -d\rho, \qquad \frac{\partial U}{\partial n} = -\frac{\partial \Phi}{\partial \rho}, \qquad \frac{\partial V}{\partial n} = -\frac{d}{d\rho}\frac{1}{\rho} = +\frac{1}{\rho^2}.$$

That part of the right member of (3.15) which refers to the $\rho$-sphere becomes therefore

$$(10) \qquad\qquad \Phi \int d\Omega + \int \frac{\partial \Phi}{\partial \rho} \rho \, d\Omega.$$

The second integral in (10) vanishes for $\rho \to 0$, and the first equals $4\pi$. (Note that the factor $\Phi$ tends to the value of $\Phi$ at $\rho = 0$).

We now consider the surface integral over the $R$-sphere. The same consideration as before leads to the following expression for this integral

$$(10a) \qquad\qquad -\int \Phi \, d\Omega - \int \frac{\partial \Phi}{\partial R} R \, d\Omega.$$

But this time both integrals vanish provided $\Phi$ goes to zero with $R \to \infty$ and $\partial\Phi/\partial R$ vanishes more strongly than $1/R$. This is what we want to enforce by requiring that $\mathbf{V}$ should vanish "sufficiently strongly" at infinity. The sum of the surface integrals over both spheres becomes then $4\pi\Phi$.

On the other hand, the first member of (3.15) is, because of (3) and $\nabla^2(1/r) = 0$,

$$\int \frac{\operatorname{div} \mathbf{V}}{r} \, d\tau.$$

This finishes the proof of Eq. (4); the proof of (9) follows exactly the same lines.

**2a.** *Proof that* div $\mathbf{A} = 0$. We take the divergence of either side of equation (8)

$$\operatorname{div} \nabla^2 \mathbf{A} = -\operatorname{div} \operatorname{curl} \mathbf{V} = 0.$$

Since the operators div and $\nabla^2$ are commutative, one concludes that $u = \operatorname{div} \mathbf{A}$ satisfies the *homogeneous* equation $\nabla^2 u = 0$. We carry out the integration as before under **1a**) and notice that the surface integral over the $R$-sphere has again the limit zero, while the volume integral that appears in Green's representation of $u$ is now identically zero. This shows that the additional condition (6) for div $\mathbf{A}$ is actually fulfilled.

**3a.** *More about the uniqueness.* When the vector field **V** is not defined everywhere in space, but only inside a boundary surface $\Sigma$, additional surface integrals over $\Sigma$ appear in place of the vanishing surface integrals over the surface of the infinitely large $R$-sphere. In the calculation of $\Phi$, for instance, we have instead of (10a)

$$(12) \qquad \int_\Sigma \Phi \frac{\partial}{\partial n} \frac{1}{r} d\Sigma - \int_\Sigma \frac{1}{r} \frac{\partial \Phi}{\partial n} d\Sigma.$$

This invalidates the representation (4) as well as our uniqueness proof, and the same is true for the representation (9) and the transformation (11). In the case of a finite domain it is indeed necessary to prescribe suitable boundary conditions on the surface $\Sigma$, in order to give a definite meaning to the resolution "irrotational-solenoidal".

Note that in the case of an infinitely extended vector field that is everywhere defined the constants in (2) and (5) must cancel each other, since the sum $\mathbf{V} = \mathbf{V}_1 + \mathbf{V}_2$ is supposed to vanish at infinity, according to our assumption at the beginning of this article.

Our uniqueness theorem has a well known counterpart in two dimensions: *a function $f(x + iy)$ that is everywhere regular in the complex plane including the point $\infty$ is necessarily a constant* (Liouville's theorem). Both real and imaginary parts of such a function represent a two dimensional potential that is irrotational and solenoidal at the same time; both parts must therefore be constants, making $f$ a complex constant.

**4.** *Representation of a velocity field with given vortices.* For the sake of simplicity we assume first the entire space filled with a moving fluid and identify the vector **V** with the fluid velocity **v**. According to the condition of incompressibility, div $\mathbf{v} = 0$; hence by (4) $\mathbf{V}_1 = 0$, and $\mathbf{V} = \mathbf{v} = \mathbf{V}_2$ (the additive constant has been omitted). Upon introducing the vortex vector $\boldsymbol{\omega}$ which, for the present, we assume continuously distributed in space, we have by (9) and (5)

$$(13) \qquad 2\pi\mathbf{A} = \int \frac{\boldsymbol{\omega}}{r} d\tau, \qquad \mathbf{v} = \text{curl } \mathbf{A}.$$

*The velocity* **v** *can be represented by a superposition of the contributions of the individual vortex elements $\boldsymbol{\omega} d\tau$.* Consider now the vortex tube (cf. p. 136) associated with a certain vortex element $\boldsymbol{\omega} d\tau$, a segment of which having the length $|\Delta s|$ and cross-section $\Delta q$ is shown in Fig. 28. If the vortex strength $\mu = \omega \Delta q$ [see (18.15)] is introduced, the contribution $\mathbf{A}'$ of our vortex element to the vector potential $\mathbf{A}$ is given by

$$(14) \qquad 2\pi\mathbf{A}' = \frac{\mu}{r} \Delta\mathbf{s}.$$

The corresponding contribution $\mathbf{v}'$ to the velocity of the flow is then

(15)          $$\mathbf{v}' = \text{curl}\, \mathbf{A}' = \frac{1}{2\pi}\, \text{curl}\left(\frac{\mu}{r}\,\Delta\mathbf{s}\right).$$

Here the vector $\mu\Delta\mathbf{s}/r$ has the direction of $\Delta\mathbf{s}$ which, according to the definition of the vortex tube is identical with the direction $\boldsymbol{\omega}$. The operation curl to which this vector is subjected refers to the coordinates of the field point $P$ for which $\mathbf{v}'$ is to be computed. It is now convenient to introduce

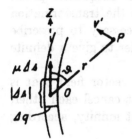

FIG. 28. The contribution of the element $\Delta\mathbf{s}$ of a vortex tube to the vector potential $\mathbf{A}$ and the flow vector $\mathbf{v}$.

an auxiliary coordinate system, taking $\Delta\mathbf{s}$ as the $z$-axis of a Cartesian system with origin $O$ at the volume element of the vortex tube under consideration.[9] With $A' = A'_z$, $A'_x = A'_y = 0$, we obtain from (15)

(16)
$$2\pi v'_x = 2\pi\, \frac{\partial A'_z}{\partial y} = -\frac{\mu}{r^2}\,\Delta s\, \frac{y}{r}$$

$$2\pi v'_y = -2\pi\, \frac{\partial A'_z}{\partial x} = +\frac{\mu}{r^2}\,\Delta s\, \frac{x}{r}, \qquad v'_z = 0.$$

These three equations can be written in condensed form:

(17)          $$2\pi\mathbf{v}' = \frac{\mu}{r^3}\,\Delta\mathbf{s} \times \mathbf{r}.$$

The direction of $\mathbf{v}'$ is orthogonal to $\Delta\mathbf{s}$ as well as to the position vector $\mathbf{r} = OP$; its magnitude is given by

(18)          $$2\pi\,|\,\mathbf{v}'\,| = \frac{\mu\,|\,\Delta\mathbf{s}\,|}{r^2}\, \sin\vartheta,$$

where $\vartheta$ is the angle between $\Delta\mathbf{s}$ and $\mathbf{r}$. The orientation of $\mathbf{v}'$ follows from the right hand rule for vector products.

The motion at the field point $P$ due to the vortex element $\mu\Delta\mathbf{s}$ is de-

---

[9]Instead of that we could also use the general formula $\text{curl}\, f\mathbf{a} = f\, \text{curl}\, \mathbf{a} + (\text{grad}\, f) \times \mathbf{a}$, which gives (17) if one puts $\mathbf{a} = \Delta\mathbf{s} = \text{const}$ and $f = \mu/r$.

termined by the law (17). The total velocity at $P$ is the vectorial sum of the elemental velocities $\mathbf{v}'$:

$$(19) \qquad\qquad \mathbf{v} = \sum \mathbf{v}'.$$

5. *The electro-magnetic interpretation of solenoidal fields.* We now compare (18) with the well known classical *law of Biot-Savart*

$$(20) \qquad\qquad |\,d\mathbf{F}\,| = \frac{i\,ds}{r^2}\sin\vartheta,$$

where $i$ is the current intensity in the element $ds$ of the conductor and $d\mathbf{F}$ the force acting on the magnetic unit pole at the field point $P$; $\vartheta$ and $r$ have the same meaning as in Fig. 28; also the oriented direction of $d\mathbf{F}$ coincides with the oriented direction of $\mathbf{v}'$, given by the vector product. rule.

This amounts to a perfect analogy between hydrodynamics and electrodynamics, which, according to a remark of Helmholtz, has helped greatly in the development of either science.

*The current intensity $i$ corresponds to the vortex strength* $\mu$, the *current density* $\mathbf{J}$ to the vortex vector $\omega$. The Biot-Savart force $d\mathbf{F}$ acting between an individual current element and the unit pole is the analogue of our elemental velocity $2\pi\mathbf{v}'$, the magnetic field strength (that is the force on the unit pole due to the total system of currents) corresponds (except for the factor $2\pi$) to the total velocity vector $\mathbf{v} = \Sigma\mathbf{v}'$. As a system of electric currents is surrounded by magnetic field lines, so a system of vortex tubes is surrounded by stream lines. In electrodynamics we speak of the field $\mathbf{F}$ as *induced* by the current density $\mathbf{J}$; it is convenient to use the same expression in hydrodynamics and to speak of the velocity $\mathbf{v}$ as *induced* by the vorticity $\omega$.

This analogy corresponds to the correlation b) on p. 110 that coordinates the flow $\mathbf{v}$ to the magnetic field strength and the rotation (previously $\phi$ and now $\omega$) to the electric field strength.

In the electromagnetic case all these things become particularly simple if one considers *linear conductors*, that is, wires of sufficiently small cross-section. Their counterparts in hydrodynamics are the *vortex filaments* introduced on p. 141 which will be treated in the following article.

6. *Boundary value problems in the finite space.* The determination of the velocity distribution in the case of the continuous flow of an infinitely extended incompressible fluid has been represented under 4) as a *summation problem*, provided the flow vector vanishes at infinity. The vector potential $\mathbf{A}$ is found by integration over the distributed vorticity which is considered as given. The scalar potential is here immaterial since it

was possible to assume the flow vector to be free of sources and sinks throughout the space. This is different when the flow is restricted to a finite domain, where we have to consider boundary conditions for the pressure (e.g. $p = 0$ at a free surface) and for the velocity (e.g. $v_n = 0$ at a rigid wall). If sources are present in the flow, they must be excluded from the domain of validity of Euler's equations and the excluding surfaces must be considered as boundary surfaces of the fluid. The simple integral representation of $\mathbf{A}$ is no longer sufficient and a scalar potential $\Phi$ appears in addition, as earlier in 3a. As a consequence of the boundary conditions we face now a *boundary value problem* which can no longer be solved by a simple superposition of sources and vortices, but calls for the application of special mathematical methods suitable for the given boundary conditions (e.g. conformal mapping in the two-dimensional case).

## 21. Straight and Parallel Vortex Filaments

In this article we consider the following two-dimensional problem: Given a number of straight vortex filaments normal to the $x,y$-plane. What can be said about their mutual interaction in the course of time? With such a question we enter the domain of vortex dynamics while we have so far only dealt with the kinematic aspect of vortex theory (for instance in our analysis of the induced velocity field). The basis of vortex dynamics is found in Helmholtz's conservation theorem of 18 where we learned about the convective character of the vorticity. Hence, if we know the velocity imparted to the particles of an individual vortex filament by the other vortex filaments, we are also informed about the velocity with which that vortex filament moves along and about its contingent change of shape. Dynamics and kinematics of vortex filaments are thus directly connected. In the two-dimensional case to which we restrict ourselves in this article, it is understood that the vortex filament remains straight.

## 1. The Single Vortex Filament

As already learned in (19.7), the velocity field of the single vortex is given by

$$(1) \qquad \Phi + i\Psi = iA \log(x + iy),$$

$$(1a) \qquad \pm v = v_\varphi = -\operatorname{grad}_\varphi \Phi = \frac{A}{r} = \frac{\mu}{\pi r}.$$

The stream lines are circles about the origin; the vortex proper, that is to say the point $O$, does not take part in the motion.

This is the place to examine the connection between the field due to a simple vortex and the Biot-Savart law of the last article. To do this we must leave the complex plane and introduce the $z$-coordinate along the vortex filament, which reaches from $-\infty$ to $+\infty$. Let $x$, $y$, $0$ and $0$, $0$, $z$ be the coordinates of the field point and of the "source" point that supplies the integration variable as before. The line element, previously $|\Delta s|$, is now $dz$, its distance from the field point, previously called $r$, is now $R = \sqrt{x^2 + y^2 + z^2} = \sqrt{r^2 + z^2}$ where the new $r$ stands for $\sqrt{x^2 + y^2}$ as in (1a). From (20.17) and (20.19) we obtain after evaluation of the vector product

$$(2) \qquad v_x + iv_y = \sum (v_x' + iv_y') = \frac{\mu}{2\pi}(-y + ix)\int_{-\infty}^{+\infty} \frac{dz}{(r^2 + z^2)^{3/2}},$$

while $v_z$ is obviously zero. To evaluate the improper integral in (2) we start with the well known formula

$$\int_0^a \frac{dz}{\sqrt{r^2 + z^2}} = \log(a + \sqrt{r^2 + a^2}) - \log r$$

and obtain by differentiation under the integral sign with respect to $r$

$$-r\int_0^a \frac{dz}{(r^2 + z^2)^{3/2}} = \frac{r}{\sqrt{r^2 + a^2}}\frac{1}{a + \sqrt{r^2 + a^2}} - \frac{1}{r}.$$

If $a \to \infty$, the first term to the right vanishes, and we have

$$\int_{-\infty}^{+\infty} \frac{dz}{(r^2 + z^2)^{3/2}} = 2\int_0^\infty \frac{dz}{(r^2 + z^2)^{3/2}} = \frac{2}{r^2}.$$

Thus Eq. (2) yields

$$(2a) \qquad v_x + iv_y = \frac{\mu}{\pi}\frac{-y + ix}{r^2}, \qquad \pm |\mathbf{v}| = v_\varphi = \frac{\mu}{\pi r}$$

in agreement with (1a) in sense and direction.

On denoting by $\mathbf{\mu}$ a vector in $\pm z$-direction* of length $|\mu|$ and by $\mathbf{r}$ the position vector in the $x,y$-plane, Eq. (2a) can be written as

$$(2b) \qquad \mathbf{v} = \frac{1}{\pi r^2}\mathbf{\mu} \times \mathbf{r}$$

The components $v_x$ and $v_y$ calculated from (2b) are, in fact, identical with the values resulting from (2a). Eq. (2b) differs from the three-dimensional formula for $\mathbf{v}'$ in (20.17) in the missing factor 2 and in the denominator of the right member which is now $r^2$ instead of $r^3$. The difference is caused by the preceding integration with respect to $z$.

*Depending on the sign of $\mu$.

## 2. Two Vortex Filaments of Equal Strength and Opposite or Equal Sense

The case of two opposite vortex filaments was dealt with in Eq. (19.12) et seq. From (19.14) or directly from Fig. 29a the following result is obtained: The velocity $v_1$ imparted to the filament $F_1$ by $F_2$ is equal and parallel to the velocity $v_2$ imparted to $F_2$ by $F_1$ :

$$(3) \qquad v_1 = v_2 = \frac{\mu}{2\pi c} = v \qquad (2c = \text{distance } F_1 F_2).$$

On the other hand, the velocity $v_0$ at the center $M$ of the segment $F_1 F_2$ , subject to the common action of $F_1$ and $F_2$ equals

$$(3a) \qquad v_0 = \frac{\mu}{\pi c} + \frac{\mu}{\pi c} = 4v.$$

Thus we see: *both vortex filaments proceed with common velocity $v$; the fluid between them moves in the same direction in which $F_1$ and $F_2$ move, but with*

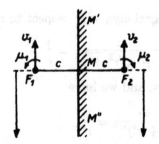

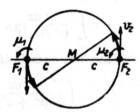

FIG. 29a. Two equally strong vortices of opposite orientation: the plane of symmetry can be replaced by a rigid surface of separation (wall).

FIG. 29b. Two equally strong vortices of the same orientation move on a circle in diametrical position.

*larger velocity.* Fluid particles in line with $F_1$ and $F_2$ but not between them move in the opposite direction; their velocity becomes very large in the immediate neighborhood of $F_1$ and $F_2$ .

Not only at $M$, but at all points $M'$, $M''$ of the bisectrix of $F_1 F_2$ the resultant velocity due to both filaments is parallel to the bisectrix. One may therefore replace it by a rigid wall since the flow along the bisectrix is just the one required by the boundary condition along a rigid wall. Hence an isolated vortex filament, originally at rest, starts to move if a wall is brought toward it; it travels parallel to the wall, the faster the closer the wall. The vortex is, as it were, pushed forward by its virtual image that is obtained by reflexion in the wall. More about that will be learned in the more complicated case of Fig. 31.

The case of *two vortex filaments of equal orientation* is also represented by (19.12), when the sign of the second term to the right is reversed. According to (3) $F_2$ induces at $F_1$ a velocity $v$ that is opposite to the velocity which $F_1$ induces at $F_2$ . While $F_1$ travels downward on a circle with radius $c$, $F_2$ travels upward by the same amount. The diameter $F_1F_2$ turns about the center $M$ which is constant in space. Both vortices travel therefore on the same circle, the one being diametrically opposite to the other.

## 3. A Theorem Concerning the "Center of Mass" of Two or More Vortices

When the strengths $\mu_1$ and $\mu_2$ of the two vortex filaments are different, the vortex points $F_1$ , $F_2$ still travel on concentric circles, but the radii are different. *The common center is the center of mass of $\mu_1$ and $\mu_2$ , when we think of $\mu_1$ and $\mu_2$ in terms of masses carried by the points $F_1$ and $F_2$ .* The center of mass is, of course, on the line through $F_1$ and $F_2$ , *inside* the segment $F_1F_2$ if $\mu_1$ and $\mu_2$ have the *same* orientation (Fig. 30a), and

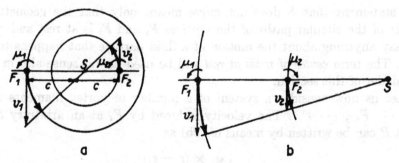

FIG. 30a (*left*), b (*right*). Two vortices move in circles about their "centroid." The centroid does (a), or does not (b), separate the vortices depending on whether the orientations of the vortices are equal or opposite.

*outside* if the orientation is *opposite* (Fig. 30b). In the latter case one of the masses must be counted negative. Figures 30a,b have been drawn under the assumption $\mu_2 > \mu_1$ and $|\mu_2| > \mu_1$ , respectively.

The common center $S$ of the circular paths described by $F_1$ and $F_2$ is found by connecting the end point of the velocity vector $v_1$ with the end point of $v_2$ (see Fig. 30a). This line intersects the line $F_1F_2$ in $S$. Since $v_1$ is due to $F_2$ and $v_2$ to $F_1$ , one has

$$(4) \qquad v_1 = \frac{\mu_2}{2\pi c}, \qquad v_2 = \frac{\mu_1}{2\pi c}.$$

For the position of the intersection point $S$ we have $\overline{SF_1}/v_1 = \overline{SF_2}/v_2$ , which yields because of (4)

$$(5) \qquad \mu_1 \, \overline{SF_1} = \mu_2 \, \overline{SF_2} \, .$$

Hence $S$ is indeed the centroid of the masses $\mu_1$ and $\mu_2$ .

The same construction is carried out in Fig. 30b for the case of opposite vortices. If here $|\mu_2| = \mu_1$ , the point $S$ is shifted to infinity: both circles degenerate into parallel lines and the first case of 2. is reobtained. The second case of 2. obviously corresponds to $\mu_1 = \mu_2$ , so that the two circles of Fig. 30a coincide.

It should be realized that the point $S$ so constructed is by no means at rest if considered as a point of the fluid. Actually the velocity at $S$ in the case of Fig. 30a is given by

$$v_s = \frac{\mu_1}{\pi SF_1} - \frac{\mu_2}{\pi SF_2} = \frac{\mu_1}{\pi SF_1} \left( 1 - \frac{\mu_2}{\mu_1} \frac{\overline{SF_1}}{\overline{SF_2}} \right).$$

The expression in parentheses is because of (5)

$$1 - \frac{\mu_2^2}{\mu_1^2} \neq 0.$$

The statement that $S$ does not move means only that the geometrical center of the circular paths of the vortices $F_1$ and $F_2$ is at rest and does not say anything about the motion of a fluid particle that happens to be at $S$. The term *center of mass at rest* will be used in this sense also in the remainder of this section.

Let us now consider a system of a number of vortex filaments $F_1$ , $F_2$ , $\cdots$ $F_i$ , $\cdots$ $F_k$ . The velocity induced by $F_i$ at an arbitrary field point $P$ can be written by means of (2b) as

$$(6) \qquad \mathbf{v} = \frac{1}{\pi} \frac{\mathbf{\mu}_i \times (\mathbf{r} - \mathbf{r}_i)}{|\mathbf{r} - \mathbf{r}_i|^2}$$

where $\mathbf{r} - \mathbf{r}_i$ is the relative position vector of $P$ with respect to $F_i$ . Let $P$ now coincide with $F_k$ and sum over *all $i$ with the exception of $i = k$* since $F_k$ is not affected by its own induction. Thus

$$(7) \qquad \mathbf{v}_k = \frac{1}{\pi} \sum_i{}' \frac{\mathbf{\mu}_i \times (\mathbf{r}_k - \mathbf{r}_i)}{|\mathbf{r}_k - \mathbf{r}_i|^2}$$

is obtained where the apostrophe at the summation sign means the exclusion of the value $k$ from the possible $i$-values. We introduce now

$$\mathbf{r}_{ik} = \mathbf{r}_k - \mathbf{r}_i \, , \, r_{ik} = |\mathbf{r}_k - \mathbf{r}_i|$$

$$(7a)$$

$$\mathbf{e}_{ik} = \frac{\mathbf{r}_k - \mathbf{r}_i}{r_{ik}} = -\mathbf{e}_{ki}$$

where $\mathbf{e}_{ik}$ is a unit vector in the $x,y$-plane. We shall further need a unit vector $\mathbf{e}$ normal to the $x,y$-plane to indicate the direction of the vector $\mathbf{\mu}$ : $\mathbf{\mu}_i = \mu_i \mathbf{e}$. With these notations, Eq. (7) transforms into

$$(7\text{b}) \qquad \mathbf{v}_k = \frac{1}{\pi} \sum_i{}' \mu_i \frac{\mathbf{e} \times \mathbf{e}_{ik}}{r_{ik}}.$$

Now we form

$$(8) \qquad \sum \mu_k \mathbf{v}_k = \frac{1}{\pi} \sum_k \sum_i{}' \mu_i \mu_k \frac{\mathbf{e} \times \mathbf{e}_{ik}}{r_{ik}}.$$

In the double sum to the right, $\mathbf{e}_{ik}$ and therefore $\mathbf{e} \times \mathbf{e}_{ik}$ is antisymmetric in $i$ and $k$ while the products $\mu_i \mu_k$ and the distances $r_{ik}$ are, of course, symmetric. Hence the terms cancel in pairs, and we have the result

$$(9) \qquad \sum \mu_k \mathbf{v}_k = 0,$$

which implies that the velocity $\mathbf{v}_s$ of the centroid of our vortex system (but only in the sense pointed out above) vanishes

$$(10) \qquad \mathbf{v}_s = \frac{\sum \mu_k \mathbf{v}_k}{\sum \mu_k} = 0.$$

## 4. The Law of Areas for a System of Vortex Filaments

In a certain sense, there exists also an analogue to the law of areas in general mechanics; as we shall see, the *total angular momentum of the system of vortices is constant*, in the absence of external forces.

We define the momentum of a single vortex filament by $\mu_k \mathbf{v}_k$ and the moment of momentum or angular momentum of the system of vortices relative to an arbitrary origin $O$ by

$$\mathbf{M} = \sum_k \mu_k \mathbf{r}_k \times \mathbf{v}_k.$$

Substituting from (7b) we obtain

$$(11) \qquad \mathbf{M} = \frac{1}{\pi} \sum_k \sum_i{}' \frac{\mu_i \mu_k}{r_{ik}} \mathbf{r}_k \times (\mathbf{e} \times \mathbf{e}_{ik}).$$

On interchanging $i$ and $k$ and considering the antisymmetric character of the $\mathbf{e}_{ik}$ we obtain

$$(11\text{a}) \qquad \mathbf{M} = -\frac{1}{\pi} \sum_k \sum_i{}' \frac{\mu_i \mu_k}{r_{ik}} \mathbf{r}_i \times (\mathbf{e} \times \mathbf{e}_{ik}).$$

We add now Eqs. (11) and (11a) and find by applying the definition of $\mathbf{e}_{ik}$ in (7a) the following form of $\mathbf{M}$:

$$(11\text{b}) \qquad \mathbf{M} = \frac{1}{2\pi} \sum_k \sum_i{}' \mu_i \mu_k \mathbf{e}_{ik} \times (\mathbf{e} \times \mathbf{e}_{ik}).$$

According to a well known vector identity

(12) $$\mathbf{e}_{ik} \times (\mathbf{e} \times \mathbf{e}_{ik}) = \mathbf{e}(\mathbf{e}_{ik} \cdot \mathbf{e}_{ik}) - \mathbf{e}_{ik}(\mathbf{e} \cdot \mathbf{e}_{ik}),$$

but the first term to the right equals $\mathbf{e}$ and the second is zero ($\mathbf{e}_{ik}$ is a unit vector, so is $\mathbf{e}$, and they are normal to each other). Thus it is seen that the right member of (11b) depends only on the vortex strengths $\mu_i$ and has the constant direction $\mathbf{e}$; it is therefore a constant vector

(13) $$\mathbf{M} = \frac{\mathbf{e}}{2\pi} \sum_i \sum_k{}' \mu_i \mu_k = \text{const.}$$

## 5. General Remarks on the Dynamics of Vortices

The dynamics of vortices which we have studied here is indeed a very peculiar one and deviates decisively from the dynamics of mass points.

To begin with, Newton's first law is altered. The isolated vortex (which is therefore not subjected to "forces") *remains in a state of rest*. A uniform rectilinear motion can only be acquired by association with a second vortex of equal strength but opposite sense of rotation or under the action of a wall at rest. Thus the relativity principle of classical mechanics according to which the state of rest and of uniform motion are equivalent is no longer valid. The reason is, of course, that the fluid to which the vortex belongs plays the role of a preferred frame of reference.

The modification of the second law is even more remarkable. The external action originating in a second vortex does not determine the *acceleration* but the *velocity*. The content of the law of motion of the mass center is shifted accordingly: not the *acceleration*, but the *velocity* of the mass center vanishes. As far as the *law of areas* is concerned the angular momentum of the vortex system is constant as in the case of a mechanical system in the absence of external forces, but the constant is entirely determined by the vortex strengths according to Eq. (13), in contradistinction to the mechanics of masses where this constant is a constant of integration that depends on initial conditions which may be chosen freely.

## 6. Atmospheric Vortices

At first thought it seems impossible to realize infinitely long vortex filaments experimentally. But, as Helmholtz noticed, in a fluid layer bounded by two parallel planes normal to the vortex filaments, the same vortex motions can be produced as in a fluid of infinite "thickness".

Such a fluid layer is realized to a certain extent by our atmosphere if one disregards the curvature of the earth. In fact, water spouts and tornadoes can be considered as vortex filaments on a tremendous scale (filament diameter 5–500m), whereas atmospheric cyclones and anti-cyclones, the horizontal dimensions of which may well extend over 1000 km, being many times larger than their vertical extensions (10-15 km), combine the properties of vortices with those of waves.

For applications in meteorology the classical theory of vortices has to be modified in two points. The normal state of the atmosphere is not a state of rest but of uniform rotation; for an observer connected with the surface of the earth the fluid in which the vortex motion takes place is not free of external forces as we have assumed so far, but under the action of the Coriolis force. In addition to this dynamical correction there is also a thermodynamical one which is due to the compressibility of the air: since the density of the air is connected with the air pressure by an equation of state $\rho = F(p, T, \cdots)$ which contains not only the pressure, but also the temperature $T$ and possibly other variables such as humidity, the expression curl $(1/\rho \operatorname{grad} p)$ does in general not vanish. That means the presence of processes which can generate or destroy vorticity.

Those are the two extensions of the classical vortex theory that are considered in the meteorological literature when more general vortex theorems are formulated. (V. Bjerknes,[10] H. Ertel[11]).

## 22. Circular Vortex Rings

Closed vortices, in particular those of circular shape, differ from the straight vortices of the preceding section in that they occur or can be easily made to occur on a small scale. As smoke rings, they are well known to smokers, but they may also be seen above chimney tops. For the purpose of demonstration the following device will serve well: A circular hole is made in the front wall of a cardboard box, the rear wall is removed and a piece of somewhat elastic cloth is stretched over that end. Into the box is put a dish of hydrochloric acid and another of ammonium hydroxide; thus a dense smoke of ammonium chloride is produced. Smoke is ejected through the hole by a sharp tap against the rear wall and curls up in the form of a circular vortex as it passes the edge of the aperture. Fig. 31a illustrates this process. The vortex ring thus created moves straight forward for a distance of several yards with considerable speed.

According to the fundamental theorem of vortex theory (cf. p. 132) the vortex adheres to the fluid, that is, the air particles are carried along with the vortex. This becomes apparent when a candle is put in the way of the vortex: The flame is blown out by the "whirlwind" with a soft whistle.—Decreasing the aperture in the box wall makes for a *smaller*

[10]V. Bjerknes, Meteorolog. Z. 1900, pp. 97 and 145, also 1902, p. 96.
[11]H. Ertel, Physikal. Z. 1942, p. 526, also Meteorol. Z. 1942, pp. 277 and 385.

radius of the vortex ring and at the same time for *larger* speed. When the circular aperture is replaced by a rectangular one the vortex, rectangular when it emerges, develops regular pulsations that indicate a tendency toward the circular shape; the latter is the *stable* one, while the rectangular shape is *unstable*.—The circular vortex increases its radius when approaching a wall parallel to the plane of the ring (Fig. 32).

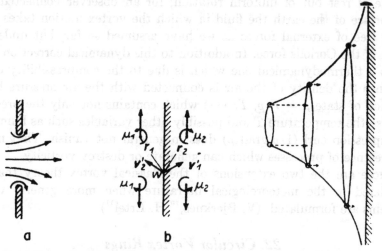

a

b

FIG. 31. a. Formation of a vortex ring when an air current is ejected through a hole. b. The vortex ring approaches a plane wall.

FIG. 32. The vortex ring widens when it approaches a wall parallel to its plane; its translatory velocity decreases.

The following experiment, not easily performed with our device, presents little difficulty to a smoker practiced in the art of blowing rings: A small ring is made to follow a bigger one. Moving with greater speed, it catches up with the bigger vortex ring and slips through it. Thereupon the smaller ring grows larger while the bigger one contracts. It has by now become the quicker of the two and will pass through the other, and, theoretically, the game should repeat over and over again. For this experiment, vortices in a liquid are more suitable than smoke rings. One uses colored liquid which is dropped upon the surface of a water reservoir from a pipette. When a drop strikes the surface, it pulls apart in the shape of a ring which penetrates into the water and expands at the same time. The next drop, or rather the vortex ring which it becomes, is initially smaller and overtakes the larger ring, etc. This arrangement permits one to observe the passing-through phenomenon several times in succession.

Turning now to the interpretation of these experiments, we start with the discussion of some aspects of the problem that can be dealt with by elementary means. This is the case if no account is taken of the *inter-*

*action of the vortex ring with itself.* The vortices, therefore, will for the present be considered as *infinitely thin* filaments.

Eq. (20.17), the equivalent of the Biot-Savart law, serves as a starting point and gives the velocity contribution **v′** induced by a single vortex element $\mu\Delta s$ at given field point. For the midpoint $M$ of the vortex ring in particular, all contributions become equal in amount and direction since in this case (20.17) becomes

$$v' = \frac{\mu\Delta s}{2\pi a^2}$$

where $a$ denotes the radius of the ring. By summation over all $\Delta s$ the velocity $v_M$ at the midpoint is obtained:

(1)     $$v_M = \frac{\mu}{2\pi a^2} \sum \Delta s = \frac{\mu}{a}$$

*Thus the velocity at M is seen to increase with decreasing a.* As we observed before and shall prove later, the translatory velocity of the vortex ring increases also with decreasing $a$.

We take now as field point an arbitrary point $A$ on the axis of symmetry of the ring, denote its distance from the center $M$ by $z$ (the distance from any point of the ring being $r = \sqrt{a^2 + z^2}$), and consider the components of **v′** parallel and normal to the axis. The normal components due to any two diametrically opposite vortex elements will cancel each other. The axial component of **v′** is $v'a/r$. By summation over all $\Delta s$, Eq. (20.17) leads to

(2)     $$v_A = \frac{\mu}{2\pi r^3} \frac{a}{r} \sum r\,\Delta s = \frac{\mu a^2}{r^3} = \mu \frac{a^2}{(a^2 + z^2)^{3/2}}.$$

Hence the velocity decreases along the axis on either side of $M$ and goes to zero as $z$ grows indefinitely.

For a field point in general position the summations that lead from **v′** to **v** are no longer quite so elementary, but require integrations of elliptic type. The same is true, as is well known, for the mathematically analogous calculation of the magnetic field induced by a circular current loop.

On the other hand, it is now possible to interpret some of the phenomena described in a *qualitative* way.

Fig. 31a indicates schematically the formation of a vortex as the air flows through the aperture, Fig. 31b illustrates the conditions that prevail when the vortex ring meets a wall parallel to its own plane. At the wall, the boundary condition is $v_n = 0$. It can be satisfied by employing the *method of images*, a procedure of general usefulness in problems of this character. To the physically real vortex on the left side of the wall one adds a *virtual* vortex on the right side, in this way extending to infinity the hydrodynamic field that, physically, is bounded by the wall. An air particle $W$ adjacent to the wall receives the velocity contribution $\mathbf{v}_1'$ from the real vortex $\mu_1$ and $\mathbf{v}_2'$ from the virtual vortex $\mu_2$ ; according to Eq. (20.17), $\mathbf{v}_i'$ is perpendicular to the plane subtended by the vector $\mathbf{r}_i$ and the axis of the vortex element $\mu_i$ . Evidently, $\mathbf{v}_1'$ and $\mathbf{v}_2'$ add up to a motion *parallel* to the wall, thus *satisfying the condition* $v_n = 0$. This, obviously, is true not only for the two pairs of vortex elements in the plane of the diagram, but for any pair of elements in image position.

We now wish to find out about the influence exerted by the image vortex upon the real vortex,[12] e.g. upon the upper vortex element $\mu_1$ in Fig. 31b. The strongest action is to be expected from the closest neighbor, that is, the upper element $\mu_2$ . Again according to (20.17), this action consists in an induced velocity directed *vertically upward*, indicated in the diagram by a vertical arrow at the upper left element. All the other elements $\mu_2$ will partly aid in this action or partly, but in a smaller degree, counteract it (e.g. the lower element $\mu_2$). The same argument applied to the lower element $\mu_1$ leads to the vertical downward arrow at the place of that element. In the diagram the influence of the real vortex upon its image is also indicated.

We see then that the real vortex is *enlarged* when it approaches the wall because of the hydrodynamic action of the image vortex. The same is true for the image vortex. As the radius of the ring grows, so the forward motion directed to the wall attenuates, since the translatory motion of the vortex ring is essentially in inverse proportion to the radius. (This has been mentioned before but has not been proved.) Hence the vortex growing ever larger and weaker has evanesced to infinity before it reaches the wall.

Using the same qualitative methods we can also understand the mutual slipping through of a pair of vortices (Fig. 33a, b, c, d). In Fig. 33a the radii and speeds of the two vortices are equal. Having been generated one after the other by the same device, their sense of rotation is the same, in contrast to Fig. 31a in which the senses of rotation were opposite. This time, the interaction of the two vortices consists in the enlargement of the more advanced vortex and, simultaneously, in the diminution of the rear vortex as indicated by the vertical arrows in Fig. 33a. In Fig. 33b and c the vortices are shown before and after passing which becomes possible on account of the greater speed of the rear vortex (cf. the horizontal arrows in Fig. 33b). In Fig. 33c, the interaction of the vortex pair at

---

[12]Speaking physically, the influence is, of course, all due to the wall, but we have just convinced ourselves that this influence is correctly described by the assumption of a virtual image vortex.

this stage is again indicated by vertical arrows: The more advanced vortex grows, the rear vortex contracts, hence the first is decelerated, the second accelerated. The result is Fig. 33d, a repetition of Fig. 33a: The game is ready to start anew.

But these qualitative methods are no longer sufficient for problems that involve the *interaction of the vortex with itself*. The translatory motion

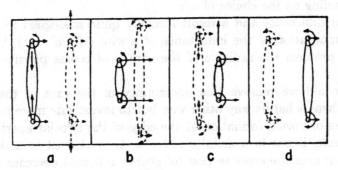

FIG. 33. The mutual threading of two vortex rings.

of a circular vortex is of that kind. Now the cross section of the vortex ring must be assumed finite, e.g. as a circle of radius $c$ very small in comparison with the ring radius $a$. Within the circular cross section one might perhaps try a finite (e.g. constant) rotation $\omega$; the vortex strength $\mu$ that so far has been used only for *filamentous* vortices is now connected with $\omega$ by

$$\mu = \pi c^2 \omega.$$

The most convenient approach is through the stream function $\Psi$ (19.24); within the vortex cross section the differential equation for $\Psi$ is not Eq. (19.26) but the inhomogeneous equation

$$\frac{\partial^2 \Psi}{\partial \rho^2} - \frac{1}{\rho}\frac{\partial \Psi}{\partial \rho} + \frac{\partial^2 \Psi}{\partial z^2} = 2\omega.$$

The representation of $\Psi$ (valid outside the cross section) by complete elliptical integrals of the first and second kind has already been given by Helmholtz. It serves as basis for the computation of the translatory velocity $v_T$, a computation that is mathematically involved and at the same time physically unsatisfactory. It will be omitted here. The result is[13]

$$(3) \qquad v_T = \frac{\mu}{2\pi a}\left(\log \frac{8a}{c} - \frac{1}{4}\right).$$

---

[13]For a critical review of literature see A. E. H. Love, Encykl. d Mathem. Wiss. IV. 3. p. 118

The term written here as $\frac{1}{4}$ is uncertain; it depends on the assumed $\omega$-distribution over the cross-section. Our previous statement, repeatedly used in the discussion of the experiments, *that $v_T$ increases with decreasing a and decreases with increasing a* is confirmed by Eq. 3, but the physically undetermined quantity $a/c$ makes the result appear rather unrealistic. A comparison between (3) and (1) shows that $v_T$ may be larger or smaller than $v_M$ , depending on the choice of $a/c$.

Similar difficulties and a result formally quite analogous to (3) occur in the computation of the inductance of a wire (cf. vol. III), but in the electrical problem the radius $c$ of the wire is of course physically determined.

How could we remove this uncertainty in the case of the circular vortex? There is hardly any other way but to investigate more completely the process of vortex formation at the edge of the circular aperture. This should result, if not in a definite radius $c$, then at least in a definite $\omega$-distribution over the cross section (of course, $\omega$ should decrease radially). However, such an investigation could hardly be carried out and would certainly not be worth the labor; hence we have to leave the attractive and successful theory of vortex rings incomplete in an essential point.

A more direct grasp of the translatory motion of the vortex may be achieved by comparison with the motion of the straight vortex pair in 21, Fig. 29a. A vortex pair may be obtained from a vortex ring by restricting one's attention to two diametrically opposite ring elements. Lengthening of these two elements makes up, in a certain way, for the disregard of the other elements and finally leads to a pair of parallel straight filaments. To be sure, the similarity of the two vortex arrangements is only superficial, but their translatory motions are remarkably analogous. If we replace in Eq. (21.3) the letter $c$, which denotes half the distance of the vortex pair, by the ring radius $a$, the translatory velocity of the vortex pair becomes

$$v = \frac{\mu}{2\pi a} .$$

This is indeed the first factor on the right side of our Eq. 3 for $v_T$ . The direction of $v$ was perpendicular to the plane of the vortex pair; our $v_T$ is perpendicular to the plane of the ring, as it should be.

A peculiar vortex motion may be observed in rowing. At the places where the oar breaks the surface of the water just previous to being lifted, small depressions of the surface appear—we could call them "dimples"—that run along on the surface. They are the endpoints of a vortex arc that has been escorting the submerged contour of the oar while it was pulled through the water. The dimple-like shape has its cause in the combined action of centrifugal force and gravity. The rotating water surface must be normal to the resultant force in the same way as it is normal to gravity if no rotation is present. Now, while the surface form of a mass of water rotating with constant $\omega$ is the well known paraboloid, in our case $\omega$ diminishes quickly in radial direction and vanishes at a finite distance from the vortex center; the depression is thus restricted to the immediate neighborhood of the end of the vortex filament. The running along of a pair of dimples is illustrative of the translatory motion $v_T$ of the vortex arc connecting the two dimples underneath the water surface.

Since it has become certain through optical investigations (Zeeman effect) that the solar matter in the sun spots is engaged in vigorous vortex motion, one would conclude that, there too, a vortex filament has emerged to the surface, coming out of the interior of the sun; and one would assume that to each sun spot there belongs a companion spot with converse sense of rotation, the two being connected by a vortex filament of roughly semi-circular shape that passes through the interior of the sun. On the other hand, one should be aware of the fact that the solar matter is not a perfect inviscid fluid. Thus it appears rather a bold step to apply our ideas about vortex motion literally to the interpretation of solar phenomena.

## CHAPTER V

# THEORY OF WAVES

Ever since waves were studied, water waves have served the natural scientist as a model for wave theory in general, although they are much more complicated than acoustical or optical waves. As *surface waves* they are bound to the common surface of two media, while the ordinary acoustic and optical waves are *three dimensional waves.*

There is this fundamental difference between vortices and waves: vortices pull the matter along in their own motion while in a wave the average locomotion of an individual fluid particle vanishes; it is not *matter* that travels but *energy* and *phase.*

We shall discuss in this chapter waves with different symmetry characteristics, such as *plane waves, circular waves, ship waves,* and *Mach waves,* starting out with the simplest type, the *plane progressive waves.* According to the nature of the restoring force we distinguish *gravity waves* and *capillary waves.* Gravity waves are the large, conspicuous waves which one usually has in mind in talking of water waves.

## 23. Plane Gravity Waves in Deep Water

We assume the wave as a completely periodic phenomenon and express the time dependence as on p. 98 in the form $e^{-i\omega t}$; waves of a more general time dependence can be obtained by superposition of partial waves having different circular frequencies (cf. 26).

We further assume that the wave motion is generated *out of the state of rest*, say, by a gust, a mechanical disturbance or the like (in problem VI, 3 we shall investigate under what circumstances an air current that grazes along a horizontal water surface can produce a wave motion). Since the fluid can be considered as *inviscid*, and since we shall consider in this and in the following article only the *potential field* of gravity, it follows from the conservation law of 18 that the motion possesses a *velocity potential.* This can depend in the case of a *plane* wave only on *two* spatial coordinates $x$ and $y$, where $x$ is the direction in which the wave progresses and $y$ the depth coordinate. The problem is independent of the third spatial coordinate $z$ which is horizontal and orthogonal to the direction of propagation. $\Phi$ is thus a two-dimensional potential as far as the space coordinates are concerned, and, according to 19, has the form $f(x \pm iy)$ in an *incompressible* fluid.

Let $y$ be counted positive downward; since we wish to obtain a train of waves advancing in the positive $x$-direction, the $x$-dependence of $f$ must be of a *trigonometrical* form, or, of the form of an imaginary exponential when written in this more convenient way. This leads to the following possibilities

(1) $$\Phi = f(x + iy)\, e^{-i\omega t} = A\, e^{i(kx-\omega t)}\, e^{-ky},$$

or

(1a) $$\Phi = f(x - iy)\, e^{-i\omega t} = B\, e^{i(kx-\omega t)}\, e^{+ky}.$$

Here, $k$ is again the wave number, and there is

(2) $$k = \frac{2\pi}{\lambda}, \qquad \omega = \frac{2\pi}{\tau},$$

where $\lambda$ is the wave length and $\tau$ the period.

We first assume the water *infinitely deep*, that is, the $y$-coordinate of the ground $y = h$ should be very large compared to $\lambda$.

At the ground,

(2a) $$ky = 2\pi\frac{h}{\lambda} \to \infty.$$

This shows that the potential (1a) is not usable since it would yield infinite velocity amplitudes at the ground, but the potential (1) satisfies all conditions that have been imposed so far.

The representation (1) contains three parameters, $A$, $\omega$, and $k$. $A$ determines the amplitude of the wave for $y = 0$; both $A$ and the (circular) frequency $\omega$ depend on the particular form of the excitation. While these two quantities can be chosen freely, the wave number $k$ must be determined in its ratio to $\omega$, for, according to (2), the ratio

(3) $$\frac{\omega}{k} = \frac{\lambda}{\tau} = V,$$

is the *velocity of propagation* of the waves.

For the determination of $k$ we must utilize the *condition for the free surface*:

(4) $$p = 0,$$

(the atmospheric pressure is taken as zero). Eq. (4) follows from (12.19) if capillarity is neglected; it introduces a dynamic element into our theory while our argument so far has been wholly of a kinematic nature.

The pressure $p$ and the potential $\Phi$ are connected by Euler's equations, which we shall use in the integrated form of Bernoulli's equation. Since our problem involves a velocity potential variable in time, we have to take Bernoulli's equation in the form (11.15). We shall, however, neglect the quadratic term $(\nabla \Phi)^2$, since we consider the amplitude factor $A$ as a *small quantity* (the usual procedure in mechanics of small oscillations). The abridged form of Bernoulli's equation is then

$$(5) \qquad -\frac{\partial \Phi}{\partial t} + \frac{1}{\rho}\,(p + U) = \text{const.}$$

As explained on p. 89, the constant in (5) is independent of the space coordinates, but in general dependent on time. The only function of time which in our case does not upset the periodicity and the uniform advancement of the wave is

$$\text{const} = F(t) = 0.$$

Under these circumstances (5) assumes the simple form

$$(5a) \qquad \frac{\partial \Phi}{\partial t} = \frac{U}{\rho}\,.$$

Here $U$ is the gravity potential per unit of volume taken at the surface. Since $y$ is counted positive downward, we have in general

$$(5b) \qquad U = -\rho g y.$$

Let the equation of the surface profile be $y = \eta$, where $\eta$ is a function of $x$ and $t$. A positive $\eta$ means a depression, a negative $\eta$ an elevation, of the surface. From (5a) we have

$$(6) \qquad \frac{\partial \Phi}{\partial t} = -g\eta.$$

The function $\eta$ must have again the form of a progressive wave, like the one we have set up for the velocity potential $\Phi$. Thus

$$(7) \qquad \eta = a\,e^{i(kx-\omega t)}.$$

The constant $a$ introduced here is in general complex since it includes amplitude *and* phase of the surface function; also, $a$ has the character of a small quantity like $A$. If we substitute (7) and (1) in (6), we obtain after cancellation of the common exponential factor

$$(8) \qquad i\omega A e^{-k\eta} = ga.$$

We now expand $e^{-k\eta}$ in powers of $k\eta$ and neglect the products $A\eta$, $A\eta^2$, etc. as small quantities of higher order; Eq. (8) then simplifies to

$$(9) \qquad i\omega A = ga.$$

This is a relation between $A$ and $a$, but it is not the relation between $k$ and $\omega$ we require. The latter is obtained by introducing a further *kinematic* condition: We stipulate *that the motion of the surface must coincide at any time with the motion of those fluid particles that happen to be at the surface at this time.* That such a condition must be satisfied is rather obvious; we specify, however, that the components of the two motions taken in the normal direction $n$ of the surface should be equal, since a motion of the fluid particles in the tangential plane does not change the shape of the surface, hence is immaterial for our problem. On denoting the velocity of the surface with $\mathbf{V}$ and the particle velocity with $\mathbf{v}$ as usual, our condition reads

$$(10) \qquad\qquad V_n = v_n .$$

The component $v_n$, expressed by the velocity potential $\Phi$, is

$$v_n = -\frac{\partial \Phi}{\partial n} .$$

However, if $A$ is sufficiently small (the wave sufficiently flat), we can replace within a "cosine error", that is, with disregard of terms of second order

$$(10a) \qquad\qquad \frac{\partial \Phi}{\partial n} \quad \text{by} \quad \frac{\partial \Phi}{\partial y} .$$

$\mathbf{V}$ is treated correspondingly: we replace

$$(10b) \qquad\qquad V_n \quad \text{by} \quad \frac{\partial \eta}{\partial t} ,$$

that is by the "sinking speed" of the surface. With these simplifications, condition (10) reads

$$(11) \qquad\qquad \frac{\partial \eta}{\partial t} = -\frac{\partial \Phi}{\partial y} .$$

Substituting for $\eta$ and $\Phi$ from (7) and (1), we obtain, again after cancellation of the exponential on both sides,

$$(12) \qquad\qquad -i\omega a = kA.$$

Now, the comparison of (9) and (12) yields at once

$$(13) \qquad\qquad \frac{A}{a} = \frac{g}{i\omega} = -\frac{i\omega}{k} .$$

Our conclusions from (13) are:

1. There is a phase difference of $\pi/2$ between the $a$-wave and the $A$-

wave. If $A$ is chosen a real quantity, which is permissible, $a$ becomes purely imaginary; or, employing real representation, we can write for $\Phi$ in accordance with (1)

$$\text{(14)} \qquad \Phi = A \cos(kx - \omega t)e^{-kv},$$

and Eq. (7) becomes now [cf. (13)]

$$\text{(14a)} \qquad \eta = -\frac{\omega}{g} A \sin(kx - \omega t)$$

or

$$\text{(14b)} \qquad \eta = -\frac{k}{\omega} A \sin(kx - \omega t).$$

2. The relation between $k$ and $\omega$ is given by Eq. (13):

$$\text{(15)} \qquad \omega^2 = gk.$$

Introducing here the velocity of propagation, we obtain [cf. (3)]

$$\text{(16)} \qquad V^2 = \frac{\omega^2}{k^2} = \frac{g}{k} = \frac{g\lambda}{2\pi}, \qquad V = \sqrt{\frac{g\lambda}{2\pi}}.$$

*The velocity of propagation depends on the wave length; long waves travel faster than smaller ones.*[1]

When the propagation velocity of a wave depends on the wave length as in Eq. (16), we speak of dispersion, using an expression borrowed from

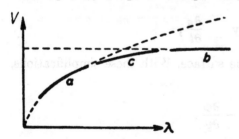

Fig. 34. The phase velocity $V$ as a function of the wave length $\lambda$. The diagram gives the dispersion of gravity waves for $\lambda \ll h$ (segment $a$), $\lambda \gg h$ (segment $b$), $\lambda \cong h$ (segment $c$); $h$ = depth of the water.

optics. The dispersion in a medium is *normal* when longer (red) waves have larger velocities (smaller index of refraction) than shorter (violet) waves. The behavior of *gravity waves in deep water thus corresponds to the case of normal dispersion in optics.*

---

[1] We should like to recommend to readers who are irritated by the complex representation of periodic phenomena, to carry out the preceding calculation once more with real quantities and to convince themselves that they obtain again Eq. (16) although in a somewhat more cumbersome way. One would have to replace (1) by (14) and (7) by $\eta = |a| \cos(kx - \omega t + \alpha)$, where $\alpha$ is the phase difference between $\varphi$ and $\eta$ that is to be determined.

Fig. 34 should make this clearer. $V$ is represented by the upper half of an ordinary parabola which has the $V$-axis as a tangent at $\lambda = 0$. Only the middle part $a$ of the parabola has been drawn as a solid line, this being the region for which our assumptions are actually valid. For, if the wave length $\lambda$ keeps increasing, it finally becomes of the same order of magnitude as the depth of the water $h$, and our assumption (2a) is no longer valid. This case will be taken up in the next article. On the other hand, if one goes to very small values of $\lambda$, gravity is no longer the decisive dynamic parameter, but *surface tension* takes the lead; this brings about a fundamental change of the dispersion law (see 25).

## 24. Plane Gravity Waves in Shallow and Moderately Deep Water

If we now assume the depth $h$ as finite, the second form of the potential (23.1a) need no longer be rejected, but is as good as the form (23.1); hence a linear combination of (1) and (1a) may be taken for the velocity potential $\Phi$:

(1)
$$\Phi = e^{i(kx - \omega t)}\{Ae^{-ky} + Be^{+ky}\}.$$

The new boundary condition at the bottom is: $v_y = 0$, or, $\partial\Phi/\partial y = 0$ for $y = 0$. This is, by (1), equivalent to

$$-Ae^{-kh} + Be^{+kh} = 0,$$

a condition for $A$ and $B$ which is more conveniently handled, if we introduce a constant $C$ defined by

$$\frac{1}{2}C = Ae^{-kh} = Be^{+kh}$$

so that

(1a)
$$A = \frac{1}{2}Ce^{+kh}, \qquad B = \frac{1}{2}Ce^{-kh}.$$

Then the potential $\Phi$ assumes the form

(2)
$$\Phi = e^{i(kx - \omega t)}\frac{C}{2}\{e^{k(h-y)} + e^{-k(h-y)}\}$$

$$= e^{i(kx - \omega t)}C \cdot \cosh k(h - y).$$

The expression (23.7) for the surface depression remains unchanged; we rewrite it here:

(3)
$$\eta = ce^{i(kx - \omega t)}.$$

Likewise, the dynamic and kinematic surface conditions (23.6) and (23.11) remain valid. They give for $y = 0$, because of (2) and (3),

(4) $$i\omega C \cosh kh = gc,$$

(5) $$-i\omega c = Ck \sinh kh.$$

We first discuss the case of *shallow* water, that is,

$$kh \ll 1, \qquad \cosh kh = 1, \qquad \sinh kh = kh.$$

Eqs. (4) and (5) yield in this case

$$i\omega C = gc, \qquad -i\omega c = Ck^2 h,$$

or

(6) $$\frac{C}{c} = \frac{g}{i\omega} = -\frac{i\omega}{k^2 h}.$$

The formula for the velocity of propagation is now

(7) $$V^2 = \frac{\omega^2}{k^2} = gh, \qquad V = \sqrt{gh}.$$

(One is reminded of the definition of the velocity head $h = v^2/2g$ on p. 45, but it is merely a formal analogy, due to dimensional reasons.)

*In shallow water the velocity of propagation is independent of the wave length; there is no dispersion.*

On the basis of Eq. (7) we can now complete Fig. 34 for large $\lambda$ values. The curve should approach a horizontal asymptote for $\lambda > h$ in the distance $\sqrt{gh}$ from the $\lambda$-axis, as indicated by the branch $b$ of this line.

When $\lambda$ and $h$ are of the same order of magnitude (*moderately deep water*), Eqs. (4) and (5) still govern the behavior of the wave. In this case, they supply directly

(8) $$\frac{C}{c} = \frac{g}{i\omega \cosh kh} = -\frac{i\omega}{k \sinh kh},$$

from which we infer

(9) $$\tanh kh = \frac{\omega^2}{gk}.$$

On introducing $x = kh$ and $d = \omega^2 h/g$, we obtain

(9a) $$\tanh x = \frac{d}{x}.$$

The solution of this transcendental equation can be obtained graphically: We plot the equilateral hyperbola $d/x$ and the curve $\tanh x$ above $x$ as abscissa (the latter is a monotonic increasing function with a horizontal asymptote at the distance 1 from the $x$-axis). There is then only one intersection point of the two curves, and its abscissa $x = x_0$ is the root of Eq. (9a); the required value of $k$ is $x_0/h$. The *velocity of propagation in moderately deep water* is therefore

$$(10) \qquad V = \frac{\omega}{k} = \frac{\omega h}{x_0}.$$

We have obtained an implicit representation of the dispersion law which should be supplemented by a chart or table of the solutions of Eq. (9a) in function of the parameter $d$. We also see in what way Fig. 34 is to be completed: the gap between segment $a$ ($h \gg \lambda$) and segment $b$ ($h \ll \lambda$) is bridged by segment $c$ ($h \cong \lambda$) corresponding to Eq. (10). Since this

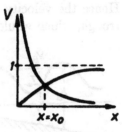

Fɪɢ. 35. Graphical solution of the transcendental Eq. (9a).

part of the curve increases, too, one sees that *the dispersion is normal throughout.*

The transition from $a$ to $c$ can also be read from the following formula which is a direct consequence of (9)

$$(10a) \qquad V^2 = \frac{\omega^2}{k^2} = \frac{g\lambda}{2\pi} \tanh \frac{2\pi h}{\lambda}.$$

One obtains from (10a)

$$\text{for} \quad h \gg \lambda: \quad V^2 \to \frac{g\lambda}{2\pi}, \qquad \text{Eq. (23.16)},$$

$$\text{for} \quad h \ll \lambda: \quad V^2 \to gh, \qquad \text{Eq. (24.7)}.$$

A remark on *surf formation* may be inserted here which is of a purely qualitative nature and includes perhaps an over-interpretation of our equations. Surf is bound to a region of very small $h$ and, consequently, small $kh = x$ so that $\tanh x \cong x$. The transcendental Eq. (9a) becomes then algebraic: $x = d/x$. The root is

$$x_0 = \sqrt{d} = \omega \sqrt{\frac{h}{g}},$$

the velocity of propagation according to (10) is therefore

(11)
$$V = \sqrt{gh},$$

which is our formula (7) for shallow water. Now, our analysis is only valid for water of *uniform* depth, but we may try to apply it to the case of decreasing depth in the vicinity of the shore (cf. Fig. 36). Besides, our results apply only to *small amplitudes* since the amplitude squares have been omitted; nevertheless, we use them now for finite amplitudes, assuming *two* values for $h$, viz., $h_C$ characterizing the crest and $h_T$ characterizing the trough of a wave (cf. Fig. 36), and obtain from (11)

(11a)
$$\frac{V_C}{V_T} = \sqrt{\frac{h_C}{h_T}} > 1.$$

Hence the velocity of propagation of the crest is larger than that of the trough, which would modify the shape of the wave as indicated in Fig. 36

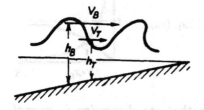

FIG. 36. Surf formation.

by the crest to the right. This offers an explanation of the surf phenomenon on the basis of our equations, although it is doubtful whether the conclusions are legitimate.

We turn to the very attractive problem of the path of an individual particle in wave motion which shall be discussed under the assumption of moderately deep water. Let $x$, $y$ be the particle coordinates in water at rest and $x'$, $y'$ the same in water disturbed by the wave motion. The displacement from the rest position is then

(12)
$$\xi = x' - x, \; \eta = y' - y;$$

where $\eta$ no longer denotes the surface disturbance as before, but the disturbance at an arbitrarily specified depth $y$, (similarly $\xi$). The particle velocity according to (2) is

(13)
$$v_x = \dot{\xi} = -\frac{\partial \Phi}{\partial x} = -ike^{i(kx-\omega t)}C \cosh k(h - y),$$

$$v_y = \dot{\eta} = -\frac{\partial \Phi}{\partial y} = \quad ke^{i(kx-\omega t)}C \sinh k(h - y).$$

It will be noticed that $\Phi$ is here differentiated with respect to $x$, $y$, while it should be differentiated with respect to the variable coordinates $x'$, $y'$; also, we should write in (13) $x'$, $y'$ or $x + \xi$, $y + \eta$ instead of $x$, $y$. The difference, however, would be of second order in the quantities $C$, $\xi$ and $\eta$ and may therefore be neglected.

Integrating (13) with respect to $t$, which amounts to a division by $-i\omega$, we obtain

(14)

$$\xi = \frac{k}{\omega} e^{i(kz-\omega t)} C \cosh k(h - y),$$

$$\eta = i \frac{k}{\omega} e^{i(kz-\omega t)} C \sinh k(h - y).$$

where the constants of integration must be taken as zero because of the periodicity of the displacement. We introduce the notations (they will justify themselves presently)

(15) $$a = \frac{kC}{\omega} \cosh k(h - y), \qquad b = \frac{kC}{\omega} \sinh k(h - y)$$

and write the *real part* of (14), using these abbreviations:

(16) $$\xi = a \cos (kx - \omega t), \qquad \eta = -b \sin (kx - \omega t).$$

To find the equation of the particle path we have to eliminate $t$ from (16). By squaring $\xi$ and $\eta$ we obtain

(17) $$\frac{\xi^2}{a^2} + \frac{\eta^2}{b^2} = 1,$$

which is the equation of an ellipse. The reason why we had to change back to the real representation becomes now apparent: the complex writing preserves its physical significance only in linear operations!

The ratio of the minor to the major axis follows from (15) as

(17a) $$\frac{b}{a} = \tanh k(h - y) = \begin{cases} 0 & \text{for} \quad y = h, \text{ bottom} \\ \tanh kh & \text{for} \quad y = 0, \text{ surface.} \end{cases}$$

The excentricity of the ellipses is the same at any depth:

(17b) $$e = \sqrt{a^2 - b^2} = \frac{kC}{\omega}.$$

Fig. 37a represents the position and shape of the ellipses in the general case, Fig. 37b in the case of deep water, Fig. 37c in the case of shallow

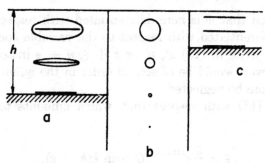

FIG. 37. The path curves of water particles (a) in moderately deep water, (b) in deep water, (c) in shallow water.

water. The ellipses of Fig. 37a are *confocal* if they are coaxially superimposed upon each other. In the limiting case $kh \to \infty$ we have by (17a)

$$\frac{b}{a} = \tanh(\infty) = 1.$$

The ellipses thus become *circles* as in Fig. 37b. The radii are determined by computing $a$ or $b$ from (15) and the first equation (1a):

$$a = \frac{kC}{\omega} \cosh k(h-y) = \frac{k}{\omega} A \, e^{-kh}(e^{k(h-y)} + e^{-k(h-y)})$$

$$= \frac{k}{\omega} A(e^{-ky} + e^{-k(2h-y)}) \to \frac{k}{\omega} A \, e^{-ky},$$

$$b = \frac{kC}{\omega} \sinh k(h-y) = \frac{k}{\omega} A \, e^{-kh}(e^{k(h-y)} - e^{-k(h-y)})$$

$$= \frac{k}{\omega} A(e^{-ky} - e^{-k(2h-y)}) \to \frac{k}{\omega} A \, e^{-ky}.$$

The radii of the circles are seen to decrease very quickly with increasing depth. In the other limiting case where $kh$ and, *a fortiori*, $k(h-y)$ go to zero we obtain from (17a)

$$\frac{b}{a} = \tanh 0 = 0.$$

The ellipses degenerate into horizontal *straight segments* (cf. Fig. 37c); in shallow water there is, so to speak, no space left for a vertical displacement. The amplitude of the horizontal oscillation, $e$, is independent of the depth $y$ [cf. (17b)].

The stream lines of the velocity field are no less interesting than the path curves of the particles. Since in the present case of plane waves we have a two-dimensional velocity field, the stream function $\Psi$ is obtained by (19.5) from the following equations

$$\frac{\partial \Psi}{\partial x} = - \frac{\partial \Phi}{\partial y}, \qquad \frac{\partial \Psi}{\partial y} = \frac{\partial \Phi}{\partial x}.$$

On substituting the general[2] expression for $\Phi$ from (2), we obtain

$$\frac{\partial \Psi}{\partial x} = ke^{i(kx-\omega t)}C \sinh k(h - y),$$

$$\frac{\partial \Psi}{\partial y} = ike^{i(kx-\omega t)}C \cosh k(h - y).$$

The first or second of these expressions is now integrated over $x$ or $y$ respectively and the integration constant set equal to zero as before; this leads to

$$\Psi = -ie^{i(kx-\omega t)}C \sinh k(h - y).$$

To find the stream lines we have to consider this equation at a specified instant $t$. Let us then take the real part for $t = 0$:

(18)                    $$\Psi = C \sin kx \cdot \sinh k(h - y).$$

For comparison, we write down the surface depression $\eta$ according to (3), express $c$ by $C$ through (6), and take again the real part at $t = 0$:

(19)                    $$\eta = - \frac{\omega}{g} C \sin kx.$$

A glance at (18) shows: the streamline $\Psi = 0$ requires $y = h$: the bottom is therefore a part of the streamline, which is fairly obvious; but for $kx = 0$, $\pm\pi$, $\pm 2\pi$, we have again $\Psi = 0$, whatever $y$ is. At the same abscissas, $\eta$ vanishes likewise, as is seen from (19). Hence the flow pattern is divided into rectangular domains which contain the stream lines. Furthermore we conclude from (18) that $\partial\Psi/\partial x = 0$ if $\cos kx = 0$, whatever $y$ is. This happens at the abscissas $kx = \pi/2, 3\pi/2, \cdots$, for which $\eta$ has a maximum or a minimum; at these points all streamlines are horizontal.—Altogether this is sufficient information to form an idea of the general pattern of the streamlines (cf. Fig. 38).

The patterns of Fig. 38 and 37a are convincingly verified by actual photographs of path curves and streamlines. For the purpose of photography the fluid is kept in a narrow glass container with parallel walls, and small light absorbing particles such as metal filings mixed in; one

---

[2]In the case of shallow water the stream lines are obviously horizontal.

photographs across the container and obtains the *streamlines* by a *short exposure*. In viewing the photographs the eye involuntarily connects the line elements traced by the dark particles and recognizes the continuous arcs of the stream lines, and also their position relative to the fluid surface which appears moderately sharp in the picture. With a *longer exposure*, e.g., as long as a half period, and a less dense interspersion of metal particles, the *path curves* can well be distinguished as elliptical arcs, but the surface does not show up distinctly. This second picture gives also evidence

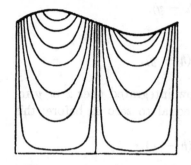

FIG. 38. The streamlines of gravity waves in water of finite depth.

of the propagation of the phase, if one compares the starting points (or end points) of the arcs left and right in the photograph.

The particles remain essentially at their places; this, at least, is true if the amplitude is sufficiently small, as assumed in our calculation, but is no longer quite correct for a finite amplitude. Therefore it is only the phase that travels with the velocity $V$ which we have computed; phase velocity would thus be a more accurate designation for $V$.

## 25. Plane Capillary Waves and Combined Capillary-Gravity Waves

In the following discussion we shall first disregard gravity entirely, taking only the *surface tension* $T$ into account. The surface is no longer free of forces, but under the action of the normal pressure $N$ caused by $T$. Accordingly, the condition for the hydrodynamic pressure at the surface is not $p = 0$ as in (23.4), but, in the case of a plane wave, according to (17.10)

$$(1) \qquad p = T \frac{\partial^2 \eta}{\partial x^2}.$$

(Note that $u$ of (17.10) must be replaced by the *surface elevation* $-\eta$.) The value of $p$ given by (1) is to be introduced in Bernoulli's equation (23.5) together with $U = 0$ (gravity is disregarded for the time being). Hence we obtain

$$(2) \qquad \frac{\partial \Phi}{\partial t} = \frac{T}{\rho} \frac{\partial^2 \eta}{\partial x^2}.$$

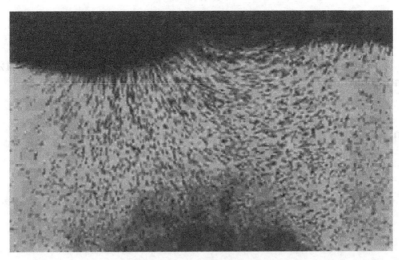

Fig. 39a. A photograph of the streamlines (short exposure).

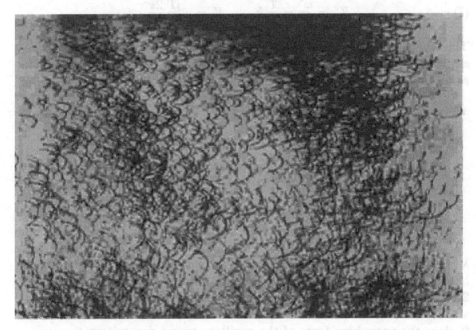

Fig. 39b. A photograph of the path curves (longer exposure).

The kinematic considerations at the beginning of 23 have dealt with the character of a plane wave in general and are consequently still valid. We therefore can take over the expressions (23.1) for $\Phi$ and (23.7) for $\eta$. Eq. (2) becomes after cancellation of the exponential and omission of second order terms

$$(3) \qquad\qquad i\omega A = \frac{T}{\rho}\, k^2 a.$$

Also the kinematic condition (23.11) is in force; it gives, as in (23.12), a second equation

$$(4) \qquad\qquad -i\omega a = kA.$$

From (3) and (4) we have

$$(5) \qquad\qquad \frac{A}{a} = \frac{T}{\rho}\frac{k^2}{i\omega} = -\frac{i\omega}{k}$$

and, therefore, the following relation between $\omega$ and $k$

$$(6) \qquad\qquad \omega^2 = \frac{T}{\rho}\, k^3.$$

The velocity of the capillary waves is

$$(7) \qquad\qquad V^2 = \frac{\omega^2}{k^2} = \frac{T}{\rho}\, k, \qquad V = \sqrt{\frac{T}{\rho}\frac{2\pi}{\lambda}}.$$

This then is a case of *anomalous dispersion* since $V$ increases with *decreasing* $\lambda$, contrary to the gravity waves which show *normal* dispersion at all finite wave lengths. In Fig. 40, we have plotted $V$ as a function of

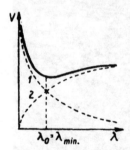

Fig. 40. Superposition of the phase velocities of capillary and gravity waves as functions of $\lambda$.

$\lambda$, obtaining the descending segment 1 (broken line). For comparison we have added the characteristic of the gravity waves from Fig. 34 as branch 2 (broken line). Let the intersection point which exists under all circumstances have the abscissa $\lambda = \lambda_0$. For $\lambda < \lambda_0$ the curve 1 is above curve 2, since the propulsive force of the capillary waves depends on the curvature of the surface profile which, for a given amplitude, is the larger the smaller the wave length. (This is also the reason why in Fig. 34 the segment $a$ was not continued beyond a certain lower limit.) For $\lambda > \lambda_0$ the curve 2 is above 1. The unimportance of capillarity for large wave lengths which expresses itself in the small $V$-values justifies our use of the velocity potential for *deep water* throughout the present article. Even at a moderate depth $h$ the quotient $h/\lambda$ is large for capillary waves, and so is *a fortiori* $hk$.

It is now easy to treat the general case of *combined action of capillarity and gravity*. We have only to introduce in Bernoulli's equation (23.5) the value $p$ from (1) and the value for the gravity potential from (23.5b) $U = -\rho g\eta$. In place of (3) we obtain

$$(8) \qquad i\omega A = \left(\frac{T}{\rho} k^2 + g\right)a.$$

From (4) and (8) follows now, instead of (5),

$$(9) \qquad \frac{A}{a} = \frac{1}{i\omega}\left(\frac{T}{\rho} k^2 + g\right) = -\frac{i\omega}{k},$$

and therefore, instead of (6) and (7),

$$(10) \qquad \omega^2 = \frac{T}{\rho} k^3 + gk,$$

$$(11) \qquad V^2 = \frac{\omega^2}{k^2} = \frac{T}{\rho} k + \frac{g}{k} = V_1^2 + V_2^2 .$$

Here $V_1$ is the velocity of the capillary wave according to (7), and $V_2$ is the velocity of the gravity wave according to (23.16); the velocity of the combination wave is found by a *quadratic superposition* formula.

Let us first compute from (11) the minimum of $V$ as a function of $\lambda$. We obtain

$$2V \frac{dV}{d\lambda} = \frac{d}{dk}\left(\frac{T}{\rho} k + \frac{g}{k}\right)\frac{dk}{d\lambda} = 0,$$

and therefore

$$(12) \qquad \frac{T}{\rho} = \frac{g}{k^2}, \qquad \lambda_{\min} = \frac{2\pi}{k} = 2\pi \sqrt{\frac{T}{\rho g}} .$$

But $\lambda_{\min}$ is at the same time the abscissa of the intersection point of the branches 1 and 2, which has been denoted by $\lambda_0$ in Fig. 40. In fact, one obtains from (7) and (23.16) the following equation for $\lambda_0$

$$\sqrt{\frac{T}{\rho}} \frac{2\pi}{\lambda_0} = \sqrt{\frac{g\lambda_0}{2\pi}},$$

hence $\lambda_0 = \lambda_{\min}$. The corresponding value of $V_{\min}$ is found as

$$(13) \qquad V_{\min}^2 = 2V_1^2 = 2V_2^2 = 2\sqrt{\frac{Tg}{\rho}} .$$

The general value (11) of $V^2$ can be put in the elegant form

(14) $$\frac{V^2}{V_{min}^2} = \frac{1}{2}\left(\frac{\lambda_{min}}{\lambda} + \frac{\lambda}{\lambda_{min}}\right),$$

which is easily verified.

We finally turn to the *numerical* values which in the present problem are not without interest. The constant of surface tension between water and air[3] is

$$T = 72\,\frac{dynes}{cm} = 72\,\frac{gr}{sec^2}.$$

From (12) and (13) one obtains with $\rho = 1$, $g = 981$:

$$\lambda_{min} = 2\pi\sqrt{\frac{72}{981}} = 1.73 \text{ cm}, \qquad V_{min} = \sqrt{2\sqrt{72 \times 981}} = 23.2\,\frac{cm}{sec}.$$

*Water waves with a velocity of less than 23 cm/sec do not exist; waves of larger and of smaller wave length than* $\lambda_{min} = 1.73$ *cm travel with greater velocity than* 23 *cm/sec.* Lord Kelvin proposed for waves $\lambda < \lambda_{min}$ the name "ripples"; one sometimes observes big gravity waves the faces of which are covered with fine capillary ripples.

## 26. The Concept of Group Velocity

At the end of 24 the term phase velocity was proposed in place of velocity of propagation $V$ and it was pointed out that the *phase* of the wave advances with the velocity $V$. The phase is given in our representation [(23.1), (23.7), (24.1), etc.] by the exponent $i(kx - \omega t)$; if it is kept constant, that is, if one looks for the locus of equal phase as time proceeds, one obtains the condition

(1) $$k\,dx - \omega\,dt = 0, \qquad \text{therefore} \qquad \frac{dx}{dt} = \frac{\omega}{k} = V.$$

For waves of a given invariable frequency (or, borrowing an expression from optics, for *monochromatic* waves), this is the only velocity that can come into consideration.

That is no longer true if waves of different frequency are superposed on each other. We speak then of a *group of waves*, particularly in the case where the frequencies are closely together. (Today one prefers, particularly in wave mechanics, the term "wave packet", an expression,

[3]Note that the most accurate method to determine $T$ consists in the measurement of $\lambda$ of standing capillary waves excited by a tuning fork. Elevation measurements in capillary tubes tend to be inaccurate because of impurities along the walls of the tube.

which is perhaps not quite so elegant as appropriate.) The wave group progresses with the *group velocity* $U$ which, in general, is different from the phase velocity $V$. It is convenient to write the expression for $U$ in the following way that will be justified later:

$$(2) \qquad U = \frac{d\omega}{dk}.$$

If the waves proceed *without dispersion*, that is, if $V$ is independent of the wave length $\lambda$ (or of the wave number $k$) one obtains from (1) $d\omega = V dk$ and therefore

$$(2a) \qquad U = V.$$

*Group and phase velocity coincide only for wave processes free of dispersion.*

In general, however, we have

$$d\omega = V \, dk + k \frac{dV}{dk} \, dk,$$

or, by (2),

$$(2b) \qquad U = V + k \frac{dV}{dk} = V - \lambda \frac{dV}{d\lambda}.$$

*For normal dispersion* $(dV/d\lambda > 0)$, *the group velocity is smaller than the phase velocity, for anomalous dispersion* $(dV/d\lambda < 0)$ *the converse statement holds.*

For gravity waves in deep water we have by (23.17)

$$V = \sqrt{\frac{g\lambda}{2\pi}}, \qquad \frac{dV}{d\lambda} = \frac{1}{2\lambda} V,$$

and by (2b)

$$(2c) \qquad U = V - \frac{1}{2} V < V.$$

On the other hand, for pure capillary waves (25.7) yields

$$V = \sqrt{\frac{T}{\rho} \frac{2\pi}{\lambda}}, \qquad \frac{dV}{d\lambda} = -\frac{1}{2\lambda} V,$$

hence

$$(2d) \qquad U = V + \frac{1}{2} V > V.$$

For the combined capillary-gravity waves we have according to Fig. 40 for $\lambda < \lambda_{\min}$ anomalous dispersion, therefore $U > V$; for $\lambda > \lambda_{\min}$ the dispersion is normal and $U < V$. For $\lambda = \lambda_{\min}$, $U = V$, corresponding to the horizontal tangent in Fig. 40 and to the vanishing factor $dV/d\lambda$ in (26).

The possibility of $U$ and $V$ having opposite signs has also been discussed.[4]

The concept of group velocity is originally a hydrodynamical one (Stokes 1876), but has proved of fundamental importance in optics (Lord Rayleigh). In wave mechanics phase velocity has formal significance while the group velocity is the important physical quantity, viz. the velocity of the particle represented by the wave packet (L. de Broglie 1924).

Let us first reproduce here the usual elementary derivation of formula (2), following Stokes. We superpose two waves that advance in positive $x$-direction, having the same amplitude, but slightly different frequency, hence slightly different wave number:

$$(3) \qquad \eta = a\{\sin (k_1 x - \omega_1 t) + \sin (k_2 x - \omega_2 t)\}.$$

The wave group so produced has the character of a *beat*: at points where the difference of two phases is equal to $2n\pi$, there is reinforcement; when the phase difference is $(2n + 1)\pi$ there is neutralization. This can be seen if (3) is written in the following form

$$(3a) \qquad \eta = 2a \cos \left(\frac{k_1 - k_2}{2} x - \frac{\omega_1 - \omega_2}{2} t\right) \sin \left(\frac{k_1 + k_2}{2} x - \frac{\omega_1 + \omega_2}{2} t\right).$$

The cosine term vanishes or equals $\pm 1$, depending on whether the phase difference of the two partial waves is an uneven or even multiple of $\pi$. Introducing

$$\frac{k_1 + k_2}{2} = k_0 , \qquad \frac{\omega_1 + \omega_2}{2} = \omega_0 , \qquad k_1 - k_2 = \Delta k, \qquad \omega_1 - \omega_2 = \Delta \omega$$

we write for (3a)

$$(3b) \qquad \eta = C \sin (k_0 x - \omega_0 t), \qquad C = 2a \cos \frac{1}{2} (\Delta k x - \Delta \omega t).$$

The introduction of the "amplitude-factor" $C$ for the cosine factor should suggest that it is a *slowly variable* quantity. We call it the group amplitude. The phase velocity of the compound wave as it follows from the sine factor in (3b) is

$$V = \frac{\omega_0}{k_0}$$

---

[4]Cf. Sir Arthur Schuster, Theory of Optics, London, 1924, p. 330.

and thus not sensibly different from the phase velocities of the component waves

$$V_1 = \frac{\omega_1}{k_1} \quad \text{and} \quad V_2 = \frac{\omega_2}{k_2} .$$

On the other hand, the velocity of propagation of the "amplitude" $C$ is found by setting

$$\Delta k x - \Delta \omega t = \text{const},$$

which yields when differentiated

(3c) $\qquad \Delta k \, dx - \Delta \omega \, dt = 0, \qquad \frac{dx}{dt} = \frac{\Delta \omega}{\Delta k} .$

In the limit $\Delta k \to 0$ we therefore obtain our formula (2).

We can go beyond the very special assumption (3) when we consider an arbitrary group of waves whose frequencies are spread over a small frequency band. Since it is now convenient to return to complex representation, we put $\eta$ in the form

(4) $\qquad \qquad \eta = \int_{k_0-\epsilon}^{k_0+\epsilon} a(k) e^{i(kx-\omega t)} \, dk.$

The amplitude of the partial wave is now $a(k)dk$ and the small band width of the whole group $2\epsilon$, if measured in wave number units, so as to be concentrated about the central wave number $k_0$ . We rewrite the exponent in (4) accordingly:

$$kx - \omega t = k_0 x - \omega_0 t + (k - k_0)x - (\omega - \omega_0)t.$$

If this is done, (4) assumes the form

(4a) $\qquad \eta = C e^{i(k_0 x - \omega_0 t)}, \qquad C = \int_{k_0-\epsilon}^{k_0+\epsilon} a(k) e^{i[(k-k_0)x-(\omega-\omega_0)t]} \, dk.$

To find the velocity of propagation of the amplitude $C$ in this more general case, we have to set the exponential in $C$, which alone contains $x$ and $t$, constant. This yields for the whole wave group the sensibly constant value

$$\frac{dx}{dt} = \frac{\omega - \omega_0}{k - k_0} = \frac{\Delta \omega}{\Delta k}$$

in agreement with Eq. (2), if $\Delta k$ is sufficiently small.

It should be noticed that the more general formula (4) as well as the expression (3) constitute a superposition of infinitely long wave trains, but do not represent an isolated hill or a few crests and troughs. To express an isolated wave group (a wave packet) by analytical means would require the use of Fourier integrals; the integration in (4) would have to be carried out over all wave numbers from $-\infty$ to $+\infty$ instead of over the narrow range $2\epsilon$.

A wave packet can proceed *without change of shape* only if there is no dispersion. In the general case the group dissolves on account of the dispersion since its partial waves do not keep together; this tendency is of particular importance for the particle concept of wave mechanics. The conservation of the packet in the absence of dispersion follows directly from d'Alembert's solution (13.11), at least in the case of a sufficiently small amplitude, when this solution applies to dispersion-free hydrodynamics as well as to acoustics. An arbitrary initial disturbance $\eta = F(x)$ is then given by $\eta = F(x - Vt)$ at any later time if the initial velocity $\partial\eta/\partial t$ has been suitably chosen. This means that the wave can advance *without change of shape.*

In Fig. 41a and b a simple construction of the group velocity is shown that uses Eq. (2b) and the relation $\tan \alpha = \pm \, dV/d\lambda$.

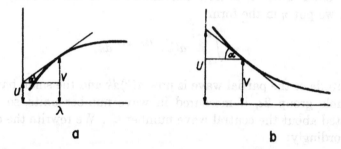

FIG. 41. Construction of the group velocity for a given phase velocity when the dispersion curve is given; (a) normal, (b) anomalous dispersion.

Reynolds[5] and, simultaneously, Lord Rayleigh[6] have pointed out the relation between the group velocity and the *transport of energy* and thus enhanced the physical understanding of the concept of group velocity. In their theory the ratio of group velocity $U$ to phase velocity $V$ appears as the ratio of the energy flow $S$ through a specified cross-section of the wave during the time $\tau$ to the energy $E$ contained in the fluid region ahead of this cross-section per length $V\tau$:

$$(5) \qquad \frac{U}{V} = \frac{S}{E}.$$

[5] Reynolds, O., Papers on math. and phys. Subj., Cambridge 1901, Vol. I, p. 198, from Nature 46 (1877) 343.

[6] Lord Rayleigh, Scientific papers, Cambridge 1901, Vol. I, p. 322 and Theory of sound, Vol. I, appendix from Proc. London Math. Soc. 9, 21.

It is convenient for what follows, to identify the time $\tau$ with the period of oscillation, so that $V\tau$ is equal to the wave length $\lambda$. The energy $E$ is more accurately defined as the energy *surplus* of the oscillating fluid over the fluid at rest.

Let us examine these relations for deep water waves of sufficiently small amplitude. Since we are concerned with energy, that is with a quadratic function of the rates of displacement, we cannot use the complex representation. Taking the real equations (23.14) and (23.14b)

(6)
$$\Phi = A \cos (kx - \omega t)e^{-ky},$$

$$\eta = - \frac{k}{\omega} A \sin (kx - \omega t),$$

we can obtain the potential energy of gravity contained between the undisturbed surface $y = 0$ and the wave surface $y = \eta$ at a specified time $t$, per length $\lambda$ in $x$-direction and per width 1 in $z$-direction:

(7)
$$E_{\text{pot}} = g\rho \int_0^\lambda dx \int_0^\eta y \, dy$$

$$= g\rho \frac{k^2}{\omega^2} \frac{A^2}{2} \int_0^\lambda \sin^2 (kx - \omega t) \, dx = g\rho \frac{k^2}{\omega^2} \frac{A^2}{4} \lambda.$$

Using the dispersion formula (23.15), we rewrite this result:

(7a)
$$E_{\text{pot}} = \frac{\rho}{4} k A^2 \lambda = \frac{\rho}{2} \pi A^2.$$

This is at the same time the difference between the potential energy of the oscillating fluid and the fluid at rest for the *entire* depth which reaches from $y = \eta$ to $y = \infty$ for the moving fluid and from $y = 0$ to $y = \infty$ for the fluid at rest.

The kinetic energy contained in this region has the same magnitude, as is always the case for small oscillations; this will be discussed in detail in problem V, 1. The total energy, or more accurately, the energy excess over the fluid at rest becomes therefore

(8)
$$E = E_{\text{kin}} + E_{\text{pot}} = \rho \pi A^2.$$

In order to determine the energy flow $S$ through the cross-section at $x = \lambda$, which may be replaced by the cross-section $x = 0$, we have to know the pressure $p$ at this cross-section. This is found from Bernoulli's equations (23.5) and (23.5b) as

(9)
$$p = \rho \frac{\partial \Phi}{\partial t} + \rho g y + \text{const.}$$

Since $p$ acts normal to the plane $x = 0$, its product with the $x$-component of the velocity, multiplied with $dy\, dt$ expresses the work $dW$ done at the surface element $dy \cdot 1$ during $dt$:

(9a)        $$dW = p v_x\, dy\, dt = - p \frac{\partial \Phi}{\partial x}\, dy\, dt.$$

The work performed at the cross-section $x = 0$ during the time $\tau$ represents the energy flow through this cross-section in this time. It is by (9a) and (9)

(10)        $$S = - \int_0^\tau dt \int_\eta^\infty dy\, p \frac{\partial \Phi}{\partial x} = - \int_0^\tau dt \int_\eta^\infty dy \left( \rho \frac{\partial \Phi}{\partial t} + \cdots \right) \frac{\partial \Phi}{\partial x}.$$

The terms $\rho g y + const$ not written out in (10) have the periodic factor $\partial \Phi / \partial x$. Since they are independent of $t$, their contributions cancel when integrated over the period $\tau$. Substitution from (6) gives

$$S = \rho \omega k A^2 \int_0^\tau dt \sin^2 \omega t \int_\eta^\infty e^{-2kv}\, dy$$

or, after integration,

(11)        $$S = \frac{\rho}{2} \pi A^2 e^{-2k\eta} = \frac{\rho}{2} \pi A^2.$$

Here, $e^{-2k\eta}$ has been replaced by 1 which means omission of terms of higher order than $A^2$.

Comparing (11) and (8), one obtains

(12)        $$\frac{S}{E} = \frac{1}{2}.$$

This is indeed the value of the ratio $U/V$ for deep water waves obtained in (2c). We have thus confirmed in this simplest case the general energetic definition (5) of the group velocity. It also holds in the considerably more complicated case of moderately deep water where Eqs. (2b) and (24.10a) lead to

(13)        $$\frac{U}{V} = \frac{1}{2} \left( 1 + \frac{2kh}{\sinh 2kh} \right).$$

Let us now have a look at this result from a more qualitative point of view; we again consider the wave train of length $\lambda$ ending at $x = 0$. If no dispersion were present, the entire energy content of the wave would flow through the cross-section during the period $\tau$; in other words, we should have $S = E$ and $U = V$. Actually, only a part of this energy is transported in the case of normal dispersion (half of it with deep water waves), whereas the other part remains in the oscillating fluid ahead of the cross-section. This makes the case of anomalous dispersion appear rather peculiar since more energy passes through the cross-section than is present in the $\lambda$-train ahead of it; the energy remaining in the train would be negative. To maintain the wave motion in such a case requires expending more energy than travels with the wave; in the first case less excitation energy is required for the same amount of energy transported. *Energy transport is thus easier in a medium of normal dispersion and more difficult in a medium of anomalous dispersion.*

## 27. Circular Waves

The waves that are produced when a stone is thrown into water form a series of concentric circular crests and troughs; their amplitudes are not constant, nor are the distances between the crests. What one observes is a sharp decrease of the amplitude and an increase of the distance between two subsequent crests, which seems to follow a peculiar law. This problem that appears so simple requires for its solution a considerable mathematical apparatus: we not only need Bessel functions and Fourier integrals, but we should have to use the method of steepest descent if we were to treat it in full accuracy. A systematic treatment of these devices will be given in Vol. VI. At this point we shall limit ourselves to such explanations as are directly required here.

For our analysis we shall replace the stone that hits the water surface by a "standard disturbance": at $r = 0$ a cylindrical piston of radius $r_0$ is immersed in the water to a distance $a$ from the surface and suddenly withdrawn at the time $t = 0$. If we again denote the surface depression by $\eta$, the initial state is given by

(1)
$$\eta = a, \qquad r < r_0 .$$
$$\eta = 0, \qquad r > r_0 .$$

In preparation for the problem of a single disturbance we first consider the much simpler case of a periodic excitation.

## 1. The Periodic Case. Introduction of Bessel Functions

The excitation we have in mind works in a similar way as the device that produces waves in a swimming pool: a straight board subjected to a periodic motion excites plane progressive gravity waves advancing normal to the board. We have investigated waves of this type in 23 for deep water and wish now to transfer our previous results to the present circular symmetric problem.

Introducing cylindrical polar coordinates $r$, $\varphi$, $y$ ($y$ positive downward), we write the condition of incompressibility for the velocity potential $\Phi$ (cf. problem I, 3) in the form

(2)
$$\frac{\partial^2 \Phi}{\partial r^2} + \frac{1}{r}\frac{\partial \Phi}{\partial r} + \frac{1}{r^2}\frac{\partial^2 \Phi}{\partial \varphi^2} + \frac{\partial^2 \Phi}{\partial y^2} = 0.$$

If we at first assume circular symmetry, $\Phi$ does not depend on $\varphi$ and can be written as follows

(3)
$$\Phi = A f(r) e^{-ky} e^{-i\omega t}.$$

Now the following differential equation results for $f(r)$:

(4)
$$\frac{d^2 f}{dr^2} + \frac{1}{r}\frac{df}{dr} + k^2 f = 0.$$

We substitute $kr = \rho$ and obtain

(4a)
$$\frac{d^2 f}{d\rho^2} + \frac{1}{\rho}\frac{df}{d\rho} + f = 0.$$

Let $J_0(\rho)$ be the solution of this equation which is regular at $\rho = 0$ and assumes there the value 1. The power series expansion of this function can be easily found from (4a) by the method of undetermined coefficients:

(5)
$$J_0(\rho) = 1 - \left(\frac{\rho}{2}\right)^2 + \frac{1}{2!^2}\left(\frac{\rho}{2}\right)^4 - \frac{1}{3!^2}\left(\frac{\rho}{2}\right)^6 + \cdots .$$

If, on the other hand, $\Phi$ is not independent of $\varphi$, then we replace $f(r)$ by $f_n(r)e^{in\varphi}$ and obtain instead of (4) and (4a)

(6)
$$\frac{d^2 f_n}{dr^2} + \frac{1}{r}\frac{df_n}{dr} + \left(k^2 - \frac{n^2}{r^2}\right)f_n = 0$$

and

(6a)
$$\frac{d^2 f_n}{d\rho^2} + \frac{1}{\rho}\frac{df_n}{d\rho} + \left(1 - \frac{n^2}{\rho^2}\right)f_n = 0.$$

Let $J_n(\rho)$ denote that solution of (6a) which is regular at $\rho = 0$ and admits the following expansion,

(7)
$$J_n(\rho) = \frac{1}{n!}\left(\frac{\rho}{2}\right)^n - \frac{1}{1!(n+1)!}\left(\frac{\rho}{2}\right)^{n+2} + \frac{1}{2!(n+2)!}\left(\frac{\rho}{2}\right)^{n+4} - \cdots$$

which may be considered as a generalization of (5). The functions $J_n$ are known as *Bessel Functions* of $n^{\text{th}}$ order; they are entire transcendental functions.

The following relation between $J_1$ and $J_0$

(8) $$J_1(\rho) = -\frac{d}{d\rho} J_0(\rho),$$

can be immediately verified from (5) and (7).

On writing Eq. (4a) in the form

$$\frac{d}{d\rho}\left(\rho \frac{dJ_0}{d\rho}\right) + \rho J_0(\rho) = 0,$$

where $J_0(\rho)$ has been written for $f$, one has because of (8),

$$\rho J_0(\rho) = \frac{d}{d\rho}(\rho J_1(\rho)).$$

This relation, integrated to an arbitrary upper limit, yields the following formula

(8a) $$\int_0^{\rho_0} \rho J_0(\rho)\, d\rho = \rho_0 J_1(\rho_0).$$

We shall also need the integral representation

(9) $$J_0(\rho) = \frac{1}{2\pi} \int_{-\pi}^{+\pi} e^{i\rho\cos\alpha}\, d\alpha.$$

It is not difficult to verify that this expression satisfies Eq. (4a) and has the value 1 for $\rho = 0$, wherefore it must be identical with (5). The same result can also be obtained, though less directly by series expansion of the exponential function in (9). Obviously, (9) can also be written in the form

(9a) $$J_0(\rho) = \frac{1}{\pi} \int_0^{\pi} e^{i\rho\cos\alpha}\, d\alpha.$$

We now substitute $f = J_0(kr)$ in the formula for the potential (3) and obtain

(10) $$\Phi = A J_0(kr) e^{-ky} e^{-i\omega t}.$$

This formula satisfies the condition for deep water: $\Phi \to 0$ for $y \to \infty$, but it should also satisfy the conditions (23.6) and (23.11) for the surface $y = 0$. The analytical form of the surface depression is assumed in correspondence to the form of Eq. (10)

(10a) $$\eta = a J_0(kr) e^{-i\omega t}.$$

Our expression for $\Phi$ and $\eta$ leads, by exactly the same steps as in Eqs. (23.9), (23.12) and (23.13), after cancellation of $J_0(kr) e^{-i\omega t}$, to

(11)
$$\frac{A}{a} = \frac{g}{i\omega} = -\frac{i\omega}{k}.$$

Thus we reobtain the dispersion law (23.15)

(11a)
$$\omega = \sqrt{gk}$$

and the final expression for the potential becomes

(12)
$$\Phi = \frac{ag}{i\omega} J_0(kr)e^{-ky-i\omega t}.$$

## 2. Single Disturbance. The Fourier-Bessel Integral

The content of the Fourier integral theorem is this: an arbitrary function $F(x)$ (provided it is not "too irregular") can be represented by superposition of trigonometric functions $\genfrac{}{}{0pt}{}{\sin}{\cos} kx$ or by the equivalent exponential functions $e^{ikx}$ in the form

(13)
$$F(x) = \frac{1}{2\pi} \int_{-\infty}^{+\infty} dk\, e^{ikx} \int_{-\infty}^{+\infty} d\xi\, F(\xi)e^{-ik\xi}, \quad -\infty < x < +\infty.$$

This theorem has an analogue in the following representation of $F$ by Bessel functions

(13a)
$$F(r) = \int_0^\infty k\, dk\, J_0(kr) \int_0^\infty \xi\, d\xi\, F(\xi)J_0(k\xi), \quad 0 < r < +\infty.$$

Both theorems will be proved in Vol. VI, where (13a) will be obtained as a consequence of (13).

We now apply (13a) to the initial state (1). Our function $F(r)$ is then given by

$$F(r) = \eta = \begin{cases} a & \text{for} \quad 0 < r < r_0, \\ 0 & \text{for} \quad r_0 < r < \infty. \end{cases}$$

Its representation according to (13a) is

(14)
$$\eta_{t=0} = a \int_0^\infty k\, dk\, J_0(kr) \int_0^{r_0} \xi\, d\xi\, J_0(k\xi).$$

For the evaluation of the inner integral Eq. (8a) is used, which gives, with $k\xi = \rho$ and $kr_0 = \rho_0$,

$$\int_0^{r_0} \xi\, d\xi\, J_0(k\xi) = \frac{1}{k^2} \int_0^{\rho_0} \rho\, d\rho\, J_0(\rho) = \frac{r_0}{k} J_1(kr_0).$$

Instead of the double integral (14) we have got

(14a) $$\eta_{t=0} = ar_0 \int_0^\infty J_0(kr) J_1(kr_0)\, dk.$$

Now we can show that $\eta$ and $\Phi$ at any later instant $t > 0$ have the following forms:

(15) $$\eta = ar_0 \int_0^\infty J_0(kr) J_1(kr_0)\, \exp\left(-i\,\sqrt{gk}\; t\right) dk,$$

(16) $$\Phi = -iagr_0 \int_0^\infty J_0(kr) J_1(kr_0)\, \exp\left(-ky - i\,\sqrt{gk}\; t\right) dk / \sqrt{gk}.$$

For a proof, we observe first that (15) takes the value of (14a) if $t = 0$. Now, the expressions (15) and (16) are obtained by applying the same "operator"

$$r_0 \int_0^\infty J_1(kr_0)\, dk \cdots$$

to the periodic solutions (10a) and (12), with due regard to the dispersion law (11a). Since (10a) and (12) satisfy the differential equation of our problem and the boundary conditions for $y = 0$ and $y = \infty$, the same must be true for our expressions (15) and (16). Furthermore (15) and (16) satisfy the initial condition for $t = 0$ and are thus the required solutions.

## 3. Integration with Respect to $k$. The Method of the Stationary Phase

The quantities of physical interest in circular waves are all connected with the surface function $\eta$ to which we therefore limit the following discussion. To evaluate $\eta$ we first replace $J_0$ in (15) by its integral representation (9a) and reverse the order of integration, obtaining the double integral

(17) $$\eta = \frac{ar_0}{\pi} \int_0^\pi d\alpha \int_0^\infty dk\, J_1(kr_0)\, \exp\left(ikr \cos\alpha - i\,\sqrt{gk}\; t\right).$$

We denote the inner integral by $K$, put $2\tau = t\,\sqrt{gr_0}/r$ and introduce a new integration variable by $p = \sqrt{kr_0}$; with these notations we have

(18) $$K = \frac{2}{r_0} \int_0^\infty p\, dp\, J_1(p^2)\, \exp\left[i(p^2 \cos\alpha - 2p\tau)r/r_0\right].$$

Here $r/r_0$ is a very large number, the exponential is, therefore, a rapidly varying function of $p$. In general, the positive and negative oscillations of the integrand will cancel except at those places where the factor of $ir/r_0$ is *slowly* variable. If we abbreviate this factor by

$$(19) \qquad f(p) = p^2 \cos \alpha - 2p\tau,$$

this will happen when

$$(19a) \qquad f'(p) = 0, \qquad p = p_0 = \frac{\tau}{\cos \alpha}.$$

This observation leads to a method of estimating the value of the integral known as the *method of the stationary phase*. Since the "phase" of the quickly oscillating exponential function becomes stationary at $p_0$, one limits the range of integration to the neighborhood of this value and thus achieves the integration by elementary means. It was particularly Lord Kelvin who applied this method expertly to many problems of hydrodynamics and optics. The mathematically exact form of this device[7] is called the *method of steepest descent* which we mentioned before.

Eq. (19) can be written in the form

$$f(p) = \cos \alpha \{(p - p_0)^2 - p_0^2\}$$

The integral (18) then becomes

$$(20) \qquad K = \frac{2}{r_0} \exp\left(-ip_0^2 \cos \alpha \cdot r/r_0\right) \int_{p_0-\epsilon}^{p_0+\epsilon} J_1(p^2) p \exp\left[i \cos \alpha (p - p_0)^2 r/r_0\right] dp.$$

where the range of integration is now an $\epsilon$-neighborhood of $p_0$ : The factors $J_1(p^2)$ and $p$ in the integrand are *slowly variable* compared with the exponential function and may be replaced by $J_1(p_0^2)$ and $p_0$. We finally substitute in (20)

$$(20a) \qquad s = q(p - p_0), \qquad q^2 = \frac{r}{r_0} \cos \alpha,$$

and obtain

$$(21) \qquad K = (2/r_0) \exp\left(-ip_0^2 \cos \alpha \cdot r/r_0\right) J_1(p_0^2) \frac{p_0}{q} \int_{-q\epsilon}^{+q\epsilon} e^{is^2} ds.$$

As to the position of the critical value $p = p_0$, (19a) gives, since $\tau$ is positive,

$$p_0 > 0 \qquad \text{for} \qquad 0 < \alpha < \frac{\pi}{2},$$

$$p_0 < 0 \qquad \text{for} \qquad \frac{\pi}{2} < \alpha < \pi.$$

---

[7]The problem of the circular waves has been treated according to that method in the Munich thesis of H. Widenbauer, Z. angew. Mathem. u. Mech. 14, 321 (1939). Our Fig. 42 has been taken from this paper.

In the first case $p_0$ and, consequently, the value $s = 0$ fall in the range of integration $0 < k < \infty$; *not* so in the second case. Thus the representation (21) is valid only for $\alpha < \pi/2$. Otherwise we have

$$(21a) \qquad\qquad K = 0; \quad \frac{\pi}{2} < \alpha < \pi,$$

which means complete extinction by "interference".

The final evaluation of (21) is achieved by means of the well known formula

$$(22) \qquad\qquad \int_{-\infty}^{+\infty} e^{-t^2} \cdot dt = \sqrt{\pi}.$$

On replacing $t$ by

$$t = e^{-i(\pi/4)}s, \quad t^2 = -is^2, \quad dt = e^{-i(\pi/4)}\, ds,$$

we are led[8] to

$$(22a) \qquad\qquad \int_{-\infty}^{+\infty} e^{is^2}\, ds = e^{i(\pi/4)}\, \sqrt{\pi}.$$

The left member is our integral in (21) whose limits become $\pm\infty$ if $r/r_0$ is made to approach infinity at constant $\epsilon$ and $\cos\alpha \neq 0$. If we introduce the result (22a) in (21) we obtain

$$(23) \qquad K = (2\sqrt{\pi}/r_0)\exp\left(-ip_0^2\cos\alpha \cdot r/r_0 + i\pi/4\right)J_1(p_0^2)\frac{p_0}{q}.$$

Now the value of $p_0$ following from (17a) and (19a),

$$p_0 = \sqrt{\frac{gr_0 t^2}{4r^2}}\,\frac{1}{\cos\alpha},$$

is very small, provided

$$(24) \qquad\qquad \frac{gr_0 t^2}{4r^2} \ll 1$$

(but we must exclude a small finite neighborhood of $\alpha = \pi/2$). The Bessel function $J_1$ is, for small $p_0$, sufficiently well approximated by the first term of (7):

$$J_1(p_0^2) = \frac{p_0^2}{2}, \quad p_0 J_1(p_0^2) = \frac{p_0^3}{2} = \frac{1}{2}\left(\frac{gr_0 t^2}{4r^2}\right)^{3/2}\frac{1}{\cos^3\alpha}.$$

---

[8]It will be noticed that the original path of the integration in (22a) is the bisectrix of the 1st and 3d quadrant in the complex $s$-plane which forms also the limit of convergence of the integral; this path may be replaced by the real axis and two circular arcs; the contributions of the latter vanish if $R \to \infty$.

Taking the value of $q$ from (20a) we obtain finally for $K$

$$(25) \qquad K = \sqrt{\frac{\pi}{\cos^7 \alpha}} \frac{r_0}{r^2} \left(\frac{gt^2}{4r}\right)^{3/2} \exp\left(-igt^2/4r \cos \alpha + i\pi/4\right).$$

## 4. Integration with Respect to $\alpha$. Discussion of a Limiting Case

We return now to (17), where we may restrict the integration interval to $0 < \alpha < \pi/2$ because of (21a). On introducing the volume displaced by the initial impulse, $V_0 = \pi r_0^2 a$, (17) is transformed into

$$(26) \qquad \eta = \frac{V_0}{r^2}\left(\frac{gt^2}{4\pi r}\right)^{3/2} \int_0^{\pi/2} \frac{d\alpha}{\cos^{7/2}\alpha} \exp\left(-igt^2/4r \cos \alpha + i\pi/4\right).$$

We shall finally have to make $r_0 \to 0$. In order to obtain a finite effect in the limit, $V_0$ must be kept constant, that is to say, the depth of immersion $a$ must approach $\infty$ in a definite way. We shall come back to this eventually.

The representation of $\eta$ by (26) depends essentially on the variable

$$(27) \qquad u = \frac{gt^2}{4r}.$$

Our aim is to determine the asymptotic behavior of $\eta$ if $u \to \infty$. Now for large $u$ the exponent in (26) becomes once more a rapidly varying function of $\alpha$, so that the method of the stationary phase can again be applied. Similarly as before, we denote the factor of $u$ in the exponent of (26) by $f(\alpha)$ and have

$$f(\alpha) = \frac{1}{\cos \alpha}, \qquad f'(\alpha) = \frac{\sin \alpha}{\cos^2 \alpha}.$$

The critical $\alpha$-value is thus

$$\alpha = \alpha_0 = 0, \qquad f(\alpha_0) = 1, \qquad f(\alpha) = 1 + \frac{\alpha^2}{2}, \qquad \frac{1}{\cos^{7/2}\alpha_0} = 1,$$

and consequently

$$(28) \qquad \eta = \frac{V_0}{r^2}\left(\frac{u}{\pi}\right)^{3/2} \exp\left(-iu + i\pi/4\right) \int_0^\epsilon \exp\left(-iu\alpha^2/2\right) d\alpha.$$

We need the value of the integral in (28) for large $u$. Keeping $\epsilon$ constant, we apply the same argument that led us from (21) to (23). This time the integral in (28) transforms as follows:

$$\int_0^\epsilon \exp\left(-iu\alpha^2/2\right) d\alpha = \sqrt{\frac{2}{u}} \int_0^\infty \exp\left(-is^2\right) ds = \sqrt{\frac{\pi}{2u}} \exp\left(-i\pi/4\right).$$

By substitution in (28) we finally obtain

$$(29) \qquad\qquad \eta = \frac{V_0}{\sqrt{2\pi r^2}}\, u e^{-iu}.$$

The dimensions in this equation are easily checked: $u$ is a number according to (27), $V_0/r^2$ is a length and so is $\eta$.

Our method seems to break down at the upper limit $\pi/2$ of the integral in (26) because of the denominator $\cos \alpha$, but according to (21a) this is at the same time the lower limit at which $K$ vanishes. Thus one is probably right to assume that a special investigation of the behavior of (26) at the upper limit is unnecessary.

The real part of (29) is the surface equation in which we are interested; we have

$$\eta = \frac{V_0}{\sqrt{2\pi r^2}}\, u \cos u.$$

$\eta$ becomes infinite for $r = 0$, which is quite understandable since the depth of immersion $a$ has become infinite in the limiting process $r_0 \to 0$ for fixed $V_0$. The amplitudes of the crests decrease according to $u/r^2$ or $r^{-3}$, the crests follow each other at the distance

$$\Delta r = \frac{8\pi}{g}\frac{r^2}{t^2},$$

as is easily seen, if the phases of neighboring crests $u = 2\pi n$, $u = 2\pi(n + 1)$ are compared for constant $t$. Hence the wave length is no longer a constant as in our previous examples of wave motion, but increases at

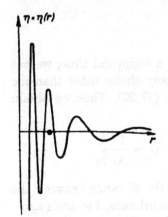

FIG. 42. Shape of the water surface $t$ seconds after the ring waves were excited; $t$ must satisfy the condition $u \gg 1$.

constant $t$ with $r^2$ and decreases at constant $r$ with $t^2$. Fig. 42 is a diagram of the surface profile $\eta$. Its appearance agrees well with that of a water surface disturbed by the fall of a small object like a stone or a rain drop.

Our problem was set as a prize problem by the academy of Paris in 1816 and solved by Cauchy. About this and a later paper of Poisson cf. the report of H. Burkhardt.[9]

## 28. Ship Waves (Kelvin's Limit Angle and Mach's Angle)

The wave pattern that is left behind by a ship at sea consists of a system of waves that envelopes the hull lengthwise and is interwoven with a system of cross waves. The two systems advance with the boat so as to be stationary relative to it. The beauty of the pattern is most impressive when viewed from an airplane or from the top of a high cliff, but the same phenomenon on a more modest scale develops behind a duck swimming in a pond.

In our analysis the object that produces the waves will be considered as a point. The problem can then be formulated in the following way: The instantaneous location of the boat is the origin of a system of circular waves; this origin is in uniform rectilinear motion, its velocity being the speed of the boat $v$. Our task is to find *the result of the superposition of the successive circular waves*. That it will be stationary relative to the boat is evident, but the detail structure of the wave pattern is surprising enough and can only be unraveled by a careful analysis.

With a view to Fig. 43, let $O$ be the location of the boat at the time $t = 0$, and $Q$ its location $t$ seconds earlier so that $QO = vt$. We wish to find the ordinate of the water surface $\eta$ at the field point $P$. It is to be compounded of all ordinates $\eta_t$ that were produced at earlier instants $t$ by means of the formula

$$(1) \qquad \eta = \beta \int_{-\infty}^{0} \eta_t \, dt.$$

The factor $\beta$ in (1) must have the dimension of a reciprocal time; we put it equal to $v/l$. For the length $l$ there is hardly any choice other than the cube root of the initial displacement $V_0$ in Eq. (27.29). Thus we obtain from (1) and (27.29)

$$(2) \qquad \eta = C \int_{-\infty}^{0} \frac{1}{r_t^2} u_t \exp(-iu_t) \, dt, \qquad C = \frac{V_0^{2/3} v}{\sqrt{2\pi}}$$

where $r_t$ is the distance $QP$ in Fig. 43, that is, the distance between the field point $P$ and the location of the source of disturbance, $t$ seconds ago.

Now let $P$ have the polar coordinates $r$ and $\vartheta$ relative to the pole $O$; $r$ and $\vartheta$ are therefore independent of $t$. According to Fig. 43 we obtain

[9]Jahresbericht der deutschen Math. Ver. Vol. X, p. 429 (1908).

(3) $$r_t^2 = r^2 + v^2 t^2 + 2rvt \cos \vartheta,$$

where $t$ is negative. As in (27.27) we put

(4) $$u_t = \frac{g}{4} \frac{t^2}{r_t} = f(t).$$

The representation (27.29) which we have applied in (2) was computed under the assumption $u \gg 1$. With this condition, $f(t)$ becomes again a rapidly varying function so that the method of stationary phase can be applied. We then have to find the roots of the equation $f'(t) = 0$. From (4) and (3) we obtain

(5) $$\frac{4}{g} f'(t) = \frac{2t}{r_t} - \frac{t^2}{r_t^3} (v^2 t + rv \cos \vartheta) = \frac{t}{r_t^3} (v^2 t^2 + 3rvt \cos \vartheta + 2r^2).$$

Hence the roots of $f'(t) = 0$ are

$$t = -\frac{3}{2} \frac{r}{v} \cos \vartheta \pm \sqrt{\frac{9}{4} \frac{r^2}{v^2} \cos^2 \vartheta - 2 \frac{r^2}{v^2}}.$$

If $\vartheta$ is an acute angle, the roots $t_1$ and $t_2$ are both negative

(6)
$$t_1 = -\frac{3}{2} \frac{r}{v} \left( \cos \vartheta - \sqrt{\cos^2 \vartheta - \frac{8}{9}} \right),$$

$$t_2 = -\frac{3}{2} \frac{r}{v} \left( \cos \vartheta + \sqrt{\cos^2 \vartheta - \frac{8}{9}} \right).$$

Now, in order to fall in our integration interval $-\infty < t < 0$, the roots not only have to be negative, they must also be real. This implies

$$\cos^2 \vartheta > \frac{8}{9}, \qquad |\vartheta| < \vartheta_0,$$

where $\vartheta_0$ denotes the limiting angle

(7) $$\cos^2 \vartheta_0 = \frac{8}{9} \quad \text{or} \quad \text{tg } \vartheta_0 = \frac{1}{\sqrt{8}}, \qquad \vartheta_0 = 19°28'.$$

This angle was first determined by Lord Kelvin. For $|\vartheta| > \vartheta_0$ there is no such $t$-value as would make the phase stationary, that is to say, *the whole wave pattern is bounded on either side by a straight line forming the angle $\vartheta_0$ with the direction of motion of the boat.* This is shown in Fig. 44 which is taken from Lamb's[10] Hydrodynamics.

The interference pattern itself can be understood on the basis of the integral (2) which essentially reduces to the two contributions of the neigh-

---

[10]H. Lamb, Hydrodynamics, Cambridge, 5th edition, p. 409 ff.

borhoods of $t_1$ and $t_2$ . These contributions contain the phase factors $\exp\,[-if(t_1)]$ and $\exp\,[-if(t_2)]$. By putting $f(t_1)$ and $f(t_2)$ constant, one obtains the two systems of curves, the lengthwise and transverse waves mentioned before. In Fig. 44 the successive crests of the two systems have been drawn. We shall discuss these curves by giving the field point $P$ a

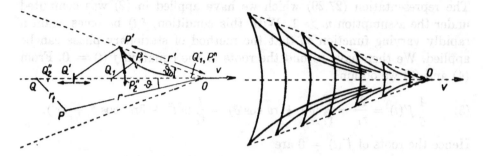

FIG. 43 (*left*). Illustrating the generation of ship waves. $O$ is the present position of the ship, $Q$, $Q'$ $\cdots$ are previous positions of the ship, $P$, $P'$ $\cdots$ are specified positions of the field point.

FIG. 44 (*right*). Lengthwise and transverse ship waves.

variety of positions, having in mind that the disturbance at $P$ at the time $t = 0$ is essentially caused by two annular waves that issued from the source in the positions $Q'$ and $Q''$ at the times $t_1$ and $t_2$ respectively.

Let us start with a field point position $P = P'$ on the limiting line $\vartheta = \vartheta_0$ . The two values $t_1$ and $t_2$ in (6) coincide: The common source position associated with this case is the point $Q'$ in Fig. 43. Its distance from $O$ is by (6)

$$OQ' \,=\, -vt_1' \,=\, -vt_2' \,=\, \frac{3}{2}\,r'\,\cos\vartheta_0\,,$$

where $r'$ is the distance of $P'$ from $O$. The direction of the curves of uniform phase $f(t_1) = f(t_2) = \text{const}$ in $P'$ is given by a circular arc element with center $Q'$ and radius $Q'P'$. Both curves of constant phase pass through $P'$ in the same forward direction. Considered as one curve they form a *cusp* at $P'$.

If now $\vartheta$ is decreased and $r$ so chosen that $f(t_2)$ remains constant the source position associated with $t_2$ moves to the *left* from $Q'$ (see Fig. 43) since $-vt_2$ increases according to (6). When $\vartheta$ reaches the value zero, $Q'$ reaches the position $Q_2''$, so that

$$OQ_2'' \,=\, -vt_2 \,=\, \frac{3}{2}\,r\Big(1 + \frac{1}{3}\Big) \,=\, 2r.$$

Here $r$ is the distance between $O$ and the field point $P_2''$ at which the transverse wave intersects the course of the boat. The direction of the phase curve is obviously perpendicular to the course; it coincides with the circular element of radius $Q_2''P_2'' = r$ about the center $Q_2''$. This gives a general idea of the shape of the phase curves $f(t_2) = $ const.

If, on the other hand, we choose the $r$-values for decreasing $\vartheta$ so that $f(t_1)$ remains constant, $-vt_1$ decreases [see (6)] and the point $Q'$ in Fig. 43 moves *right* to the new position $Q_1$. The slope of the lengthwise wave through $P'$ becomes flatter, as is shown by the circular arc about $Q_1$ with radius $Q_1 P_1$. The distance $OQ_1''$ becomes for very small $\vartheta$

$$-vt_1 = \frac{3}{2}r\left(1 - \frac{\vartheta^2}{2} - \sqrt{\frac{1}{9} - \frac{\vartheta^2}{2}}\right) = r\left(1 + \frac{3}{8}\vartheta^2\right)$$

so that $Q_1''$ nearly coincides with the field point $P_1''$ when both are close to $O$. The lengthwise waves should become tangential to the direction of travel at $O$ if our method were still valid in the neighborhood of $O$. This, however, is not the case: Our me hod of stationary phase breaks down for a short running time $t$. Nevertheless the general shape of the lengthwise waves has thus been clarified.

The phase curves $f(t_1) = $ const, $f(t_2) = $ const are essentially identical with the actual wave pattern observed. Along either curve the contribution of the integral whose phase is *not* constant shows up as a secondary ripple, at least at some distance from $O$. For a comprehensive analytical representation of both families of curves see Lamb loc. cit. and L. Hopf in his Munich thesis quoted on p. 120; there the evaluation is carried out by complex integration.

Looking back at our result it appears strange that there should exist a limiting angle $\vartheta_0$ *independent of the traveling speed v*. This seems to disagree with the well known photographs of projectiles moving with supersonic speed, first obtained by the Austrian philosopher and physicist Ernst Mach (1838-1916) with the so-called schlieren method. The projectile in Mach's beautiful theory is shrunk to a moving point, just as the steam boat before, from which compression waves originate continually. At the time of observation the wave that has originated $t$ seconds ago has now spread over a spherical surface of radius $r = ct$ where $c$ is the sound velocity. In the meantime the projectile has traveled a distance $x = vt$. The spherical shells so produced have an enveloping circular cone, the *Mach cone*. Half its apex angle is called the *Mach angle* and given by

$$\sin \vartheta_0 = \frac{r}{x} = \frac{c}{v}.$$

The Mach angle is approaching zero with increasing $v$, in contrast to the limiting angle $\vartheta_0 = 19°18'$ in the case of the ship waves.

The reason for this different behavior is found in the *dispersion*. The sound waves travel at fixed velocity $c$ without dispersion. The deep water

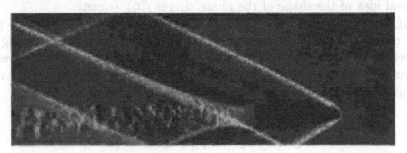

FIG. 45. A projectile moves with supersonic velocity between two parallel walls, the Mach cone is reflected at the walls.

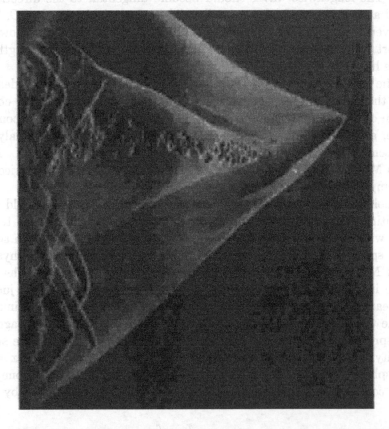

FIG. 45a. The projectile has pierced a thin wooden wall. The wood splinters also produce Mach cones of varying apex angles, according to their velocities.

waves follow the dispersion law $V = \sqrt{g/k}$. With the velocity thus depending on the wave length, there exist waves which at any given speed of the boat run along with the boat, while in the Mach phenomenon all waves are overtaken by the projectile. Thus the fact that $\vartheta_0$ is independent of $v$ becomes understandable.

We could substantiate this interpretation if we treated the *ship waves in shallow water*, in the sense of 24. According to (24.7) there is no dis-

FIG. 45b. The projectile passes a cylinder with circular slots. The spherical waves originating at the openings are tangential to the Mach cone of the projectile.

persion in this case, and Mach's purely geometrical argument may be applied. If $v > V$ the limiting angle should go to zero with increasing $v$. This problem, however, in which we would have to assume a flat disturbance (a flat-bottomed boat) seems to us rather artificial. However, the classical investigations of annular waves on which the theory of ship waves is based (Euler 1759, Laplace 1776, Lagrange 1787) assumed shallow water.

We close this chapter with some of the less well known pictures of the Mach phenomenon which were taken in the ballistic laboratory of

C. Cranz. Fig. 45 shows a projectile flying between two walls: the Mach cones suffer regular reflection at the walls, also, on this and the other photographs a tail of eddies is shed from the rear end of the projectile. Fig. 45a shows a shot through a thin wooden wall which is located at the left border of the picture: the projectile itself which has somewhat deviated from its path after piercing the wall and the numerous splinters flying along show their Mach cones. Note that the splinters possess Mach cones with larger apex angles than the projectile in accordance with their smaller velocities which, however, are still larger than the sound velocity $c$. Fig. 45b is very instructive: the projectile was shot through a cylinder furnished with circular slots and is just emerging from the other end. Spherical waves come from the slots whose envelops fit the Mach cone of the projectile accurately. Here we see a direct experimental realization of the mental picture we have of the formation of the Mach cone as the envelop of individual spherical waves. At the same time the photograph is a beautiful visualization of *Huygen's principle*, which likewise operates with envelopes of spherical waves.

## CHAPTER VI

# FLOW WITH GIVEN BOUNDARIES

## 29. Flow Past a Plate

In this and the following articles we restrict ourselves to two-dimensional flow and may, therefore, make use of the powerful tool of the theory of complex functions.

We take up again our discussion of elliptical coordinates $\xi$, $\eta$ in 19. The connection with Cartesian coordinates $x$, $y$ is given by Eq. (19.15)

$$(1) \qquad x + iy = c \cosh (\xi + i\eta).$$

Here $c$ is half the focal distance (cf. Fig. 27) and, at the same time, half the length of the projection of our plate in the $x,y$-plane; the plate is supposed to be infinite in $\pm z$-direction.

In elliptic coordinates the front and back sides of the plate are simultaneously given by

$$\xi = 0, \qquad -\pi < \eta < +\pi,$$

corresponding to the infinitely narrow ellipse of Fig. 27. For these values the right member of Eq. (1) is real, hence $y = 0$ and $-c < x < c$.

On the other hand, for

$$\eta = \pm \frac{\pi}{2}, \qquad 0 < \xi < \infty$$

the right member of (1) becomes a positive or negative purely imaginary number, hence $x = 0$ and $y \gtrless 0$, as can also be seen from Fig. 27.

Consider now the analytic function

$$(2) \qquad \Phi + i\Psi = \text{const} \sinh (\xi + i\eta)$$

already anticipated in (19.20). Let the constant in (2) be equal to $iqc$, where $q$ is a real quantity having the dimension of a velocity:

$$(2a) \qquad \text{const} = iqc; \qquad q, c > 0.$$

Now, $\Phi + i\Psi$ is not only an analytic function of $\xi + i\eta$, but through (1) an analytic function of $x + iy$ likewise. Consequently $\Phi$ and $\Psi$ may be interpreted as *velocity potential* and *streamfunction* (19) provided they satisfy such boundary conditions in the $x,y$-plane as are required by the problem under consideration. The values of $\Phi$ and $\Psi$ along certain lines of the $x,y$-plane have been listed in the following table, the auxiliary variables $\xi$ and $\eta$ playing the role of parameters:

| $\xi$, $\eta$-plane | $x$, $y$-plane | values of $\Phi$, $\Psi$ |
|---|---|---|
| $\xi = 0$ | $x = c \cos \eta, \quad y = 0$ | $\Psi = 0, \quad \Phi = -qc \sin \eta$ |
| $\eta = -\dfrac{\pi}{2}$ | $x = 0, \quad y = -c \sinh \xi < 0$ | $\Psi = 0, \quad \Phi = qc \cosh \xi > 0$ |
| $\eta = +\dfrac{\pi}{2}$ | $x = 0, \quad y = +c \sinh \xi > 0$ | $\Psi = 0, \quad \Phi = -qc \cosh \xi < 0$ |
| $\xi \to \infty$ | $x^2 + y^2 = \dfrac{c^2}{4} e^{2\xi} \to \infty$ | $\Psi = +qx, \quad \Phi = -qy$ |

The first three lines of this table are explained in our previous remarks, if one takes into account that the right member of (2) becomes real for $\xi = 0$ as well as for $\eta = \pm\pi/2$ because of (2a). As regards the last line, the effect of the limit $\xi \to \infty$ on $x$, $y$, etc. can be easily seen if one separates real and imaginary parts in (1) and (2). One obtains for large $\xi$

$$x = \frac{c}{2} e^{\xi} \cos \eta, \qquad y = \frac{c}{2} e^{\xi} \sin \eta,$$

(3)

$$\Phi = -\frac{qc}{2} e^{\xi} \sin \eta, \qquad \Psi = \frac{qc}{2} e^{\xi} \cos \eta,$$

in agreement with the last line of the table.

According to the $\Phi$ and $\Psi$ values in the last line, the flow field in the neighborhood of the point infinity of the $x,y$-plane is uniform, viz.

(3a)        $$v_y = -\frac{\partial \Phi}{\partial y} = q, \qquad v_x = -\frac{\partial \Phi}{\partial x} = 0.$$

The *first three lines* of the table contribute the following additional information about the flow: the streamline $\Psi = 0$ coincides with the negative $y$-axis (line 2) and with the positive $y$-axis (line 3). According to line 1, a singularity appears at $y = 0$.

In its total extension, $\Psi = 0$ may be described as follows: the branch $\eta = -\pi/2$ corresponding to the negative $y$-axis splits into two branches at $y = 0$ (center of the front side of the plate). They follow the surface of the plate and are obtained according to the first line, when $\eta$ passes from $-\pi/2$ to $\pi/2$ either by adding or subtracting $\pi(-3\pi/2$ being equivalent to $\pi/2)$. The two branches join at the center of the back side and, from

there on, $\Psi = 0$ coincides with the positive $y$-axis.[1] This streamline together with the known flow pattern at infinity forms a frame for the
family of streamlines $\Psi =$ const, that makes the diagram of Fig. 46
acceptable.

The point or, rather, the points $O$ on the front and back side of the
plate are of particular interest. They are called *points of bifurcation* with

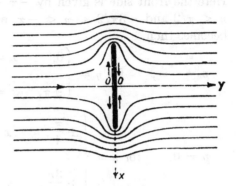

FIG. 46. Plate at right angle to flow.
The theoretical flow pattern downstream
is a mirror image of the upstream pattern.

reference to the division of the streamline $\Psi = 0$, or *branch points* with
reference to the mapping function. Physically speaking, they are *stagnation points* since the velocity of flow vanishes there ($v_x = v_y = 0$).
According to Bernoulli's theorem they are points of *maximum pressure*:
From (11.9) we have for steady flow and in the absence of external forces
($U =$ const)

$$(4) \qquad p = \text{const} - \frac{\rho}{2} v^2.$$

If we put the pressure at infinity equal to zero, so that $p$ is the "hydrodynamic" pressure due to the acceleration of the fluid in the presence of
the plate, then *const* $= \rho/2 \, q^2$ [$q$ is the flow velocity at infinity according
to (3a)] and

$$(4a) \qquad p = \frac{\rho}{2} (q^2 - v^2).$$

This shows immediately that $p_{\text{max}}$ is larger than any other possible $p$-value.

The pressure distribution along the plate is found from our formulas
as follows: From

$$\Phi = -qc \sin \eta, \qquad x = c \cos \eta,$$

---

[1] If the $\xi$, $\eta$-strip in Fig. 27a which is in one-to-one correspondence with the $x$, $y$-plane
is shifted downward by $\pi/2$, the streamline $\Psi = 0$ is represented by an $E$-shaped figure,
the upper and lower horizontal of which correspond to the same ray $y > 0$, $x = 0$.

we find

$$v = -\frac{\partial \Phi}{\partial \eta}\frac{d\eta}{dx} = q\,\frac{\cos \eta}{\sin \eta},$$

and by (4a)

$$p = \frac{\rho}{2}\,q^2\!\left(1 - \frac{\cos^2 \eta}{\sin^2 \eta}\right).$$

Here the front side is given by $-\pi < \eta < 0$ and the back side by $0 < \eta < \pi/2$ and $-3\pi/2 < \eta < -\pi$, or simply, $0 < \eta < \pi$. The pressure becomes then

$$p = -\infty, \qquad \text{for } \eta = \left\{\begin{array}{ll} 0, & x = +c \\[2mm] \pm\pi, & x = -c \end{array}\right\} \text{ at the edges of the plate}$$

$$p = 0, \qquad \text{for } \eta = \left\{\begin{array}{ll} \pm\dfrac{\pi}{4}, & x = +\dfrac{c}{\sqrt{2}} \\[4mm] \pm\dfrac{3\pi}{4}, & x = -\dfrac{c}{\sqrt{2}} \end{array}\right\} \begin{array}{l}\text{at a distance } 0.29c \text{ from} \\ \text{either edge toward the} \\ \text{center}\end{array}$$

and, as we have found before $p = p_{\max} = \rho q^2/2$ for $\eta = \pm\pi/2$, $x = 0$, center of plate.

*This pressure distribution is impossible in a perfect fluid because of the occurrence of a negative pressure* (which at the edges becomes even infinite); *the fluid would separate under a negative pressure.*[2]

A second consequence which is no less paradoxical, is this: The resultant pressure force exerted by the stream upon the plate is zero, (the resultant moment vanishes, too) since the pressure at any surface element on the front side is canceled by the corresponding pressure on the back side. In other words, *if the fluid is transformed to rest at infinity* (by superposition of the constant velocity field $\Phi = qy$), *so that the plate moves through the fluid with uniform velocity $q$* no work would have to be expended in such a process. Once the plate together with the surrounding fluid has been set in motion, it will keep on moving with constant velocity.

This paradoxical assertion in a generalized form, in which it applies to rigid bodies of arbitrary shape, is attributed to d'Alembert or Euler.[3]

---

[2]Here the emphasis lies on the fact that the pressure becomes *arbitrarily* strongly negative; if it were only negative, but would stay bounded, we could make good for that by choosing the pressure at $\infty$ sufficiently positive.

[3]In the introduction to his book, Hydrodynamik, Leipzig, Akademische Verlagsgesellschaft 1927, C. W. Oseen quotes Spinoza as an even earlier authority. One could say, in a sense, that Oseen's book has as its aim the analysis of the paradox. For instructive detail observations in two dimension see G. Hamel, Z. angew. Math. u. Mech. 15 (1935) 52

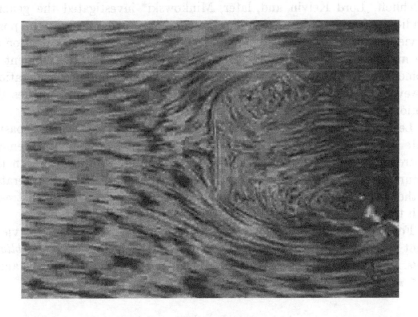

FIG. 47a. The vertical plate in real flow. Stagnation point at the upstream center of the plate.

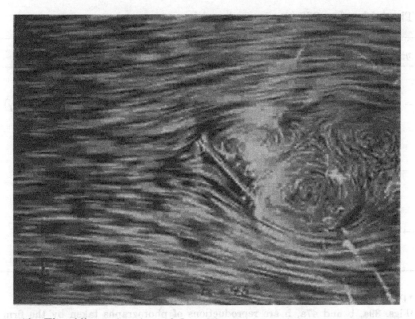

FIG. 47b. The oblique plate in real flow. The stagnation point is shifted toward the upstream edge of the plate.

Kirchhoff, Lord Kelvin and, later, Minkowski* investigated the general conditions for a steady motion like the one in our example from the point of view of abstract hydrodynamics. It was found that the direction of the momentum would have to coincide with the axis of the moment of momentum to make the motion possible. In the case under consideration, however, the occurrence of arbitrary large negative pressures makes the motion physically impossible.

Let us now compare Fig. 46 with the photograph of a real flow past a plate in Fig. 47a.[4] The two Figs. 46 and 47a are in very good agreement upstream of the plate but have nothing in common downstream. In the picture of the real flow eddies occur behind the plate that are generated in the process of passing the edges of the plate and move downstream with the fluid. We shall investigate this phenomenon in 30 and 32.

For the present, however, we wish to maintain our theoretical viewpoint incomplete though it is, and investigate the flow past an *oblique plate*. We rotate the $x,y$-system with respect to the $\xi,\eta$-system by the angle $\gamma < \pi/2$ and change Eqs. (1) and (2) into

(5)
$$x + iy = ce^{i\gamma} \cosh (\xi + i\eta),$$

(6)
$$\Phi + i\Psi = iqc \sinh (\xi + i\eta + i\gamma).$$

The table of p. 208 is now replaced by:

| $\xi, \eta$-plane | $x, y$-plane | values of $\Phi, \Psi$ |
|---|---|---|
| $\xi = 0$ | $\begin{cases} x = c \cos \eta \cos \gamma \\ y = c \cos \eta \sin \gamma \end{cases}$ | $\Psi = 0, \quad \Phi = -qc \sin (\eta + \gamma)$ |
| $\eta = \mp \dfrac{\pi}{2}$ | $\begin{cases} x = \pm c \sinh \xi \sin \gamma \\ y = \mp c \sinh \xi \cos \gamma \end{cases}$ | $\begin{cases} \Psi = \pm qc \sinh \xi \sin \gamma \\ \Phi = \pm qc \cosh \xi \cos \gamma \end{cases}$ |
| $\xi \to \infty$ | $x^2 + y^2 = \dfrac{c^2}{4} e^{2\xi} \to \infty$ | $\Psi = qx, \quad \Phi = -qy$ |

*Preuss. Akad. Ber., 1888, 1095.

[4] Figs. 39a, b and 47a, b are reproductions of photographs taken by the firm of Ahlborn in Hamburg.

According to the table, the front and back sides of the plate represented by $\xi = 0$, $-\pi < \eta < 0$, and $0 < \eta < \pi$ are still the streamline $\Psi = 0$. In the neighborhood of the point infinity of the $x,y$-plane, that is for $\xi \to \infty$, we have a uniform field of flow with constant velocity $q$ as before; the streamlines in this region are the lines $x = $ const, but the lines $\eta = \mp\pi/2$ are no longer streamlines. Geometrically, they represent a pair of rays that coincide with the two halves of the bisectrix of the plate; the center column of our table gives for these lines the equation $y = -x \cot \gamma$. Altogether, there are no rectilinear streamlines in the finite domain.

In order to analyze the flow pattern we first determine the *stagnation points*. They are the branch points of the function that maps the $\Phi + i\Psi$-plane on the $x + iy$-plane. Now, the general condition for a branch point is:

$$(7) \qquad \frac{d(\Phi + i\Psi)}{d(x + iy)} = 0.$$

The differential quotient is calculated from (5) and (6) as follows:

$$(7a) \qquad \frac{d(\Phi + i\Psi)}{d(\xi + i\eta)} \bigg/ \frac{d(x + iy)}{d(\xi + i\eta)} = \frac{iq \cosh(\xi + i\eta + i\gamma)}{e^{i\gamma} \sinh(\xi + i\eta)}.$$

Condition (7) leads therefore to

$$(7b) \qquad \cosh(\xi + i\eta + i\gamma) = 0,$$

or, on separating into real and imaginary parts, to

$$(7c) \qquad \cosh \xi \cos(\eta + \gamma) = 0, \qquad \sinh \xi \sin(\eta + \gamma) = 0.$$

Since $\cosh \xi$ is never zero, and $\sinh \xi$ vanishes only for $\xi = 0$, the solution of (7c) is

$$(8) \qquad \eta = \pm\frac{\pi}{2} - \gamma, \qquad \xi = 0.$$

Thus there are again two stagnation points one of which is in the front, the other in the back of the plate:

$$(8a) \qquad \xi = 0, \quad \eta = -\frac{\pi}{2} - \gamma \quad \text{and} \quad \xi = 0, \quad \eta = +\frac{\pi}{2} - \gamma.$$

They are not in the center of the plate, but shifted up and down by the same amount $|x + iy| = c \sin \gamma$, as follows from (5). In Fig. 48 they have been denoted by $O_1$ and $O_2$. At these points we have again $v_x = v_y = 0$ and the pressure is, according to Bernoulli's equation (4a), a maximum: $p_{stag} = p_{max} = \rho q^2 / 2$. From our derivation it follows that the stagnation points are points of bifurcation as before. The streamline that meets the plate at $O_1$ and divides there into two branches is normal to the plate, and so is the stream line issuing from $O_2$.

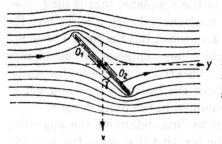

FIG. 48. Flow past an oblique plate. The theoretical flow pattern downstream is obtained by rotating the upstream pattern about the center of the plate by 180°.

The differential quotient computed in (7a) yields quite generally the velocity of flow in the complex form

(9)
$$\frac{d(\Phi + i\Psi)}{d(x + iy)} = -v_x + iv_y .$$

This is a direct consequence of the form of $f'(z)$ in (19.4a) together with the Cauchy-Riemann equations (19.5). If we take the absolute value in (9) and substitute (7a), we obtain

(10)          $$|\, \mathbf{v} \,| = q \left| \frac{\cosh (\xi + i\eta + i\gamma)}{\sinh (\xi + i\eta)} \right| .$$

In particular for $\xi = 0$, that is, on the plate itself, we have on either side

(10a)          $$|\, \mathbf{v} \,| = q \left| \frac{\cos (\eta + \gamma)}{\sin \eta} \right| .$$

The pressure distribution is found from (10a) by substitution in Bernoulli's equation (4a):

(11)          $$p = \frac{\rho}{2} q^2 \left( 1 - \frac{\cos^2 (\eta + \gamma)}{\sin^2 \eta} \right) .$$

This leads again to

$$p = -\infty \quad \text{for} \quad \eta = \begin{cases} 0 \\ \pm\pi \end{cases} ,$$

that is at the edges of the plate, and the point where the pressure vanishes is found from $p = 0$ as

(12)          $$\cos^2 (\eta + \gamma) = \sin^2 \eta$$

which can be written $\cos(\eta + \gamma) = \pm \cos(\pi/2 - \eta)$, and therefore

either $\qquad\qquad \eta + \gamma = \dfrac{\pi}{2} - \eta, \qquad \eta = \eta_1 = \dfrac{\pi}{4} - \dfrac{\gamma}{2},$

or $\qquad\qquad \eta + \gamma = -\dfrac{\pi}{2} - \eta, \qquad \eta = \eta_2 = -\dfrac{\pi}{4} - \dfrac{\gamma}{2}.$

The value $\eta_1$ is positive, the corresponding point lies therefore on the backside; the converse is true for $\eta_2$. The zone of negative pressure is on the back side between $\eta = 0$ and $\eta = \eta_1$, on the front side between $\eta = -\pi$ and $\eta = \eta_2$. Since arbitrarily large negative pressures are not permissible, our solution in the case of the oblique plate has as little physical reality as in the special case of a normal plate.

The comparison of Fig. 48 with the photograph of a real flow in 47b illustrates what we mean. Ahead of the plate the two stream patterns are rather similar, in particular, there is a stagnation point in front which has the expected position. The downstream flow, however, is entirely different. Instead of an antisymmetric repetition of the upstream pattern we meet again the characteristic eddies in alternating positions.

In this article we have only dealt with the *steady* flow past a plate. A plate problem occurs also in acoustics, when a light circular disk is suspended in the *non-steady* field of a sound beam. This so-called Rayleigh disk plays a significant part in the measurement of sound intensities; the analytic methods discussed so far and developed in the following are of importance for the theory of this instrument, as Lord Rayleigh[5] has shown; for this reason our statement about the unrealistic character of the obtained solutions should be restricted to the case of steady flow.

## 30. The Problem of the Wake; Surfaces of Discontinuity

"There is nothing in the nature of a liquid that should prevent two adjacent fluid layers from sliding past each other with finite relative velocity, provided we consider the fluidity as perfect, that is, exclude all friction. Our assertion is certainly true for those properties of the liquid which are considered in the hydrodynamic equations, viz. the conservation of mass in each volume element and the equality of the pressure in all directions; these two factors do not preclude the possibility of a finite difference of the tangential velocities on either side of a surface in the interior of the fluid. The components of the velocity normal to the surface and the pressure, however, must be equal on both sides of such a surface".[6]

---

[5]Lamb, Hydrodynamics, Cambridge, 5th ed., Chap. IV, Art. 77.
[6]Helmholtz, Preuss. Akad. Ber., 1868.

Helmholtz further points out that a sharp edged obstacle always produces discontinuity surfaces in the flow and that in this way the negative pressures that would otherwise be present are avoided. He also observes that the discontinuity surface which originally is conceived as a geometrical surface curls up under the influence of friction and takes the form of a sequence of eddies (Helmholtz speaks of vortex filaments). All this will be discussed in 32; for the present we maintain the assumption of a perfect inviscid fluid.

With this in mind we again take up the two-dimensional problem of the plate, assuming the plate normal to the flow for the sake of simplicity, as at the beginning of 29. Two lines of discontinuity—if we now use two-dimensional terminology—issue from the end points of the segment that represents our plate. They are symmetric to the $y$-axis, but of unknown shape (cf. Fig. 49, where the edges of the plate are denoted by $A$ and $B$).

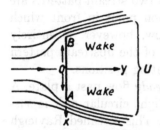

FIG. 49. Helmholtz flow past a vertical plate $AB$. The wake is delimited by the discontinuity lines $AU$ and $BU$ that originate at the edges of the plate.

The wake is enclosed between the two discontinuity lines and the plate, and we imagine the state of the wake to be dead water in the literal sense: its velocity is assumed as zero, and the pressure is everywhere constant, say, equal to $p_0$. According to Helmholtz, we then have also for the fluid that *flows along* the discontinuity lines the following boundary conditions for the pressure and the normal component of the flow:

(1)
$$p = p_0, \qquad v_n = 0.$$

The tangential component $v_s$ follows now from Bernoulli's equation (29.4a) which yields because of (1)

(2)
$$|\mathbf{v}|^2 = v_s^2 = q^2 - \frac{2}{\rho} p_0.$$

The tangential component is thus constant along the lines of discontinuity. Infinite velocities and negative pressures no longer occur.

Kirchhoff[7] worked out Helmholtz's idea in detail and applied it among other things to the plate problem. He considers in addition to the two complex variables

---

[7] G. Kirchhoff, Crelles Journ. 70 (1869) and Vorlesungen über Mechanik, Chap. XXII.

$$z = x + iy \quad \text{and} \quad f = \Phi + i\Psi$$

a third variable

(3)
$$\zeta = \frac{1}{-v_x + iv_y}$$

connected with $z$ and $f$ by the relation

(4)
$$\zeta = \frac{dz}{df}$$

as in (29.9), and introduces the assumption that *the wake reaches to infinity* (we have tried to indicate that in Fig. 49). The pressure $p_0$ then equals the pressure at infinity which we defined as zero in (29.4a), so that now $p_0$ in Eq. (2) must also be put equal to zero. From (2) we obtain

(5)
$$| -v_x + iv_y | = v_s = q$$

all along the lines of discontinuity. Consequently we have for $\zeta$ the boundary condition

(5a)
$$|\zeta| = \frac{1}{q}.$$

We wish to emphasize, however, that the assumption of discontinuity lines reaching to infinity is too restricted. Indications for this will be found at the end of this article.

Let us now look for the image of the discontinuity line, which separates the moving fluid from the wake in the plane of the complex variable $\zeta$ (cf. Fig. 50).

FIG. 50. Mapping of the $z$-plane on the $\zeta$-plane.

Starting with the point infinity of the $z$-plane, we have there

$$v_y = q, \quad v_x = 0, \quad \text{and by (3)} \quad \zeta = -\frac{i}{q}.$$

The image point $U$ in the $\zeta$-plane is therefore on the negative imaginary axis. The images of the two discontinuity lines must pass through the point $U$, in agreement with (5a). We draw now a semi-circle of radius $1/q$ and center zero; it passes through $U$ and terminates at the real axis in the points $\zeta = \pm 1/q$, which are the images of the points $A$ and $B$ in Fig. 49 (edges of the plate). All along the plate one has $v_y = 0$ and therefore $\zeta$ real; this holds also at the edges.

We finally consider the stagnation point $O$ in Fig. 49. There

$$v_x = v_y = 0, \qquad \text{and by (3)} \qquad \zeta = \infty.$$

The image of $O$ is the point $\infty$ of the $\zeta$-plane. This determines completely the mapping of the two halves of the plate: Since $\zeta$ is real along the plate, as was just pointed out, one half of the plate maps on the segment $OA$, the other on the segment $OB$ of the real axis in Fig. 50.

In describing the circuit $UAOBU$ in Fig. 49, the region of the "live" flow[8] remains to the left, and the region of the dead water to the right. Accordingly, the marked region in Fig. 50 to the left of the image circuit corresponds to the region of flow, the unmarked region to the wake. Yet that does not determine the analytic solution of our mapping problem since the shape of the lines of discontinuity is still unknown.

This part of the problem calls for a discussion of the mapping of the $z$-plane on the $f$-plane, the characteristic points of which are listed in the following table. As before we assign the equation $\Psi = 0$ to the "symmetric" stream line which separates at the stagnation point, whereupon the entire boundary of the dead water becomes real in the $f$-plane. The zero level of the velocity potential $\Phi$ is now assigned to the stagnation point $O$; this makes $\Phi$ negative along $OA$ and along $OB$. We denote the unknown value of $\Phi$ at $A$ and $B$ by $f_0$. (Note that $f_0$ is at the same time the value of $f$ at $A$ and $B$ since $\Psi = 0$; $\Phi$ must decrease from $O$ to $A$ or from $O$ to $B$, hence is negative along $OA$ and $OB$.) $f_0$ is negative and given by

$$(6) \qquad f_0 = -\int_0^{+c} v_x \, dx = -\int_0^{-c} v_x \, dx.$$

We have thus found a correspondence between the total boundary of the wake and the negative real axis of the $f$-plane, but it is not a one-to-one

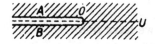

Fig. 51. Mapping of the $z$-plane on the $f$-plane.

correspondence, since, for instance, the plate front $AOB$ is mapped on the double segment $f_0 - 0 - f_0$, cf. Fig. 51. This can be helped by making a cut along the negative real axis. The two borders of the cut form the image of the wake boundary, while the positive real axis has already been

---

[8]If we had used the reciprocal variable $\zeta' = 1/\zeta$, instead of $\zeta$, the region of flow would map on the *interior* of the semi-circle and the circuit $UAOBU$ on the closed boundary of the semi-circle. We have retained Kirchhoff's original definition of $\zeta$ in (3) to make a comparison with the original paper easier, although the map of the flow in the plane $\zeta'$, being simply the *hodograph* of the flow, has a stronger physical appeal.

in one-to-one correspondence with the stream line $\Psi = 0$. The marking of Fig. 51 indicates that the cut $f$-plane is the image of the flow field in Fig. 49. The wake is mapped on a second sheet of the $f$-plane connected with the sheet of Fig. 51 across the branch cut (the two sheets form one Riemann surface).

| | $z$ | $\zeta$ | $f$ | angular change |
|---|---|---|---|---|
| $A$ | $+c$ | $-\dfrac{1}{q}$ | $f_0 < 0$ | $\dfrac{1}{2}$ |
| $B$ | $-c$ | $+\dfrac{1}{q}$ | $f_0 < 0$ | $\dfrac{1}{2}$ |
| $O$ | $0$ | $\pm\infty$ | $0$ | $\dfrac{1}{2}$ |
| $U$ | $\infty$ | $-\dfrac{i}{q}$ | $\infty$ | |

Let us now try to find the analytical form of the mapping of the $\zeta$-plane on the $f$-plane, omitting the $z$-plane and comparing the Figs. 50 and 51. (It will be noticed that this is feasible without knowing the discontinuity lines, since the plane of the physical flow matters no longer.) If the mapping were everywhere conformal, it would be given by a bilinear relation between $\zeta$ and $f$. There are, however, points where the conformality breaks down, though not in the interior of the marked regions of the two figures, but on the boundary; they have been indicated in the last column of our table. The number $\frac{1}{2}$ going with $A$ and $B$ means that the angle $\pi$ occurring at the boundary points $A$ and $B$ in the $f$-plane appears as the angle $\pi/2$ in the $\zeta$-plane; the number $\frac{1}{2}$ going with $O$ should likewise tell us that the angle $2\pi$ at $O$ in the $f$-plane appears in the $\zeta$-plane as $\pi$. From this we infer that the relation between $\zeta$ and $f$ must contain

$$f \text{ in the form } \sqrt{f}, \qquad \zeta \text{ in the form } \left(\zeta \pm \frac{1}{q}\right)^2.$$

We therefore consider the following bilinear relation of these quantities

$$\frac{\left(\zeta + \dfrac{1}{q}\right)^2}{\left(\zeta - \dfrac{1}{q}\right)^2} = \frac{\sqrt{f_0} - \sqrt{f}}{\sqrt{f_0} + \sqrt{f}}$$

or, written more conveniently,

(7)
$$\frac{(q\zeta + 1)^2}{(q\zeta - 1)^2} = \frac{1 - \sqrt{\dfrac{f}{f_0}}}{1 + \sqrt{\dfrac{f}{f_0}}}.$$

This function fulfills the requirements of the table as can be shown by expanding (7) in power series at the points $A$, $B$, $O$. We shall come back to this in the next article in a more systematic way.

Let us now compute $\zeta$ explicitly. Eq. (7) can be rearranged in the form

(7a)
$$(q\zeta)^2 + 2\sqrt{\frac{f_0}{f}}\, q\zeta + 1 = 0;$$

the roots of this equation are

(8)
$$\zeta = -\frac{1}{q}\left(\sqrt{\frac{f_0}{f}} \pm \sqrt{\frac{f_0}{f} - 1}\right).$$

The salient point in this analysis is now that it is possible to eliminate the auxiliary variable $\zeta$ by (4) and to obtain a *differential equation between the two original variables z and f*, viz.

(9)
$$\frac{dz}{df} = -\frac{1}{q}\left(\sqrt{\frac{f_0}{f}} \pm \sqrt{\frac{f_0}{f} - 1}\right).$$

Its integration determines the mapping function.

It is convenient for this purpose to introduce the parameter $\alpha$ by

(10)
$$\sqrt{\frac{f}{f_0}} = \sin \alpha, \qquad df = 2f_0 \sin \alpha \cos \alpha \, d\alpha.$$

$$\sqrt{\frac{f_0}{f}} = \frac{1}{\sin \alpha}, \qquad \sqrt{\frac{f_0}{f} - 1} = \frac{\cos \alpha}{\sin \alpha}.$$

Choosing the upper sign in (9), we obtain

$$dz = -\frac{2f_0}{q}(\cos \alpha + \cos^2 \alpha)\, d\alpha$$

which gives integrated

(11)
$$z = -\frac{2f_0}{q}\left\{\sin \alpha + \frac{1}{2}(\sin \alpha \cos \alpha + \alpha)\right\}.$$

In this integration the constant must be set equal to zero, since $f = 0$ for $\alpha = 0$ according to (10), and $z$ is supposed to be zero if $f = 0$ according to our table.

For $\alpha = \pi/2$, $f$ equals $f_0$ by (10); according to our table this characterizes the edge of the plate $A$, or $z = c$. Thus we have

$$(12) \qquad c = -\frac{2f_0}{q}\left(1 + \frac{\pi}{4}\right), \qquad \frac{f_0}{q} = \frac{-2c}{\pi + 4}$$

which determines the hitherto unknown quantity $f_0$. We substitute this value in (11) and obtain by separation of real and imaginary parts the following parametric representation of the plate:

$$(13) \qquad x = \frac{4c}{\pi + 4}\left\{\sin\alpha + \frac{1}{2}(\sin\alpha\cos\alpha + \alpha)\right\},$$

$$y = 0,$$

traced by the point $x$, $y$ when $\alpha$ varies according to

$$(13a) \qquad -\frac{\pi}{2} < \alpha < +\frac{\pi}{2}.$$

Now we allow $\alpha$ to take complex values, starting from $\pi/2$:

$$(13b) \qquad \alpha = \frac{\pi}{2} - i\beta, \qquad 0 < \beta < \infty.$$

Then one has

$$(13c) \qquad \sin\alpha = \cosh\beta, \quad \cos\alpha = +i\sinh\beta.$$

If $\alpha$ changes according to (13b), $f$ remains real by (10) and (12), hence (13b) represents the continuation of the stream line $\Psi = 0$ beyond the end point $A$, that is to say, the lower branch of the discontinuity line in Fig. 49. Its parametric representation is obtained from (11) by separation of the real and imaginary part in the following form:

$$(14) \qquad x = \frac{4c}{\pi + 4}\left(\cosh\beta + \frac{\pi}{4}\right),$$

$$y = \frac{2c}{\pi + 4}(\cosh\beta\sinh\beta - \beta).$$

This is a transcendental curve; its numerical determination makes no difficulty. For large $\beta$-values it can be approximated by a parabola.

Eqs. (10), (11) and (12) contain the general solution of our flow problem. When $\alpha$ takes arbitrary complex values we obtain

$$(15) \qquad f = \Phi + i\Psi = -\frac{2qc}{\pi + 4}\sin^2\alpha,$$

$$(16) \qquad z = x + iy = \frac{4c}{\pi + 4}\left\{\sin\alpha + \frac{1}{2}(\sin\alpha\cos\alpha + \alpha)\right\}.$$

An explicit representation of $\Phi$ and $\Psi$ in terms of $x$ and $y$ such as would be obtained by elimination of the parameter $\alpha$ is neither feasible nor necessary.

To check our results we may compare the value of $f_0$ in (12) with its original definition in (6). According to (6) we ought to have

$$(17) \qquad f_0 = - \int_0^c v_r \, dx = \int_0^c \frac{\partial \Phi}{\partial x} \, dx = \Phi_c - \Phi_0 .$$

By (15) we have the $\Phi$-values

$$\Phi_c = - \frac{2cq}{\pi + 4} \qquad \text{for} \qquad \alpha = \frac{\pi}{2} ,$$

$$\Phi_0 = 0 \qquad \text{for} \qquad \alpha = 0.$$

Thus (17) and (12) are in agreement.

Contrary to 29, the results of which led us to the d'Alembert-Euler paradox, our present theory gives a reasonable value for the *total pressure* $P$ acting on the plate per unit of length under the influence of the streaming fluid.

From Bernoulli's theorem (29.4a) we have

$$(18) \qquad P = \int_{-c}^{+c} p \, dx = \frac{\rho}{2} q^2 \int_{-c}^{+c} \left( 1 - \frac{v^2}{q^2} \right) dx.$$

Along the front of the plate $v$ equals $v_x$, hence $v^2 = 1/\varsigma^2$ by (3), and (18) takes the form

$$(19) \qquad P = \frac{\rho}{2} q^2 \int_{-c}^{+c} \left( 1 - \frac{1}{q^2 \varsigma^2} \right) dx.$$

This integral can be evaluated in the following elegant manner: along the plate, $dx$ equals $dz$; thus we obtain from (4)

$$(19a) \qquad dx = \varsigma \, df, \qquad P = \frac{\rho}{2} q \int \left( q\varsigma - \frac{1}{q\varsigma} \right) df.$$

From (8), where we choose again the upper sign, we have

$$- q\varsigma = \sqrt{\frac{f_0}{f}} + \sqrt{\frac{f_0}{f} - 1} .$$

The reciprocal value of this expression may be written

$$- \frac{1}{q\varsigma} = \sqrt{\frac{f_0}{f}} - \sqrt{\frac{f_0}{f} - 1} ,$$

so that

$$q\zeta - \frac{1}{q\zeta} = -2\sqrt{\frac{f_0}{f} - 1},$$

a result that could have been read from the quadratic equation (7a). Upon introducing the last expression in (19a), we obtain

(20) $$P = -2\rho q \int_0^{f_0} \sqrt{\frac{f_0}{f} - 1}\, df.$$

As regards the limits of the integral and the factor 2, note that the pressure distribution is symmetric with respect to the center, and therefore the contributions of the upper and the lower half equal.

If we now introduce the parameter $\alpha$ from (10) we obtain

(21) $$P = -4\rho f_0 q \int_0^{\pi/2} \cos^2 \alpha\, d\alpha = -\rho f_0 q \pi$$

and, on substitution of $f_0$ from (12),

(22) $$\frac{P}{2c} = \frac{\pi}{\pi + 4}\, \rho q^2 = 0.88\, \frac{\rho}{2}\, q^2.$$

*The pressure per unit of surface is thus proportional to the square of the flow velocity at infinity* in agreement with the general hydraulic experience (cf., e.g., p. 80); we could formulate our result also in this way: *the pressure is proportional to the kinetic energy per unit of volume of the undisturbed flow,* or, in other words, *proportional to the stagnation pressure at 0.* The numerical coefficient, however, does not agree too well with the results of careful measurements. In the case of the infinitely long plate the actual coefficient is about 2.0 according to Prandtl (Strömungslehre, p. 165)[9] instead of our value 0.88. The reason is—also according to Prandtl[10]—that the real wake does not reach to infinity and the two separate regions of the flow join at some distance behind the plate. Besides, the pressure downstream of the plate is smaller than the pressure of the undisturbed flow because of the eddies present in the wake. This leads to a "suction" on the back side that adds to the pressure on the front side.

The last remarks throw a good deal of doubt on the basic assumption of Kirchhoff's wake theory which has been widely accepted in literature. An essential feature of the theory is the assumption $p_0 = 0$ introduced in connection with Eq. (2). It is identical with the hypothesis of an infinitely extended wake as explained on p. 217. Attempts at a general treatment

---

[9]Cf. also Hunsaker and Rightmire, *op. cit.*, p. 198.

[10]See C. Schmieden, Z. für Luftfahrtforschung 17, (1940) 37 and M. Kolscher, *ibid.* p. 154.

of the problem which drop this assumption (see, e.g., footnote p. 223) are too complicated to be presented in this book.

Let us finally have a look at the interesting question of stability; suppose the plate is not kept rigidly fixed, but is allowed to revolve about an axis through its center, what will be its position relative to the flow? If the preferred orientation were parallel to the flow, the total pressure $P$ would vanish. The experiment shows, however, that the plate orients itself normal to the flow.

A complete investigation of this question would require knowledge of the flow pattern and of the discontinuity surfaces also in the case of the oblique plate. We can do without that if we trust the stream pattern in Fig. 48 as far as the qualitative position of the stagnation point on the front side of the plate is concerned which, in Fig. 47b, is confirmed by the experiment. We take it then that the stagnation point travels toward the leading edge of the plate whenever the plate turns from the normal to an oblique position. The pressure maximum then travels in the same direction. When the plate, as in Fig. 48, is turned counterclockwise out of the normal position, the pressure produces a clockwise moment which tends to restore the normal position. *This is therefore the stable position of the plate.* The same argument applied to the *parallel position* shows that *it is unstable.*

A weather vane likewise, would not turn into the wind but normal to it, if its axis were centrally located (actually the rear face is much larger than the front face). In fact, the Rayleigh disk (cf. p. 215) which is centrally suspended places itself normal to the sound beam.

### 31. The Problem of the Free Jet Solved by Conformal Mapping

It might appear that the mapping function given in (30.7) is the result of a lucky guess rather than of rigorous deduction. These doubts will be dispelled by the systematic derivation of (30.7) that follows here.

Let it be required to map the positive quadrant[11] of a $t$-plane on the upper half of an $s$-plane; the simplest way to do this is by the relation

$$(1) \qquad\qquad t = s^{\frac{1}{2}}.$$

The mapping is conformal everywhere in the finite domain with the exception of the origin of the two planes, where the straight angle $\pi$ formed by the real axis at $s = 0$ is mapped into the angle $\pi/2$ between the positive

---

[11]Our quadrant can be considered as a closed polygon, bounded by two straight lines and possessing two right angles at $t = 0$ and $t = \infty$ as one sees by stereographic projection. More generally, any polygon with arbitrary angles can be mapped on the half plane. There exists for this purpose a general integral formula of H. A. Schwarz and E. B. Christoffel, but we do not have to use it here.

real and the positive imaginary axes at $t = 0$. We consider now the $\zeta$-plane connected with the $t$-plane by the bilinear transformation

$$(2) \qquad t = \frac{\zeta + a}{\zeta - a}, \qquad \zeta = a\frac{t + 1}{t - 1},$$

where $a$ is real and positive. By (2) the positive-imaginary $t$-axis is mapped on the semicircular arc $|\zeta| = a$ which passes from $\zeta = -a$ through $\zeta = -ia$ to $\zeta = +a$ (bilinear transformations are known to map straight lines into circles or straight lines), while the positive real axis of the $t$-plane is mapped into the two parts of the real $\zeta$-axis $a < \zeta < \infty$ and $-\infty < \zeta < -a$.

We now substitute (2) in (1) and obtain the mapping function

$$(3) \qquad \frac{(\zeta + a)^2}{(\zeta - a)^2} = s,$$

which maps the upper half plane of the variable $s$ on the region of the $\zeta$-plane marked in Fig. 50.

Let us turn to the relation between $s$ and $f$: here we wish to map the upper half plane of $s$ on the $f$-plane cut along the negative real axis as indicated in Fig. 51. The angle $\pi$ at $s = 0$ must be doubled in the mapping, so that the angle at $f = 0$ becomes $2\pi$. The simple relation $s = \sqrt{f}$ of Eq. (1) would do this. We wish, however, to achieve at the same time the coordination of $f$-values and $\zeta$-values prescribed by our table on p. 219, and therefore subject $\sqrt{f}$ first to a bilinear transformation which we write

$$(3a) \qquad s = \frac{a\,\sqrt{f} + \beta}{\gamma\,\sqrt{f} + \delta}.$$

Combination with (3) and replacement of $a$ by $1/q$ gives

$$(4) \qquad \frac{(q\zeta + 1)^2}{(q\zeta - 1)^2} = \frac{\alpha\,\sqrt{f} + \beta}{\gamma\,\sqrt{f} + \delta}.$$

The coefficients $\alpha$, $\beta$, $\gamma$, $\delta$, determined[12] according to that table are

| | | |
|---|---|---|
| $q\zeta = -1,$ | $f = f_0$ | $\alpha\,\sqrt{f_0} + \beta = 0,$ |
| $q\zeta = +1,$ | $f = f_0$ | $-\gamma\,\sqrt{f_0} + \delta = 0,$ |
| $q\zeta = \infty,$ | $f = 0$ | $\beta = \delta,$ |
| $q\zeta = -i,$ | $f = \infty$ | $\alpha = -\gamma.$ |

---

[12]The negative sign on $\gamma$ in the second line of the table is obtained from Fig. 51 in the following way: $\sqrt{f}$ changes its sign in a transition from the upper to the lower border of the cut, if this transition is carried out without leaving the hatched sheet in Fig. 51, that is without crossing the cut. The value $\sqrt{f_0}$ in the first line that corresponds to the point $A$ is thus to be replaced by $-\sqrt{f_0}$ in the second line that corresponds to the point $B$.

Putting $\alpha = 1$, we obtain for the other constants

$$\gamma = -1, \quad \beta = \delta = -\sqrt{f_0},$$

and Eq. (4) assumes the form

(4a)        $$\frac{(q\zeta + 1)^2}{(q\zeta - 1)^2} = \frac{\sqrt{f} - \sqrt{f_0}}{-\sqrt{f} - \sqrt{f_0}} = \frac{1 - \sqrt{\dfrac{f}{f_0}}}{1 + \sqrt{\dfrac{f}{f_0}}}.$$

Thus we have achieved a more cogent derivation of Eq. (30.7) based on the known mapping properties of the bilinear function and the square root.

We now apply the same method to the classical problem connected with Torricelli's name, viz. the outflow from a large reservoir, already treated in 11, p. 87. At that time we took it for granted that the fluid leaves the orifice in the form of a jet. Now we wish to follow Helmholtz who was the first to realize that the problem of jet formation should be made the object of hydrodynamic investigation. In the paper quoted on p. 215 he actually gives the present problem as the only example to illustrate the idea of the discontinuity surface.

We shall treat here the two-dimensional equivalent of the problem of jet formation. This means replacing the orifice in the tank wall by an infinitely long slit of width $2c$ in a plane wall perpendicular to the drawing of Fig. 52. The action of gravity on the emerging jet will be disregarded.

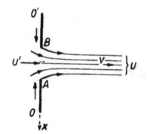

FIG. 52. Two dimensional scheme of jet contraction.

The center line of the jet is taken as $y$-axis, the wall as $x$-axis. The two discontinuity lines that issue from the sharp edges of the slit will approach each other since the water enters the orifice also from the sides; the cross-section of the jet decreases therefore after its emergence from the orifice (vena contracta). The principal aim of our investigation is to determine the degree of contraction, that is to say, the ratio of the jet cross-section at large distance to that in the immediate neighborhood of the orifice. The details of the calculation need not be presented since the problem is mathematically identical with that of 30.

Fig. 52 has been drawn in the plane $z = x + iy$. In addition we consider as on p. 217 and p. 218 the planes

(5) $$\zeta = \frac{1}{-v_x + iv_y} \quad \text{and} \quad f = \Phi + i\Psi.$$

Along the two discontinuity lines the pressure is constant; it is equal to the atmospheric pressure which we take as the zero level for $p$. From Bernoulli's equation (29.4) we then have $v^2 = $ const along the discontinuity lines, therefore

$$|\mathbf{v}| = \text{const} = q \quad \text{and from (30.3)} \quad |\zeta| = \frac{1}{q}.$$

Here $q$ represents the magnitude of $|\mathbf{v}|$ at the points $A$ and $B$, or, as we may also put it, the value of the velocity at sufficient distance from the orifice, as it is given by Torricelli's Eq. (11.10).

At the infinitely distant point $U$ of the jet, where the side component $v_x$ evidently vanishes, we obtain from (5)

(5a) $$\zeta = -\frac{i}{q}.$$

Along the inside of the wall where $v_y$ vanishes, $\zeta$ becomes real; for, we have from Fig. 52 and Eq. (5)

(5b)
$$+\infty > \zeta > \frac{1}{q} \qquad \text{at the wall } OA,$$

$$-\infty < \zeta < -\frac{1}{q} \qquad \text{at the wall } O'B.$$

The region of flow in the $\zeta$-plane of Fig. 53 has therefore the same shape as in Fig. 50. In addition, Fig. 53 shows the image of the infinitely distant point $U'$ on the upstream prolongation of the center line of the jet. (In approaching the point $U'$ from the orifice we have to assume $v_x = 0$, $v_y \to 0$, therefore $\text{Im}(\zeta) \to -\infty$.)

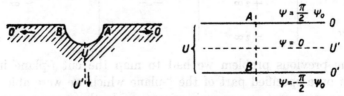

FIG. 53 (*left*). Mapping of the $z$-plane on the $\zeta$-plane.
FIG. 54 (*right*). Mapping of the $z$-plane on the $f$-plane.

Fig. 54 shows the flow region of the $f$-plane. Let the stream line $\Psi = 0$ coincide with the axis of symmetry of the flow $UU'$ (real axis of the $f$-

plane). $\Phi$ increases along this line in the direction from $U$ to $U'$, since $v = v_y$ has here the direction from $U'$ to $U$. We assign to the streamlines that run along the walls of the reservoir and the discontinuity lines the value[13] of the stream function

$$\Psi = \pm \frac{\pi}{2} \Psi_0 \,.$$

The entire region of flow is mapped on the strip of the $f$-plane bounded by these two $\Psi$-lines. If we now assign the zero level of the velocity potential $\Phi$ to the equipotential line that passes through the edges $A$ and $B$ of the slit, then the images $A$ and $B$ in Fig. 54 are located on the imaginary axis of the $f$-plane. The "lower" and "upper" tank walls are represented by the segments $AO$ and $BO'$ of the boundaries of the strip, while $UA$ and $UB$ are the images of the boundaries of the jet.

As before we have now to find the analytical relation between $\zeta$ and $f$, and use again a table that incorporates the correlations already indicated in Figs. 52-54.

|       | $z$        | $\zeta$           | $f$                              |
|-------|------------|-------------------|----------------------------------|
| $A$   | $+c$       | $\dfrac{1}{q}$    | $i\dfrac{\pi}{2}\Psi_0$          |
| $B$   | $-c$       | $-\dfrac{1}{q}$   | $-i\dfrac{\pi}{2}\Psi_0$         |
| $O$   | $+\infty$  | $+\infty$         | $\infty + \dfrac{i\pi}{2}\Psi_0$ |
| $O'$  | $-\infty$  | $-\infty$         | $\infty - \dfrac{i\pi}{2}\Psi_0$ |
| $U$   | $+i\infty$ | $-\dfrac{i}{q}$   | $-\infty$                        |
| $U'$  | $-i\infty$ | $-i\infty$        | $+\infty$                        |

In the previous problem we had to map the cut $f$-plane in its full extension on a specified part of the $\zeta$-plane which we were able to do by using the square root as a mapping function. In the present case we want to map a mere strip of the $f$-plane on the same part of the $\zeta$-plane as before. This can be achieved by the exponential function. We use Kirch-

---

[13]The factor $\pi/2$ of $\Psi_0$ simplifies some of the following formulas.

hoff's notation $e^w$ where the variable $w$ is proportional to $f$ with a scale factor[14] that depends on the choice of $\Psi_0$ :

$$(6) \qquad\qquad w = \frac{f}{\Psi_0}.$$

The mapping function (4) takes now the following form

$$\frac{(q\zeta + 1)^2}{(q\zeta - 1)^2} = \frac{ie^w - 1}{ie^w + 1}.$$

Here $\alpha$, $\beta$, $\gamma$, $\delta$ have been determined according to the foregoing table, e.g., by satisfying the conditions required for the points $B$, $A$ and $U$; the conditions at the remaining points are then automatically fulfilled. In analogy to (30.7a) we obtain for $q$ the quadratic equation

$$(q\zeta)^2 + 2ie^w q\zeta + 1 = 0$$

with the solution

$$q\zeta = -i(e^w + \sqrt{e^{2w} + 1}).$$

Because of (30.4), where $df$ is to be replaced by $\Psi_0 dw$, this yields the differential equation

$$(7) \qquad\qquad dz = -\frac{i\Psi_0}{q}(e^w + \sqrt{e^{2w} + 1})\, dw.$$

Let us now consider the three straight lines of Fig. 54

$$\Psi = 0, \qquad \Psi = \pm\frac{\pi}{2}\Psi_0.$$

The quantities $f$ and $w$ are real if $\Psi = 0$; the left member of (7) is then purely imaginary. The point $z$ travels along the $y$-axis from $-\infty$ to $+\infty$ as it is expected to do (Fig. 52). Along the two other lines $e^w$ becomes purely imaginary. On putting

$$e^w = \pm is, \qquad dw = \frac{ds}{s},$$

Eq. (7) appears in the following form:

$$(8) \qquad\qquad dz = \frac{\Psi_0}{q}(\pm s + \sqrt{s^2 - 1})\frac{ds}{s}.$$

[14]Note that Kirchhoff in his treatise Vorlesungen über Mechanik, Chap. XXII, 3, defines the velocity potential with reversed sign, chooses the coordinate axes not symmetrical to the jet, and does not set in evidence physical parameters of the problem such as $q$. This procedure, which amounts to putting $q$ equal to 1, has been generally accepted in the literature. We are of the opinion, however, that a better grasp of Kirchhoff's ingenious method becomes possible by the introduction of definite units from the start.

The square root is real for $\infty > s > 1$; for these values of $s$, $z$ is real and travels along the "upper" and "lower" tank wall $y = 0$, $x > c$ and $x < -c$.

As we are mainly interested in the *free boundaries of the jet* that issue from the slit, we must consider the range $s < 1$. This makes $dz$ in (8) complex, as it has to be in the case of curved jet boundaries. We choose the upper sign in (8) which corresponds to the lower jet boundary in Fig. 52 and put

$$s = \sin \alpha, \quad \frac{\pi}{2} > \alpha > 0.$$

Eq. (8) takes now the form

$$dz = \frac{\Psi_0}{q} (\sin \alpha + i \cos \alpha) \frac{\cos \alpha}{\sin \alpha} \, d\alpha = \frac{i\Psi_0}{q} e^{-i\alpha} \frac{\cos \alpha}{\sin \alpha} \, d\alpha.$$

This yields after integration the result

(9)
$$z = i\frac{\Psi_0}{q} \left( e^{-i\alpha} - \frac{1}{2} \log \frac{1 + \cos \alpha}{1 - \cos \alpha} \right) + \text{const},$$

which may be checked by differentiation. The constant can be determined by requiring $z = c$ for $\alpha = \pi/2$, that is, at the point $A$. Then we obtain

$$\text{const} = c - \frac{\Psi_0}{q}, \text{ and}$$

(9a)
$$z = c - \frac{\Psi_0}{q} \left( 1 - ie^{-i\alpha} + \frac{i}{2} \log \frac{1 + \cos \alpha}{1 - \cos \alpha} \right).$$

Separation into real and imaginary parts gives:

$$x = c - \frac{\Psi_0}{q} (1 - \sin \alpha),$$

(10)

$$y = \frac{\Psi_0}{q} \left( \cos \alpha - \frac{1}{2} \log \frac{1 + \cos \alpha}{1 - \cos \alpha} \right).$$

These equations are the parametric representation of one of the two jet boundaries; the other is obtained from (10) by changing the sign of $x$.

For $\alpha = 0$ we obtain from (10)

$$x = c - \frac{\Psi_0}{q}, \qquad y = \infty,$$

that is, the behavior at the infinitely distant point $U$ in Fig. 52. Half the width of the jet at $U$ is therefore

(11)
$$b = c - \frac{\Psi_0}{q}.$$

Finally, we have to find the physical meaning of the constant $\Psi_0$ which is of course connected with the velocity $q$. We write down the second Cauchy-Riemann equation (19.5)

(12)
$$\frac{\partial \Psi}{\partial x} = -\frac{\partial \Phi}{\partial y}.$$

At the cross-section $U$, that is, at large distance from the orifice, we find the velocity uniformly equal to $q$ and parallel to the $y$-axis; the right member of (12) is therefore constant. Integration of (12) between the limits $x = 0$ and $x = b$ which correspond to $\Psi = 0$ and $\Psi = (\pi/2)\Psi_0$ yields

(12a)
$$\frac{\pi}{2}\Psi_0 = qb.$$

In combination with Eq. (11) one has

$$b = c - \frac{2}{\pi}b, \qquad \frac{2+\pi}{\pi}b = c$$

so that the relative contraction of the jet is

$$\frac{b}{c} = \frac{\pi}{2+\pi} = 0.611.$$

The actual contraction coefficients used in hydraulic practice are close to our theoretical result; they serve to predict the efflux through a given orifice from its geometry.

## 32. Kármán's Vortex Street

The analysis of a given physical situation usually requires a certain amount of idealization, so that the tools of mathematics can be successfully applied; to find workable idealizations is the art of the mathematical physicist. Bearing this in mind, let us investigate a problem already presented by Helmholtz (cf. p. 216): What is the nature of the mechanism that leads to the curling up of the discontinuity surface into a sequence of vortex filaments? In the case considered in 30, the vortices originate at the sharp edges of the plate; in general vortices may be generated by any obstacle that is in the way of the flow. The photographs 47 a, b give a good idea of them. Loosely speaking, the vortices operate like ball or roller bearings distributed along the discontinuity surface that separates the streaming fluid from the dead water. This visualization gives at least the correct sense of rotation of the vortices on either side of the wake. There is however this difference: the vortices do not run along in the same way as the balls in the race of a bearing. They would move with the stream, but they influence each other; actually the mutual influence produces an *additional velocity* which is superimposed on the *principal velocity* of the flow and has in the simplest case the opposite direction as we shall see.

Kármán[15] idealized the sequence of vortices produced at an obstacle by assuming an *infinitely long straight vortex street* and disregards the origin of the vortices by transferring it to infinity. The vortices on either side of the street are equidistantly arranged: this characterizes the steady state which has developed in the (infinitely) long time that passed after they had been generated. On either side of the street the sense of rotation is different; the "upper" vortices are clockwise, the "lower" vortices counterclockwise. The principal velocity points to the right, corresponding to that of a free flow that rolls on "bearings" along a wake at rest. Kármán restricts himself to the case in which the additional velocity is parallel to the street; we shall start our analysis with the same assumption.

The first consequence is, as we shall show, that the two rows of vortices must be either opposite as in Fig. 55 or alternating as in Fig. 56. (Note

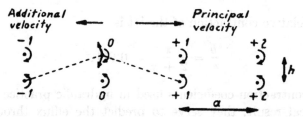

Fig. 55. Kármán's vortex street consisting of two rows of vortices of different orientation. The vortices are in opposite position.

that "vortex" means concentrated vortex filament, indicated in the figure by a dot with semicircular arrow.)

Consider Fig. 55; number the vortices from $-\infty$ to $+\infty$ and choose the vortex 0 of the upper street as "field" vortex, that is, consider this vortex as subjected to the induction of all the other vortices as in 21. There are *no contributions to its velocity* from the neighboring vortices $\pm 1$ on the *same* side of the street, since the induced velocity is normal to the line connecting with the inducing vortex and points in the sense of its rotation. The contributions of the pair $\pm k$ cancel for the same reason. The contribution of the opposite vortex 0 points in the *negative direction*

[15]Th. v. Kármán, Über den Mechanismus des Flüssigkeits- und Luftwiderstandes, Göttinger Nachr. 1911 und 1912. Cf. also Kármán and H. Rubach, Physikal. Z. 13 (1912) 49.

*of the street.* Also the contributions of the pair $\pm 1$ on the other side, which, individually, are inclined to the street, give a *resultant in the negative direction of the street,* and the same is true for the pair $\pm k$.

When the vortices are not exactly opposite to each other, the same construction will give a total induction at our field vortex that is *oblique to the street* with the one exception of the alternating case of Fig. 56. Here we compound the inductions due to the vortices 0 and 1 on the opposite side (broken lines in Fig. 56), and generally the inductions of the pairs $(-k, k + 1)$ and obtain again an additional velocity at the field vortex, pointing in the negative direction of the street. Fig. 56 gives in addition an idea of the presumable path of the fluid particles that wind in and out between the vortices, in reasonable agreement with the photographs in Fig. 47a and b.

These photographs suggest at the same time that the alternating arrangement of Fig. 56 rather than the one of Fig. 55 occurs in nature. In

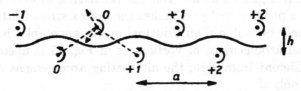

FIG. 56. Kármán's vortex street; the vortices are in alternating position.

order to explain this preference Kármán investigates the *stability* of both arrangements.

The concept of stability applied to a problem such as ours needs some comment. In Figs. 55 and 56 the locations of the vortices

$$z_k = x_k + iy_k$$

are accurately given by the positive or negative integer number $k$ so that the additional velocity can be calculated according to well known complex methods. To investigate the stability we have to impart small displacements or disturbances

$$\bar{z}_k = \bar{x}_k = i\bar{y}_k$$

to the vortices.[16] The introduction of such disturbances changes also the

_____

[16]This variation is actually too restricted since it occurs only in the $x$, $y$-plane while in the $z$-direction the displaced vortex filaments are still supposed to be straight as they originally were. A more general stability investigation which introduces variations in $z$-direction has been carried out by K. Schlayer in his Munich thesis [Z. angew. Mathem. und Mech. 8, (1928) 352]. One can defend the special variations admitted by Kármán, observing that "infinitely long" straight vortex filaments can at any rate only be produced in a flow between parallel plates at a small distance (see p. 160). The

additional velocity. For instance, the statement that the total induction of all vortices on the same side as the field vortex vanishes is no longer true. In the disturbed state the velocity of the field vortex depends on all $\bar{z}_k$ regardless on which side of the street they are. We assume the dependence as linear neglecting the higher powers of $\bar{z}_k$. The variation of the induction produced by the variation of the $\bar{z}_k$ determines the time rate of change $\bar{z}_0$ of the field vortex as a linear function of all the other $\bar{z}_k$. Thus we obtain first an infinite system of linear differential equations of first order for the $\bar{z}_k$ which can be reduced, however, to a system of only two differential equations of the same character (see later). According to the general theory of small oscillations[16a], one determines first the principal modes of the system, that is, one finds solutions with the time dependence $e^{\lambda t}$. This procedure was carried out in a very simple case in Vol. I, 3.24 and prob. IV, 2. If $\lambda$ is purely imaginary the vortex street is considered stable since oscillations of constant amplitude would be attenuated by friction which has been neglected so far. But the occurrence of even one complex $\lambda$-value with a positive real part makes the vortex street unstable, since a system can only be stable if all principal modes are stable. Kármán[17] has shown that an arrangement of vortices as in Fig. 55 is unstable for any value of $h$ different from zero; the alternating arrangement of Fig. 56 is stable if and only if

(1) $$\sinh \frac{h}{a}\pi = 1, \qquad \text{hence} \qquad \frac{h}{a} = 0.281.$$

Our Figs. 55 and 56 have been drawn with about this numerical value of $h/a$.

The complete representation of the stability investigation, which is rather an involved mathematical problem, would take up too much space here. We shall indicate only the main points of the calculation. In doing so we shall at the same time broaden the basis of our investigation by following a paper of Maue.[18] We shall assume that one row of the street

---

stability of such an experimental arrangement is indeed demonstrable by means of the special variations used in the text, provided the plate distance is sufficiently small. According to Schlayer, there is, however, always a weak instability toward general variations, and it increases with increasing plate distance.

[16a]For a discussion of the method of small oscillations see, for instance, Kármán, Th. v. and Biot, M. A. Mathematical methods in Engineering, New York, 1940, Chapters V and VI.

[17]Kármán's investigation has been supplemented by Bl. Dolaptschiew who considered terms of second order. [Z. angew. Mathem. u. Mech. 17 (1937) 313, 18 (1938) 263]

[18]A. W. Maue, Z. angew. Mathem. u. Mech. 20 (1940) 130. Eqs. (6) to (8) that go beyond Kármán's original investigation have been taken from this paper. Mr. Maue kindly contributed the appendix to the present article.

is shifted relative to the other by the amount $d < a$ (in Fig. 55, $d = 0$, in Fig. 56, $d = a/2$). This makes no formal difficulties if the origin of the $z$-plane is located at the center of the pair 0, 0 and if the points are numbered in such a way that the points $\pm k$ on the one side are diametrically opposite to the points $\mp k$ on the other side of the street, viz:

$$(2) \qquad z_k^2 = -z_k^1, \qquad z_k^1 = z_0 + ka, \qquad z_0 = \frac{1}{2}(d + ih).$$

(The superscripts 1 and 2 refer to the upper and lower rows of the street respectively, $z_0$ indicates the location of the 0-vortex belonging to the upper side of the street.)

First, we give a very simple expression for the complex potential $f = \Phi + i\Psi$ valid for arbitrary $d$, as it depends on $d$ only by $z_0$. Our starting point is Eq. (19.10) which describes the combined action of a positive and a negative vortex; the numerator and also the denominator occurring there vanish respectively at the center of the positive and negative vortex. The generalization of this equation for our vortex street is simply

$$(3) \qquad f = -\frac{i\mu}{\pi} \log \frac{\sin \dfrac{(z - z_0)\pi}{a}}{\sin \dfrac{(z + z_0)\pi}{a}}.$$

In fact, the numerator vanishes now for all points

$$z = z_0 + ka = z_k^1,$$

and the denominator for all points

$$z = -(z_0 + ka) = z_k^2$$

as required by (2). The periodicity of the sine makes each vortex act with the same strength. The factor $A$ in (19.10) that characterizes the vortex strength has been changed in accordance with (19.8a) and the signs have been adjusted in such a way as to correspond to the orientation of the vortices in Fig. 56.

By differentiation of $f$ with respect to $z$ one obtains as in the familiar relations (30.3) and (30.4)

$$(4) \qquad \frac{df}{dz} = -\mathbf{v}^* \begin{cases} \mathbf{v} = v_x + iv_y, \\ \mathbf{v}^* = v_x - iv_y. \end{cases}$$

Here $\mathbf{v}$ is the velocity of a fluid particle characterized by its complex coordinate $z$. If we wish to apply this formula to one of the vortex points, e.g., to the field vortex $z = z_0$ itself, then we must keep in mind that its induction upon itself is zero and has to be omitted in the expression for $f$. This can be done by adding to the right member of (3)

$$+ \frac{i\mu}{\pi} \log (z - z_0).$$

According to this, Eq. (4) yields, when we introduce $\mathbf{u} = u_x + iu_y$ to denote the velocity of the field vortex $z_0$ ,

$$\mathbf{u}^* = \frac{i\mu}{\pi} \frac{d}{dz} \left[ \log \frac{\sin \frac{(z - z_0)\pi}{a}}{\sin \frac{(z + z_0)\pi}{a}} - \log (z - z_0) \right];$$

the differential quotient (4) is to be taken at $z = z_0$. We therefore introduce

$$(z - z_0) \frac{\pi}{a} = \epsilon, \qquad (z + z_0) \frac{\pi}{a} = 2z_0 \frac{\pi}{a} + \epsilon,$$

and find in the limit $\epsilon \to 0$

(5) $$\mathbf{u}^* = - \frac{i\mu}{a} \operatorname{ctg} \frac{2\pi z_0}{a}.$$

This formula is valid not only for the vortex $z = z_0$ , but, because of the periodicity of the street, also for any other vortex. The condition for the "additional velocity" $\mathbf{u}$ to point in the direction of the street is simply that (5) be real; consequently the expression

$$\operatorname{ctg} \frac{2\pi z_0}{a}, \qquad \text{or because of (2)} \qquad \operatorname{ctg} \frac{\pi}{a} (d + ih)$$

must be *purely imaginary*. This occurs only if $d = 0$ or $d = a/2$. In both cases we have from (5)

(5a)
$$d = 0, \qquad \mathbf{u}^* = u_x = - \frac{\mu}{a} \coth \frac{h}{a} \pi,$$

$$d = \frac{a}{2}, \qquad \mathbf{u}^* = u_x = - \frac{\mu}{a} \tanh \frac{h}{a} \pi.$$

By means of the stability criterion stated in (1), the second line of (5a) can be written in the form

(5b) $$d = \frac{a}{2}, \qquad \mathbf{u}^* = u_x = - \frac{1}{\sqrt{2}} \frac{\mu}{a}.$$

Thus the additional velocities constructed in Figs. 55 and 56 have been found. The principal velocity indicated in the same figures—it could be denoted by $q$ as in the plate problem—has the value zero for the particular frame of reference we have chosen (center of the pair 0, 0). This can be seen from Eq. (3): If $z$ is allowed to go to infinity along some specified path lying outside the vortex street, then the quotient of the sines tends to a constant limit; thus $f$ is constant and $q = 0$.

In all other cases $\mathbf{u}$ is *oblique* relative to the vortex street. The angle of inclination $\gamma$ follows from (5) by an elementary calculation:

$$(6) \qquad \operatorname{tg} \gamma = \frac{u_y}{u_x} = \frac{\sin 2\pi \dfrac{d}{a}}{\sinh 2\pi \dfrac{h}{a}} .$$

The absolute value $u$ of $\mathbf{u}$ is also obtained from (5):

$$(6a) \qquad u = \frac{\mu}{a} \sqrt{\frac{\cosh^2 \pi \dfrac{h}{a} - \sin^2 \pi \dfrac{d}{a}}{\cosh^2 \pi \dfrac{h}{a} - \cos^2 \pi \dfrac{d}{a}}} .$$

This formula is the generalization of $u_x$ in Kármán's result, which is our $u_x$ in Eq. (5a).

If we let the vortex street be generated by a flow past a plate, as at the beginning of this article, Eq. (6) must obviously be associated with the case of an *oblique plate*. This should not imply, however, that the angle $\gamma$ in (6) is equal to the angle $\gamma$ in Fig. 48 (the former is actually smaller than the latter). Unfortunately, the photograph 47b does not show clearly the obliquity of the street produced.

We have still to discuss the central part of our problem, viz. the investigation of stability. Following Maue, we choose a slightly modified approach: instead of prescribing displacements $\bar{z}_k$ for the vortex points, we modify the sine functions occurring in the complex potential (3) in the following way:

$$(7) \qquad \sin (z - z_0) \frac{\pi}{a} - \zeta_1(z) \qquad \text{and} \qquad \sin (z + z_0) \frac{\pi}{a} - \zeta_2(z).$$

The perturbations $\zeta_1$, $\zeta_2$ are supposed to be small. They are connected to the arbitrary variations $\bar{z}_k$ by the requirement that the zeros of the functions (7) should occur at $z_k^1 + \bar{z}_k^1$ and $z_k^2 + \bar{z}_k^2$ respectively. This is achieved by putting

$$\zeta_1(z_0 + ka) = (-1)^k \frac{\pi}{a} \bar{z}_k^1 , \qquad \zeta_2(-z_0 + ka) = (-1)^k \frac{\pi}{a} \bar{z}_k^2 .$$

In this way the functions $\zeta$ have been fixed except for a function of the period $a$. The velocity of the vortex points after perturbation is now found as in Eq. (4) by differentiation of the modified function $f$. Upon expressing the same velocity by the time derivatives of the perturbations one obtains two simultaneous linear differential equations between $\zeta_1$, $\zeta_2$, their first and second derivatives with respect to $z$, and $\dot{\zeta}_1$, $\dot{\zeta}_2$. The discussion of these equations which we omit here yields the stability criterion

$$(8) \qquad\qquad \sinh \frac{h}{a}\pi = \sin \frac{d}{a}\pi.$$

*This is the direct generalization of Kármán's criterion* (1). Thus there exists also in the case of a vortex street that advances in oblique direction, a certain value $h/a$ for any given value $d \le a$ such that the vortex street maintains its arrangement after perturbation except for small oscillations that do not increase.

Fig. 57 illustrates this result and at the same time Kármán's special

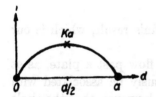

FIG. 57. Illustrating the generalized stability criterion. The width $h$ of the vortex street as a function of the mutual displacement of the two rows of vortices.

case [Eq. (1)]. It is an $h$ versus $d$ plot for all possible stable vortex streets. If the sinh in (8) were approximated by its tangent at the origin, the $h$ versus $d$-curve would become an ordinary sine curve. Since the hyperbolic sine increases monotonically with its argument, the actual $h$, $d$-dependence is a slight distortion of the sine curve appearing on the right side of (8). The maximum ordinate corresponds to Kármán's case $d = a/2$. For $d = 0$ one has $h = 0$, just as in the geometrically identical case $d = a$. In both cases the two rows of vortex points coincide point by point and cancel each other: the end points of the stability curve have no physical significance.

One can expect that an experimental investigation such as was carried out by Kármán and Rubach for straight vortex streets (cf. footnote on p. 232) will also confirm the generalized stability criterion (8) for oblique vortex streets.

# Appendix

## The Drag Problem

Kármán's real aim in his investigation of the vortex street was to establish an analytic drag formula. For this purpose he considers a body moving through the fluid in the rear of which a vortex street has developed that reaches to infinity. The calculation is performed in the $z$-plane of the street, $x$ being the direction of motion of the body. The body itself is considered as infinitely long in the direction normal to the $z$-plane. One delimits in the plane a sufficiently large rectangular region by a control surface, which includes the body and a sufficiently long stretch of the vortex street, and investigates the balance of momentum inside the control surface. The terms to be considered are: the pressure interaction between fluid and body, the increase of momentum inside the control surface due to the generation of new vortices, and the flux of momentum through the control surface. The first of these three terms is to be taken negative; its real part represents the drag, its imaginary part the side force normal to the direction of motion.

Even in the case where the direction of the vortex street coincides with the direction of motion of the body the calculation is rather involved. The more general case studied by Maue where the directions of the vortex street and that of the motion of the body form an angle with each other will probably always occur, unless there is a special relation of symmetry between the body and the direction of its motion. The drag formula is not given in Maue's paper quoted on p. 234 but has been kindly contributed by the author for use in the following discussion. It contains Kármán's formula as a particular case.

The notations used here conform to those of 30 in the plate problem. The drag $P$ corresponds to the integral of the pressure over the width of the plate in (30.22). The effective width of the body normal to the direction of motion in the $x + iy$-plane is now denoted by $2c$, while previously $2c$ was directly the width of the plate. In the present notation $q$ is the velocity of the body relative to the fluid which is at rest at infinity; the previous $q$ was the flow velocity at infinity while the plate was at rest. The dimensionless drag coefficient is denoted by $\lambda$ (previously, 0.88). The distance of two neighbored vortices is $a$, $h$ is the width of the street, and $u$ its velocity of progression as calculated in (5).

According to Maue, the drag formula that replaces (30.22) is

$$(9) \qquad \frac{P}{2c} = \lambda \frac{\rho}{2} q^2$$

where $\lambda$ is given by

(10) $$\lambda = \frac{a}{c}\left\{A\eta - B\eta^2 \sqrt{1 - C\eta^2} - 2C\eta^3\right\}.$$

Here, $A$, $B$, $C$ stand for

(11)
$$A = \frac{2\sqrt{2}}{\pi}\beta \sinh \beta, \qquad B = \frac{4}{\pi}\sinh \beta(\beta \cosh \beta - \sinh \beta),$$

$$C = \frac{1}{2}(1 - \sinh^2 \beta);$$

the numbers $\beta$ and $\eta$ are defined by

(12) $$\beta = \pi\frac{h}{a}, \qquad \eta = \frac{u}{q}.$$

The drag coefficient (10) thus depends on three ratios, viz. the two quantities appearing in (12) and $a/c$ that appears in (10).

In Kármán's case $d = a/2$; hence, with $\sinh \beta = 1$ by (8), (11) gives

(13) $$A = 0.794, \qquad B = 0.31, \qquad C = 0.$$

From (10) Kármán's drag coefficient is now obtained in the form

(14) $$\lambda = \frac{a}{c}\left(0.794\frac{u}{q} - 0.314\left(\frac{u}{q}\right)^2\right),$$

depending only on the ratios $a/c$ and $u/q$. In the papers quoted on p. 232 these ratios have been obtained from experiments carried out with a flat plate:

(15) $$a/2c = 5.5, \qquad u/q = 0.20.$$

With these numerical values (14) gives

(16) $$\lambda = 1.6,$$

a considerably larger value than Kirchhoff's coefficient 0.88 in (30.22). The latter value is too small as already mentioned in connection with the results of Prandtl and his coworkers.

W. Heisenberg[19] has made an attempt to determine the quantities (15) theoretically from certain conservation theorems that should hold in the generation of vortices. He finds, in excellent agreement with Kármán, $a/2c = 5.5$, $u/q = 0.23$, which would yield, by (14), $\lambda = 1.8$.

---

[19]Physik. Z. 23 (1922), 363 with an additional remark of Prandtl.

## 33. Prandtl's Boundary Layer

It seems unsatisfactory that in all examples of potential flows[20] discussed in this chapter the fluid was supposed to be inviscid. However, in the case of an *incompressible* fluid in two- or three-dimensional potential flow the disregard of viscosity has no bearing on the *differential equations*, although it is of decisive importance for the *boundary conditions*.

This can be seen by a glance at the Navier-Stokes equations (16.1). If curl $\mathbf{v} = 0$ (potential flow) and div $\mathbf{v} = 0$ (incompressible fluid), the friction term $\mu \nabla^2 \mathbf{v}$ can be dropped entirely from Eqs. (16.1) because of the relation

$$\nabla^2 \mathbf{v} = \text{grad div } \mathbf{v} - \text{curl curl } \mathbf{v}.$$

Hence irrotational solutions of the Euler equations are always solutions of the Navier-Stokes equations: *the viscous forces have no dynamic effect in the interior of the fluid if the flow is irrotational.*

This is no longer true at the boundary between the fluid and a solid wall. While the boundary conditions for a perfect fluid are for this case

(1a)                          $v_n = 0, \; v_s$ arbitrary,

they read for a viscous fluid according to (10.9) (condition of adherence),

(1b)                          $v_n = 0, \qquad v_s = 0.$

*It is only in the proximity of the wall that the viscous forces become dynamically important as a consequence of the modified boundary condition* (1b). For a flow with large Reynolds numbers we may assume, however, that the modification of the perfect fluid stream pattern due to viscosity does not extend very far into the interior of the fluid.

The existence of the two different types of boundary conditions (1a) and (1b) has its *mathematical* basis in the occurrence of second (spatial) derivatives in the general Navier-Stokes equations in contrast to the Euler equations, that contain only first order derivatives. In order to single out a solution one therefore needs *two* vectorial boundary conditions in the first case and only *one* such condition in the second case. On the other hand, to reconcile *physically* the contradictory conditions (1a) and (1b), a fluid layer is inserted between the presumable region of validity of the potential flow and the wall. In passing through this layer, the thickness of which becomes arbitrarily small with decreasing $\mu$, the tangential velocity $v_s$ drops rapidly to zero. For small but finite $\mu$ (air, water), the

---

[20]Kármán's vortex street is of course also a potential flow; its vorticity is concentrated in the vortex filaments, that is, in the singularities of the potential. Also the wave motions in Chap. V were treated as potential flows (see p. 168).

layer will still be reasonably thin, provided the overall dimensions of the wetted wall and the immersed bodies are large and the flow is sufficiently fast. The principal part of the flow lies then outside the layer and behaves like the potential flow of a perfect fluid. This idea forms the basis of the boundary layer theory first proposed by Prandtl in 1904.

The actual calculations by Prandtl[21] and his coworkers have the character of an approximate theory.[22] In the calculations which are mostly two-dimensional, inside the boundary layer the influence of the small velocity differences parallel to the layer on the friction terms is neglected. For small $\mu$ the thickness of the boundary layer is proportional to $\sqrt{\mu}$. Expressed in terms of the overall Reynolds number of the flow, $R$, it is proportional to $1/\sqrt{R}$. The pressure gradient normal to the boundary layer is neglected on account of its small width, and the pressure inside the layer therefore identified with the pressure of the adjacent potential flow which impresses, as it were, its own pressure on the whole boundary layer beneath. For the flow immediately adjacent to the boundary layer Bernoulli's equation applies, thus we have for a line element $ds$ in the zone where the boundary layer passes over into the free stream

$$\frac{dp}{ds} = -\rho v_s \frac{dv_s}{ds}.$$

Note that the fluid begins to flow against the pressure gradient if $dv_s/ds$ changes sign from plus to minus at some point.

Further calculation shows that $\partial^2 v_s/\partial n^2$ at $n = 0$ is negative in the region downstream from such a point. If the pressure gradient is large enough $(\partial v_s/\partial n)_{n=0}$ can also become negative, which means that the flow in the immediate neighborhood of the wall is reversed. This leads to the generation of vortices so that downstream from this point the laminar flow is replaced by flow processes of a different character.

The accuracy of this theory is quite high in spite of its approximate character, because the Reynolds numbers that occur in practical cases are usually large. In a given case the theory serves to decide whether the flow will everywhere stick to the rigid boundary and thus essentially remain a potential flow except for the thin boundary layer, or whether separation of the flow will occur, and, consequently, formation of eddies.

For very large $v_s$ (corresponding to very large $R$-values) the assumption of a laminar flow within the boundary layer is no longer correct; the flow

[21]Verh. des III. Int. Math. Kongresses in Heidelberg, 1904, Leipzig, 1905. Detailed representation in Prandtl-Tietjens, Hydro- and Aerodynamics, New York, 1934, Vol. II. and in Modern Developments in Fluid Dynamics, ed. S. Goldstein, Oxford, 1938.

[22]The discussion in the text follows a report of Th. v. Kármán in Vorträge aus dem Gebiet der Hydro- und Aerodynamik (Innsbruck 1922), Berlin 1924, Springer.

becomes turbulent even inside the narrow boundary layer. This is of importance for the overall flow pattern because the separation point may be shifted downstream quite considerably, if the boundary layer becomes turbulent. An approximate prediction of the behavior of turbulent boundary layers based on a semi-empirical theory, has likewise become feasible.

The flow past a wing is one of the problems which have been solved analytically in this sense, that is, by the application of potential flow theory. The two-dimensional wing theory, the beginnings of which go back to Kutta[23] and Joukowski,[24] represents the flow as a plane potential motion with non-vanishing circulation and uses freely the methods of conformal mapping. The mathematical form of the three-dimensional wing theory is due to Prandtl. Here the flow is represented as a three-dimensional potential flow with a discontinuity surface, that may be visualized as a two-dimensional system of vortex filaments. The theory furnishes not only the lift distribution over the span but also the (induced) drag, that is, the dynamic equivalent of the kinetic energy that remains in the flow field behind the wing.

A. Betz and his coworkers developed the theory of the finite wing into a tool for the aeronautical engineer so that today pressure distributions for wings of any plan form (also with deflected ailerons, etc.) can be computed.

---

[23]Sitzungsber. Bayer. Akad. 1910 and 1911.

[24]Z. f. Flugtechnik u. Motorluftschiffahrt, 1 (1910).

## CHAPTER VII

# SUPPLEMENTARY NOTES ON SELECTED HYDRODYNAMIC PROBLEMS

## 34. Lagrange's Equations of Motion

Lagrange's equations[1] have been referred to this supplement since they cannot compete with Euler's equations in importance as far as the setting up of hydrodynamic problems is concerned. Before we occupy ourselves with them, we wish to emphasize that Euler's equations, rather than Lagrange's, are the "unusual" equations. In ordinary mechanics we consider a mass point $m$, or some well defined part of a mechanical system, give it a certain tag $q$ (coordinate) and determine the time rate of change of momentum $p$, corresponding to $q$, in accordance with the given forces and constraints. Now, in the case of Euler's equations we consider a *volume element* $d\tau$ and identify *it* rather than the mass element $\rho d\tau$, by coordinates; we again determine the change of momentum in accordance with the given forces and constraints, but we do this at the *space-fixed* element $d\tau$ (note that the pressure plays the role of a constraint, cf. 12). The mass element that occupies $d\tau$ at the given instant is lost sight of thereafter.

On the other hand, when setting up Lagrange's equations we follow the usual procedure and ascribe to the mass element $dm$ certain tags $a$, $b$, $c$, e.g. its coordinates $x_0$, $y_0$, $z_0$ at the time $t = 0$. We ask then for the location $x$, $y$, $z$ of this element $dm$ at the time $t$. Thus $x$, $y$, $z$ and the pressure $p$ are the dependent variables, while $a$, $b$, $c$ and the time $t$ are the independent variables of Lagrange's equations.

No difference has now to be made between local acceleration $\partial \mathbf{v}/\partial t$ and material acceleration $d\mathbf{v}/dt$. The material acceleration which is the decisive parameter in the momentum balance, is now directly given by[2]

$$\frac{\partial^2 x}{\partial t^2}, \quad \frac{\partial^2 y}{\partial t^2}, \quad \frac{\partial^2 z}{\partial t^2},$$

---

[1]The usual terminology is historically not correct, as it so often happens: "Lagrange's equation" occurs actually in the papers of Euler quoted on p. 84.

[2]This notation seems to be more consistent than the notation $d^2x/dt^2$ used by Kirchhoff and sometimes found in text books.

The equations that determine the acceleration vector are:

$$(1) \qquad \frac{\partial^2 x}{\partial t^2}\frac{\partial x}{\partial a} + \frac{\partial^2 y}{\partial t^2}\frac{\partial y}{\partial a} + \frac{\partial^2 z}{\partial t^2}\frac{\partial z}{\partial a} = -\frac{\partial}{\partial a}\frac{U+p}{\rho}$$

and two corresponding equations to be formed with respect to $b$ and $c$, instead of $a$. Eq. (1) follows from (11.2) if the component equations of (11.2) are multiplied by $\partial x/\partial a$, $\partial y/\partial a$, $\partial z/\partial a$, respectively and added, provided the fluid is incompressible ($\rho =$ const) and the external force has a potential $U$ ($\mathbf{F} = -\operatorname{grad} U$). Equations (1) are non-linear, but so are Euler's equations; if Lagrange's equations were linear they would indeed present a tremendous advantage.

The derivative $\partial/\partial a$ means of course that $b$, $c$ and also the fourth variable $t$ are to be kept constant; we have e.g.

$$\frac{\partial U}{\partial a} = \frac{\partial U}{\partial x}\frac{\partial x}{\partial a} + \frac{\partial U}{\partial y}\frac{\partial y}{\partial a} + \frac{\partial U}{\partial z}\frac{\partial z}{\partial a}.$$

Note, that (1) contains no other differentiations than with respect to the four independent variables $a$, $b$, $c$, $t$ and that they apply only to the four dependent variables $x$, $y$, $z$, $p$ (and to the given function $U$).

The three equations (1) must be supplemented by the condition of incompressibility. Now the ratio of the instantaneous volume $d\tau$ to the initial volume $d\tau_0$ is (as in any point transformation) given by the functional determinant (Jacobian)

$$(2) \qquad \frac{d\tau}{d\tau_0} = \begin{vmatrix} \dfrac{\partial x}{\partial x_0}, & \dfrac{\partial x}{\partial y_0}, & \dfrac{\partial x}{\partial z_0} \\[2ex] \dfrac{\partial y}{\partial x_0}, & \dfrac{\partial y}{\partial y_0}, & \dfrac{\partial y}{\partial z_0} \\[2ex] \dfrac{\partial z}{\partial x_0}, & \dfrac{\partial z}{\partial y_0}, & \dfrac{\partial z}{\partial z_0} \end{vmatrix}$$

or, with the usual abbreviation, by

$$(2a) \qquad \frac{d\tau}{d\tau_0} = \frac{\partial(x,\,y,\,z)}{\partial(x_0,\,y_0,\,z_0)}.$$

The $x_0$, $y_0$, $z_0$ can be expressed in terms of the tags $a$, $b$, $c$; according to the multiplication theorem for determinants the right member of (2) then becomes the product of the two functional determinants

$$\frac{\partial(x,\,y,\,z)}{\partial(a,\,b,\,c)} \qquad \text{and} \qquad \frac{\partial(a,\,b,\,c)}{\partial(x_0,\,y_0,\,z_0)}.$$

The last determinant is evidently independent of the time $t$. The invariability of (2) in course of time, i.e. the condition of incompressibility, can thus be written in the form:

$$(3) \qquad \frac{\partial D}{\partial t} = 0, \qquad D = \frac{\partial(x, y, z)}{\partial(a, b, c)}.$$

It contains only differentiations of the three dependent variables $x$, $y$, $z$ with respect to the four independent variables $a$, $b$, $c$, $t$.

The set (1) and Eq. (3) are Lagrange's equations for incompressible fluids.

The corresponding relations for compressible fluids are obtained from (1), if one replaces

$$(3a) \qquad \frac{p}{\rho} \quad \text{by} \quad \mathcal{P} = \int \frac{dp}{\rho} \quad \text{and} \quad \frac{U}{\rho} \quad \text{by} \quad V,$$

where $V$ is the potential per unit of mass of the external force. In addition, (3) must be replaced by

$$(3b) \qquad \frac{\partial}{\partial t}(\rho D) = 0.$$

Kirchhoff[3] used Lagrange's equations in a rigorous proof of Helmholtz's theorems on vortex conservation and was therefore able to avoid the use of formula (18.8), which we had to use, since our proof starts from Euler's equations.

## 35. Stokes's Resistance Law

On p. 78 we discussed the very slow ("creeping") motion of a viscous incompressible fluid as an example for a flow the dynamics of which is almost exclusively determined by viscous forces; the next step would be neglecting the inertia forces altogether. According to the value of the Reynolds number (16.8)

$$R = \frac{va}{\nu}, \qquad \nu = \frac{\mu}{\rho},$$

we decide when omission of the inertia terms is justifiable, the condition being

$$(1) \qquad R \ll 1.$$

The problem which interests us here,—it is usually connected with the name of Stokes—is the steady motion of a sphere through a liquid. Let

---

[3]Vorlesungen über Mechanik, Chap. XIV, 1.

$a$ be the radius of the sphere and $v$ its velocity which has become uniform after a short initial period. The motion proceeds under the influence of an external force such as the difference of gravity and buoyancy, possibly an electric field, etc. Condition (1) can not only be satisfied by a large value of the kinematic viscosity $v$, but also by sufficiently small values of $a$ or $v$. Instead of moving the sphere through the fluid one may as well keep the sphere at rest and have the fluid flow past the sphere (inverse flow); the resultant pressure force acting on the sphere is then equal to the resistance force encountered by the sphere in the direct flow.

Stokes resistance law plays an important role in the evaluation of certain fundamental experiments in physics. Einstein made use of it in his theory of the Brownian motion which leads to a determination of Loschmidt's number (cf. p. 53). Later it has become important again in Millikan's determination of the electronic charge.

We consider the steady state of the motion. The weight of the sphere is then balanced by the resultant friction forces and cannot appear in the equations or boundary conditions. Adopting now the viewpoint of the inverse motion, we take as a starting point the *equilibrium conditions* for a viscous fluid (10.8a, b), that is, the Navier-Stokes equations with omission of the inertia terms. The body force $\mathbf{F}$ of gravity acting on the fluid can be disregarded. In other words, we have now a kinematic problem defined by a system of equations, which reads in Cartesian coordinates:

(2) $$\frac{\partial p}{\partial x} = \mu\nabla^2 u, \qquad \frac{\partial p}{\partial y} = \mu\nabla^2 v, \qquad \frac{\partial p}{\partial z} = \mu\nabla^2 w;$$

it is to be supplemented by the condition of incompressibility

(2a) $$\frac{\partial u}{\partial x} + \frac{\partial v}{\partial y} + \frac{\partial w}{\partial z} = 0$$

and by the boundary conditions of the problem. Let the origin coincide with the center of the sphere, the $z$-axis pointing in the direction opposite to the motion of the sphere (the fluid flows past the sphere in the direction of the positive $z$-axis). From (2) and (2a) we have directly

(3) $$\nabla^2 p = 0.$$

We now introduce spherical coordinates $r$, $\vartheta$, $\varphi$ referring again to the center of the sphere, so that e.g.

(4) $$\cos\vartheta = \frac{z}{r}.$$

The spherically symmetric solution of (3) is, according to (19.22),

(5) $$p = \frac{A_0}{r}.$$

but our problem is not spherically symmetric, the $z$-axis being the preferred direction in which the fluid moves toward the sphere. We consider therefore the expression

(5a)
$$p = A_1 \frac{\partial}{\partial z} \frac{1}{r},$$

which obviously is also a solution of (3). On carrying out the differentiation in (5a) and replacing $A_1$ by $A$, we obtain, because of (4),

(5b)
$$p = -\frac{A}{r^2} \cos \vartheta.$$

A much more general solution of (3) is obtained in the form of the series

(5c)
$$p = \sum_{m=0}^{\infty} A_n \frac{\partial^n}{\partial z^n} \frac{1}{r},$$

the first two terms of which coincide with (5) and (5a) respectively. Eq. (5c) represents an expansion in terms of negative powers of $r$, the coefficients being the spherical harmonics $P_n (\cos \vartheta)$. Without giving here the general definition of spherical harmonics, we note that the factors 1 and $\cos \vartheta$ appearing in (5) and (5a) are spherical harmonics in the usual normalization: $P_0 (\cos \vartheta) = 1$, $P_1 (\cos \vartheta) = \cos \vartheta$, and that the spherical harmonics with higher subscripts are polynomials in $\cos \vartheta$ of higher order. (Legendre's polynomials). Since in our boundary conditions [(12a, b), see below] only the first power of $\cos \vartheta$ appears, we conclude that in (5c) only the term with $n = 1$ lends itself to our purpose; when we put $A_0 = A_2 = A_3 = \cdots = 0$, the series (5c) reduces to our trial solution (5b).

We first have to write equations (2) in their legitimate vectorial form. This is done by means of Eq. (3.10a) which supplies, because of div $\mathbf{v} = 0$,

(6)
$$\mu \text{ curl curl } \mathbf{v} = -\text{grad } p$$

instead of (2). It is now easy to pass over to spherical coordinates. We denote the components of the velocity vector $\mathbf{v}$ taken in direction $r, \vartheta, \varphi$ by $v_r, v_\vartheta, v_\varphi$, and observe that the $\varphi$-component must vanish because of the axial symmetry of the problem; also, all differential quotients with respect to $\varphi$ must vanish, in other words,

(7)
$$v_\varphi = 0, \qquad \frac{\partial}{\partial \varphi} = 0.$$

The $r$-component of Eq. (6) is, according to the solution of problem I.3, or, according to Appendix III and IV,

(8)
$$\frac{\mu}{r^2} \left[ \frac{\partial^2 (r^2 v_r)}{\partial r^2} + \frac{1}{\sin \vartheta} \frac{\partial}{\partial \vartheta} \left( \sin \vartheta \frac{\partial v_r}{\partial \vartheta} \right) \right] = \frac{\partial p}{\partial r}.$$

Here we assume the $r$ and $\vartheta$ dependence of $v_r$ in the form

(8a) $$rv_r = R\cos\vartheta,$$

where $R$ denotes a function dependent on $r$ alone. Substituting in (8) and expressing $p$ according to (5a), we obtain, after dropping the factor $\cos\vartheta$, the ordinary differential equation for $R$

$$\frac{\mu}{r^2}\left(\frac{d^2}{dr^2}(rR) - \frac{2}{r}R\right) = \frac{2A}{r^3},$$

which may also be written

(9) $$r^2R'' + 2rR' - 2R = \frac{2A}{\mu}.$$

A particular integral of (9) is $R = -A/\mu$ and the general integral of the associated homogeneous equation

$$R = Br + \frac{C}{r^2},$$

hence the general integral of (9) is according to a well known theorem

$$R = \frac{-A}{\mu} + Br + \frac{C}{r^2}$$

and the velocity component $v_r$ is found

(10) $$v_r = \left(-\frac{A}{\mu r} + B + \frac{C}{r^3}\right)\cos\vartheta.$$

The component $v_\vartheta$ may now be found from the condition of incompressibility (see solution of problem I.3) the form of which suggests a trial separation $v_\vartheta = R_1\sin\vartheta$, where again $R_1$ depends on $r$ alone. One finds easily

$$R_1 = \frac{A}{2\mu r} - B + \frac{C}{2r^3},$$

hence

(11) $$v_\vartheta = \left(\frac{A}{2\mu r} - B + \frac{C}{2r^3}\right)\sin\vartheta.$$

The constants of integration $A$, $B$, $C$ must be determined so as to satisfy the boundary conditions of the problem. If $q$ denotes the undisturbed flow velocity at infinity, the boundary conditions are

(12a) $$r = a : v_r = v_\vartheta = 0$$

(12b) $$r = \infty : v_r = q\cos\vartheta, \qquad v_\vartheta = -q\sin\vartheta.$$

Eq. (12b) is fulfilled if we put $B = q$; then Eqs. (12a) can be solved for $A$ and $C$ and give

$$\frac{A}{\mu} = \frac{3q}{2}\, a, \qquad C = \frac{q}{2}\, a^3;$$

hence we obtain finally

(13)
$$v_r = q\left(1 - \frac{3}{2}\frac{a}{r} + \frac{1}{2}\frac{a^3}{r^3}\right)\cos\vartheta$$

$$v_\vartheta = q\left(-1 + \frac{3}{4}\frac{a}{r} + \frac{1}{4}\frac{a^3}{r^3}\right)\sin\vartheta$$

and by (5b)

(13a)
$$p = -\frac{3}{2}\,\mu q a\,\frac{\cos\vartheta}{r^2}.$$

The result shows that the choice of the special solution (5b) in place of the more general one (5c) was in order. It is true that we have not proved that (13) and (13a) are the only possible solutions of our problem. On the other hand we are convinced that every problem of mathematical physics that has been correctly formulated can have only *one* solution.

We have yet to derive the formula for the *resistance* experienced by the *moving* sphere or, which is the same, for the pressure force acting on the sphere at *rest*; it is composed of the following two parts: the component of the hydrodynamic pressure force in the direction of the positive $z$-axis and the contribution of the viscous pressure force in the same direction.

The *hydrodynamic* pressure acts *radially*. According to (13a) its component in the positive $z$-direction is at the surface of the sphere, that is for $r = a$

$$-p\cos\vartheta = \frac{3}{2}\,\mu\,\frac{q}{a}\cos^2\vartheta.$$

Integration over the sphere whose surface element is $2\pi a^2 \sin\vartheta d\vartheta$ gives

(14)
$$3\pi\mu a q \int_0^\pi \cos^2\vartheta \sin\vartheta\, d\vartheta = 3\pi\mu a q \int_{-1}^{+1} x^2\, dx = 2\pi\mu a q.$$

To calculate the *viscous pressures*, we start from Eq. (10.2) which correlates the tensors $p_{ik}$ and $\dot{\epsilon}_{ik}$ and is valid in any orthogonal coordinate system, hence also in our polar coordinates. In the present notation we have

(15)
$$p_{rr} = -2\mu\dot{\epsilon}_{rr}, \qquad p_{r\vartheta} = -2\mu\dot{\epsilon}_{r\vartheta}.$$

The displacements $\epsilon$ have been defined in curvilinear coordinates in Eqs. (4.26) and (4.28). The rates of displacement $\dot{\epsilon}$ are obtained from these equations if the $\delta q_1$, $\delta q_2$, $\delta q_3$ are replaced by the components of the velocity

$$v_r, v_\vartheta, v_\varphi, \qquad (v_\varphi = 0)$$

and the $p_1$, $p_2$, $p_3$ by $r$, $\vartheta$, $\varphi$. Since $ds^2 = dr^2 + r^2 d\vartheta^2 + r^2 \sin^2 \varphi \, d\varphi^2$, one has $g_1 = 1$, $g_2 = r$, $g_3 = r \sin \vartheta$. Hence (4.26) yields

(16) $$\dot{\epsilon}_{rr} = \frac{\partial v_r}{\partial r}$$

and (4.28)

(17) $$2\dot{\epsilon}_{r\vartheta} = \frac{1}{r}\frac{\partial v_r}{\partial \vartheta} + \frac{\partial v_\vartheta}{\partial r} - \frac{1}{r} v_\vartheta .$$

When we substitute for the velocity components according to (13), only the middle term of (17) gives a non-vanishing contribution for $r = a$, so that

$$\dot{\epsilon}_{rr} = 0, \qquad \dot{\epsilon}_{r\vartheta} = -\frac{3}{4}\frac{q}{a} \sin \vartheta.$$

According to (15), the friction pressures are then

(18) $$p_{rr} = 0, \qquad p_{r\vartheta} = \frac{3}{2}\frac{\mu q}{a} \sin \vartheta.$$

The component of $p_{r\vartheta}$ in the positive $z$-direction is

$$p_{r\vartheta} \sin \vartheta = \frac{3}{2}\frac{\mu q}{a} \sin^2 \vartheta$$

which must be integrated over the sphere:

(19) $$3\pi\mu a q \int_0^\pi \sin^3 \vartheta \, d\vartheta = 3\pi\mu a q \int_{-1}^{+1} (1 - x^2) \, dx = 4\pi\mu a q.$$

This together with the contribution (19) gives Stokes's famous resistance formula

(20) $$D = 6\pi\mu a q.$$

Analogous to the problem of the sphere in uniform translatory motion $q$ is the problem of a sphere rotating uniformly, with angular velocity $\omega$. In this case the opposing moment[4] (see problem VII.4) is

(21) $$M = 8\pi\mu a^3 \omega.$$

---

[4]Cf. Kirchhoff, Vorlesungen über Mechanik, Chap. XXVI, 3.

These results are valid for liquids and gases at ordinary pressures. For gases at low pressures one has to add a correction term to Eq. (20) which is due to Cunningham. In the corrected formula the right member of (20) appears with the denominator

$$(22) \qquad 1 + \alpha \frac{l}{a},$$

where $l$ is the *mean free path* of the gas molecules, which increases with decreasing pressure; the physical interpretation of the numerical coefficient $\alpha$ is somewhat problematical.[5] Cunningham's correction was of great importance in the determination of the electronic charge according to Millikan's method.

To introduce the concept of the mean free path means to leave behind the framework of a continuum theory and enter the domain of statistical mechanics. The continuum theory which has led us to Stokes' formula is only valid on a scale that is large in comparison with $l$. In our case, the scale is set by the radius $a$. Cunningham's correction term is subject to the same condition if it is to be considered a correction: this is no longer the case if $l \sim a$, as one sees directly from (22), and a comprehensive treatment of the problem is then no longer feasible.

Another limitation of our considerations is that they are only good for *small* Reynolds numbers [cf. inequality (1)]. When $R$ increases, the inertia terms that have been neglected so far gain in importance; the laminar character of the flow, however, is still maintained, disregarding some vortices in the wake. The resistance law, which was linear in $q$ for Stokes's case now assumes by and by a quadratic character. Full turbulence is only reached at a certain $R = R_{\text{crit}}$ at which, in addition to the wake, the entire flow field in the neighborhood of the sphere becomes turbulent. At this point the resistance law changes its character suddenly.

Let us now have a critical look at the omission of the inertia terms in the preceding analysis. According to (5b), we have the following asymptotic behavior at $r \to \infty$:

$$p \sim \frac{1}{r^2}, \qquad \text{hence grad } p \sim \frac{1}{r^3}.$$

The terms $\nabla^2 u$, $\nabla^2 v$, $\nabla^2 w$ retained in Eq. (2) have the same order of magnitude as grad $p$. The inertia terms which we neglected vanish at $r \to \infty$ partly with the third, but partly with a smaller power of $1/r$. The latter happens in the case of those inertia terms which read in Cartesian components $u$, $v$, $w$

---

[5] Th. Sexl, Zur gaskinetischen Begründung der Stokes-Cunninghamschen Formel, Ann. Physik. 81 (1926) 855.

(23)  $$\rho w \frac{\partial u}{\partial z}, \qquad \rho w \frac{\partial v}{\partial z}, \qquad \rho w \frac{\partial w}{\partial z}.$$

When these terms are calculated from (13) they are found to vanish only as $1/r^2$. It is true that the Navier-Stokes equations can be rewritten in such a way that the Reynolds number $R = qa/\nu$ appears in these terms (introduce the dimensionless coordinates $x/a$, etc. and suitably rearrange the factors $\rho$, $\mu$, and $a$, cf. 16). But although $R$ is supposed to be small, this cannot change the fact that the ratio of the neglected terms (23) to the terms retained becomes arbitrarily large if $r \to \infty$.

On account of this Oseen devised a treatment of the problem that retains the terms (23) in principle, but replaces them by their first order approximation[6]

(23a)  $$\rho q \frac{\partial u}{\partial z}, \qquad \rho q \frac{\partial v}{\partial z}, \qquad \rho q \frac{\partial w}{\partial z}.$$

The integration of the equations so augmented is by no means simple and forms one of the principal parts of Oseen's book quoted on p. 210. In the light of his result our preceding formulas appear as the first term of a power series expansion in terms of the Reynolds number $R$. In particular, the resistance law (20) has the form

(24)  $$W = 6\pi\mu a q \left(1 + \frac{3}{8} R\right),$$

when the first order term is retained.

Fortunately the validity of Stokes result which has found such wide acceptance in experimental and theoretical physics is not perceptibly affected by this correction; this is due to condition (1), the fulfillment of which we have assumed throughout our discussion of laminar motion. On account of Eq. (24) one can now comprehend why the transition from the linear to the quadratic resistance law mentioned before does not proceed suddenly but gradually. The onset of full turbulence, however, at $R = R_{\text{crit}}$ cannot be understood in this way.

## 36. The Hydrodynamic Theory of Lubrication

Lubrication today is essentially a problem of "creeping" motion. In older text books of applied mechanics the friction between journal and bearing was treated as a problem of *dry* friction. The arguments put forth had nothing to do with hydrodynamics and are only mentioned here to point out the contrast to the later developments.

---

[6] Thus an error is introduced at $r = a$, where actually $w = 0$ and not $= q$, and the error for $r \to \infty$ is diminished but that should not imply that Oseen's calculation leaves anything to be desired in the way of mathematical rigor.

In order to apply the ordinary concepts of friction theory developed in Vol. I, 14[7] one must assume direct contact between journal and bearing. In Fig. 58 the journal pressure $P$, which is referred to the unit of length of the bearing, has been drawn vertically downward. This should not imply that it is actually vertical. In general it is composed of the gravity load, possible components of the driving force and inertia effects; it will therefore be inclined to the vertical.

At the point of contact $B$ the acting forces are the friction $R$, which is opposite to the sense of rotation, and the normal pressure $N$. The re-

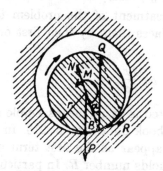 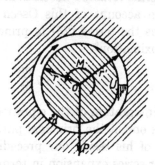

FIG. 58 (*left*). Relative position of journal and bearing according to the assumption of dry friction.

FIG. 59 (*right*). Fluid friction in the case of the centered journal.

sultant of the two, $Q$, forms the angle of friction $\alpha$ with $N$ ($\alpha$ means here the kinetic angle of friction, defined by the coefficient of kinetic friction $f$ according to $f = \tan \alpha$). The equilibrium condition requires that $Q$ and $P$ are antiparallel and equal. The remaining couple $QP$ then represents the moment of friction about the centerline of the journal which is balanced by an opposite driving moment. If $r$ denotes the radius of the journal and $r \sin \alpha$ the arm of the couple, we have

(1)        $$M = Pr \sin \alpha.$$

According to this the moment of friction should be *independent of the circumferential velocity $U$* and *proportional to the journal pressure $P$*; the point of contact $B$ should be situated *behind* the vector of the journal pressure, in reference to the sense of rotation.

The Russian engineer and officer N. Petroff was the first to look at this problem from the point of view of the viscosity of the lubricant.[8] In

[7]See e.g. Synge and Griffith, *op. cit.*, Sec. 3.2.

[8]Cf. Petroff, N.  Friction in Machines and the Effect of the Lubricant (in Russian), Eng. Jour., St. Petersb. 1883, a German translation in Ostwalds Klassiker, No. 218 (Leipzig).  In the same volume also the following papers: Reynolds, On the Theory of Lubrication etc., Phil. Trans. Roy. Soc. Vol. 177 pt. I, 1886, or Papers, Cambridge 1901,

his conception the journal is exactly centered in the bearing so that the surrounding lubricant forms a layer of uniform thickness $d$ (see Fig. 59). Throughout this layer the velocity drops from its value $U$ at the journal to the value zero at the bearing in an approximately linear way. The tangential friction pressure at the bearing is then $\mu U/d$. By summation over the circumference one obtains the moment of friction per unit of length

$$(2) \qquad\qquad M = 2\pi\mu \, \frac{U}{d} \, r^2.$$

This dependence is inverse to that of (1), since we have now *proportionality with U and independence of P*; also, the factor $\mu$ points to the lubricant and not to the materials of journal and bearing like $\sin \alpha$ in (1).

The equilibrium conditions of statics, however, are not satisfied by (2) since the journal pressure $P$ can never be balanced by the friction forces. Reynolds came therefore to the conclusion that the actual position of the journal in the bearing is not the central one; it could only be central for infinitely large circumferential velocity $U$. In addition Reynolds showed that the point of smallest layer thickness $h$ is not behind $P$ but *in front* of it. Unfortunately, Reynolds formulas are difficult to understand, since his calculations refer to bearings with only partially encased journals, as of railroad cars.

Fully encased journals with complete peripheral lubrication, the most desirable type for high speed electrical machinery, lend themselves to a comprehensive analysis, (cf. the third of the papers quoted in the footnote. Only a little later and under the influence of this paper Mitchell carried out his construction of the particular type of step bearing that has been named after him and has become important for heavy thrust bearings in turbines with vertical shafts.)

The essential point is that the journal is displaced in the direction normal to the journal pressure while it runs (see Fig. 60). The foundation of the theory of the *bearing* is obtained by an approximate calculation of the flow pattern of the *fluid wedge* in Fig. 61, where we neglect the $y$-component $v$ of the velocity in comparison with the $x$-component $u$; $w$ is exactly zero, since the problem is considered as two-dimensional. According to the fundamental equations of "creeping" motion (35.2) and (35.2a) we then have

Vol. II; Sommerfeld, Zur hydrodyn. Theorie der Schmiermittelreibung, Z. Math. Phys., 50, 1904. Mitchell, A. G. M. The Lubrication of Plane Surfaces, Z. Math. Phys., 52, 1905; Sommerfeld, Zur Theorie der Schmiermittelreibung, Z. techn. Phys., 2, 1921.— More recent literature e.g. in Hersey, M. D., Theory of Lubrication, New York 1938, Wiley.

$$p = p(x), \qquad u = u(y), \qquad \frac{dp}{dx} = \mu \frac{d^2u}{dy^2},$$

or

$$u = \frac{1}{2\mu} \frac{dp}{dx} y^2 + ay + b;$$

this yields because of the boundary condition $u = U$ for $y = 0$ (journal) and $u = 0$ for $y = h$ (bearing)

$$(3) \qquad u = U\left(1 - \frac{y}{h}\right) - \frac{1}{2\mu} \frac{dp}{dx} yh\left(1 - \frac{y}{h}\right).$$

The flux through the cross-section is

$$\int_0^h u\, dy = \frac{h}{2} U - \frac{h^3}{12\mu} \frac{dp}{dx}.$$

Although $h$ is variable in Fig. 61, the flux must be the same for all cross-sections (condition of incompressibility!). Thus, if $h_0$ denotes an assumed cross-section at which $dp/dx = 0$, we have

$$(4) \qquad \frac{h^3}{6\mu} \frac{dp}{dx} = (h - h_0)U.$$

From (3) and (4) the friction pressure per unit of area of the boundary $y = 0$ is found

$$(4a) \qquad p_{yx} = \mu \frac{du}{dy} = -\mu \frac{4h - 3h_0}{h^2} U.$$

Now transfer these results from Fig. 61 to Fig. 60, introducing the following notations: $O$, $r$ and $O'$, $r'$ are center and radius of journal and bearing, respectively, the excentricity $e =$ distance $OO'$, $d = r' - r$ is

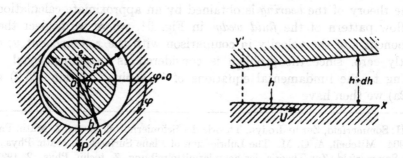

Fig. 60 (*left*). The journal is shifted normal to the direction of the journal pressure.
Fig. 61 (*right*). Simplified scheme of a lubricant film of variable thickness.

the play, $A$ is a variable point at the bearing surface, characterized by the angle $\varphi$ whose vertex is at $O$. We have correspondingly

$$OA = r + h, \; O'A = r', \qquad r + h = e \cos \varphi + r'.$$

The last (approximate) equation, obtained by projecting the segments $OO'$ and $O'A$ on the line $OA$, can be written in the form

(5) $$h = d + e \cos \varphi.$$

This meaning of $h$ is now introduced into Eqs. (4) and (4a) in order to adapt[9] them to the conditions of Fig. 60:

(6) $$\frac{1}{r}\frac{dp}{d\varphi} = 6\mu U\left(\frac{1}{h^2} - \frac{h_0}{h^3}\right),$$

(6a) $$p_{r\varphi} = -\mu U\left(\frac{4}{h} - \frac{3h_0}{h^2}\right).$$

By integration with respect to $\varphi$ between the limits 0 and $2\pi$ one has

(7) $$p(2\pi) - p(0) = 6\mu r U(J_2 - h_0 J_3),$$

(7a) $$M = -\mu r^2 U(4J_1 - 3h_0 J_2).$$

As before, $M$ stands for the friction moment transferred to the journal per unit of length; the $J$'s serve as abbreviations for

(8)
$$J_1 = \int_0^{2\pi} \frac{d\varphi}{d + e \cos \varphi} = 2\pi(d^2 - e^2)^{-1/2},$$

$$J_2 = \int_0^{2\pi} \frac{d\varphi}{(d + e \cos \varphi)^2} = 2\pi d(d^2 - e^2)^{-3/2},$$

$$J_3 = \int_0^{2\pi} \frac{d\varphi}{(d + e \cos \varphi)^3} = 2\pi\left(d^2 + \frac{1}{2}e^2\right)(d^2 - e^2)^{-5/2}.$$

If the value given for the integral $J_1$ is known, the values of $J_2$ and $J_3$ can be found according to

$$J_2 = -\frac{\partial J_1}{\partial d}, \qquad J_3 = -\frac{1}{2}\frac{\partial J_2}{\partial d}.$$

Since $p$ must have the period $2\pi$, one concludes from (7) and (8)

(9) $$h_0 = \frac{J_2}{J_3} = d\,\frac{d^2 - e^2}{d^2 + \frac{1}{2}e^2}.$$

---

[9]It would not have been difficult to derive these equations directly in polar coordinates instead of in coordinates $x, y, z$ as we did it.

With this value of $h_0$ the moment becomes

(10)
$$M = - \frac{2\pi\mu r^2 U}{\sqrt{d^2 - e^2}} \frac{d^2 + 2e^2}{d^2 + \frac{1}{2} e^2} .$$

Putting in analogy to (5)

(11)
$$h_0 = d + e \cos \varphi_0 ,$$

one obtains from (9)

(11a)
$$\cos \varphi_0 = - \frac{3}{2} \frac{de}{d^2 + \frac{1}{2} e^2} .$$

If $e = 0$ (centered position of the bearing), we have

$$\cos \varphi_0 = 0, \qquad \varphi_0 = \frac{\pi}{2} \quad \text{or} \quad \frac{3\pi}{2} ;$$

on the other hand, we obtain in the case of the largest possible excentricity $e = d$

$$\cos \varphi_0 = -1, \qquad \varphi_0 = \pi.$$

Now, the definition of $h_0$ according to (6) has been

$$\frac{dp}{d\varphi} = 0 \quad \text{for} \quad h = h_0 ,$$

therefore the two values $\varphi_0$ belonging to $h_0$ determine the points of minimum and maximum pressure. In the centered position these points are diametrically opposite to each other. At maximum excentricity, that is to say, if contact is made between journal and bearing, they coincide at the point of contact.

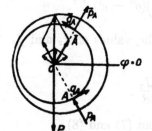

FIG. 62. Equilibrium between the hydrodynamic pressures $p$, $q$, and the bearing pressure $P$.

Our next task is to relate the pressure distribution to the direction of the vector of the journal pressure.

We notice that the pressure gradient $dp/d\varphi$ depends only on $h$ [see (6)], and is therefore an *even* function of $\varphi$, as $h$ is itself, when we count $\varphi$ from the zero line $OO'$. The pressure itself is consequently an *odd* function of $\varphi$, say $p_1$, plus an integration constant, say $p_0$. In forming the sum total of the pressure forces, the contribution of the *constant* pressures $p_0$ obviously vanishes. Denoting by $p_A$ and $p_{\bar{A}}$ the pressure values $p_1$ at two symmetrically situated points relative to $\varphi = 0$, we find the resultant action on the journal indicated in the larger of the two force parallelograms in Fig. 62: it is normal to $OO'$.

Applying the same reasoning to Eq. (6a), we see that the *friction pressures* (shear stresses) at the journal surface are *even* functions of $\varphi$; they are indicated by $q_A$, $q_{\bar{A}}$ in the figure (note, however, their tangential direction opposite to the sense of increasing $\varphi$). By construction of the corresponding parallelogram of forces at $O$ we see that the friction pressures at $A$ and $A'$ give also a resultant *normal* to $OO'$. Thus the overall resultant of the hydrodynamic forces acting on the journal is normal to $OO'$. Since this resultant balances the journal pressure $P$, it must be opposite to $P$, in other words, *the displacement of the journal is normal to the journal pressure* $P$, a result anticipated in Fig. 60.

The *value* of the displacement $e$ can now be determined. One has only to form the sum of the resultants of $p$ and $q$ and balance it against the prescribed value of $P$:

$$(12) \qquad \frac{P}{r} = \int p \sin \varphi \, d\varphi - \int q \cos \varphi \, d\varphi.$$

The first term can be transformed by partial integration; one of the two resulting terms, $[-p \cos \varphi]_0^{2\pi}$, vanishes because of the periodicity. Altogether (12) becomes

$$(12a) \qquad \frac{P}{r} = \int \left( \frac{dp}{d\varphi} - q \right) \cos \varphi \, d\varphi.$$

Here the term $q$ may be omitted since $dp/d\varphi$ outweighs $q = p_{r,\varphi}$ by a full order of magnitude of $r/h$ [cf. Eqs. (6) and (6a)]. Hence we obtain the simpler relation

$$(12b) \qquad \frac{P}{r} = 6\mu r U(J_4 - h_0 J_5)$$

where

$$J_4 = \int \frac{\cos \varphi \, d\varphi}{(d + e \cos \varphi)^2}, \qquad J_5 = \int \frac{\cos \varphi \, d\varphi}{(d + e \cos \varphi)^3}.$$

From (8) we find easily

$$J_4 = \frac{1}{e} (J_1 - d \cdot J_2), \qquad J_5 = \frac{1}{e} (J_2 - d \cdot J_3).$$

If this is substituted in (12b) and the value of $h_0$ taken from (9), an elementary calculation gives the result

(13)
$$\frac{P}{2\pi\mu U}\frac{d^2}{r^2} = \frac{3e}{\sqrt{d^2-e^2}}\frac{d^2}{d^2+\frac{1}{2}e^2}.$$

The excentricity $e$ or rather the ratio $e/d$ is here determined as a function of the left member of (13). This value of $e$, substituted in (10), leads to the determination of the moment $M$. In what follows we are only interested in the magnitude and not in the sign of the moment. Note, that a particularly simple expression for $M$ is obtained, if (10) is divided by (13):

(13a)
$$\frac{M}{P} = \frac{d^2+2e^2}{3e}.$$

In discussing this result, we first consider two limiting cases:
a) $U \to \infty$. From (13) we have $e = 0$, that is, the journal is centered; Eq. (10) yields immediately

(14)
$$M = 2\pi\mu r^2 \frac{U}{d},$$

which is Petroff's equation.
b) $U \to 0$. In this case, (13) requires $e = d$, that is, contact between journal and bearing. Eq. (13a) gives accordingly $M = Pd$ which we may also write in the form

(15)
$$M = rP \cdot \frac{d}{r}.$$

This becomes identical with our *formula* (1) *for dry friction* if we define the angle of friction in the present case by $\sin\alpha = d/r$.
c) We are also interested in the case where $M$ is a minimum for given journal pressure $P$ and given play $d$. When the differential quotient of the right member of (13a) with respect to $e$ is set equal to zero, one obtains

$$-\frac{d^2}{e^2} + 2 = 0, \qquad e = \frac{d}{\sqrt{2}}.$$

With this value of $e$ Eq. (13a) gives for the moment of friction

(16)
$$M = \frac{2\sqrt{2}}{3}Pd = 0.94\,Pd.$$

that is, only 6% less than the limit of $M$ for $U \to 0$, which has been determined in (15).

These results are collected in the diagram of Fig. 63. Here the dimensionless quantity $y = M/Pd$ has been plotted as ordinate over the abscissa $z$, which is the reciprocal of the left side of (13), viz.

(17) 
$$z = \frac{2\pi\mu U}{P} \frac{r^2}{d^2}.^{10}$$

The solid curve represents the content of the two Eqs. (10) and (13), that is to say, the result of the elimination of the displacement $e$. It is best computed from the following parametric representation which is identical with (13a) and (13) or (10) and (13):

(18)  $$y = \frac{1 + 2\beta^2}{3\beta}, \qquad z = \sqrt{1 - \beta^2} \frac{1 + \frac{1}{2}\beta^2}{3\beta}, \qquad \beta = \frac{e}{d}.$$

In the same figure formulas (1) and (2) have been plotted as broken lines; in the present variables they simply read $y = 1$ and $y = z$, the tangents

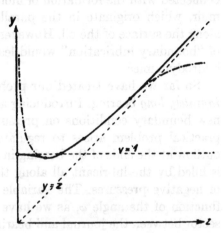

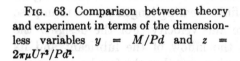

FIG. 63. Comparison between theory and experiment in terms of the dimensionless variables $y = M/Pd$ and $z = 2\pi\mu Ur^2/Pd^2$.

of the solid curve at $z = 0$ and $z = \infty$. The fact that our curve depends solely on the dimensionless quantity $z$ expresses a *similarity law* which is valid in the hydrodynamic theory of lubrication: No matter what particular values of the dimensions $r$ and $d$, the lubricant $\mu$, the speed $U$, and the load $P$ combine to give a certain $z$, it will always lead to the same $\beta = e/d$ and thus to the same value $y = M/Pd$ by (18).

When we now ask how well this theoretical result is confirmed by measurements on actual bearings, the answer is this: in the neighborhood of the minimum and to the right of it for not too large $z$-values we have not only qualitative but numerical agreement between theoretical and experimental values. One would then conclude that in this range the phenomenon of lubrication is determined entirely by the coefficient of viscosity $\mu$, and that the laminar character of the motion, which we have assumed throughout this analysis, actually occurs. But for large and for very small $z$-values deviations from the theoretical curve occur, as indicated by the

---

[10]This expression is sometimes called the Sommerfeld variable in lubrication literature.

dotted branches in the figure. Since in these regions the similarity law is no longer valid, the proper representation should consist of a family of curves $y$ vs. $z$ depending on the variation of one of the parameters $P$, $U$ etc.

The empirical curve does not rise nearly so steeply for large $z$-values as the theoretical curve, which has been interpreted to indicate that the motion of the lubricant is no longer laminar but turbulent.

For very small $z$-values the actual rise is very much steeper than it should be according to our theory. Here, qualities of the lubricant other than its viscosity become prevalent; they have been given the name of *oiliness* and refer to the adhesion of the lubricant to the metallic boundaries of the bearing. It is highly probable that the strength of adhesion be connected with the formation of monomolecular layers as studied by Langmuir, which originate in the parallel orientation of the chain molecules along the surface of the oil. However, a discussion of the physical problems of "boundary lubrication" would lead us far beyond the limits of classical hydrodynamics.

So far we have treated our problem as two-dimensional, assuming an *infinitely long* bearing. Introducing a finite bearing length means to impose new boundary conditions on pressure and velocity. For the designer the practical problem arises to regulate the inflow of the lubricant and its outflow at the ends of the bearing in such a way that the free bearing space is filled by the lubricant all along the bearing. There is also the difficulty of negative pressures. The variable part of $p$, denoted by $p_1$, is an *odd* function of the angle $\varphi$, as we have seen. While $p_1$ rises in the narrowing sector between the journal and bearing wall, it must drop again to negative values when the point of proximity is passed. There is thus the danger that the layer breaks in the region of minimum $p_1$, unless the constant part of $p_0$ of the total pressure $p$ is large enough to make $p = p_0 + p_1$ everywhere positive. The quantity $p_0$ which depends on the inflow and outflow conditions of the lubricant is in our theory an arbitrary parameter; its proper choice is left to the skill of the designer.

One cannot expect an idealizing theory to do better than to establish general rules in an eminently practical problem like ours. Yet these rules may be quite important for the practical design as is born out by the success of Mitchell's segmental bearings.

## 37. Riemann's Shock Waves. General Integration of Euler's Equations for a Compressible Fluid in One-dimensional Flow

We have so far successfully avoided dealing with the quadratic terms in the equations of hydrodynamics assuming either infinitesimal amplitudes (wave motion) or "creeping" flow, in which case those terms were dropped as part of the effect of inertia. In the present article, however, we shall finally attack the quadratic terms directly. Riemann[11] was the first to brave the challenge of this problem. Hugoniot, Hadamard and many others followed in his steps.

Riemann's method consists in transforming the *non-linear* hydro-dynamic equations by an interchange of the dependent and independent variables; this leads to *linear* equations in the case of one-dimensional states. Riemann then shows how to integrate the resulting linear differential equations, and his method has become the model for the general integration problem of linear partial differential equations of second order of the corresponding (hyperbolic) type. This will be taken up in Vol. VI.

We first present a preliminary review of the results of Riemann's investigation: two waves proceed in opposite directions from each point of the initial disturbance in a similar way as in d'Alembert's solution of the corresponding problem for infinitesimal amplitudes (see Fig. 16, 13). Let the pressure be given as function of the density, $p = \varphi(\rho)$, for example in the polytropic form of (7.1); then we have according to Riemann:[12] "The velocity of propagation of both waves relative to the gas is $\sqrt{\varphi'(\rho)}$, but in space it is increased by the velocity of the gas flow measured in the direction of the propagation of the wave. If it is supposed that $\varphi'(\rho)$ does not decrease with increasing $\rho$, an assumption that holds in reality, stronger compressions must advance with larger velocities; it follows that the waves of compression will diminish in breadth and must finally become compression shocks".

So much for an introduction. We start now with the transition from the non-linear to the linear equations. Disregarding gravity or other external forces we choose the $x$-axis parallel to the direction of motion. If differentiations are indicated by the usual subscript notation the original hydrodynamic equations for a one-dimensional gas read:

$$(1) \qquad \rho_t + u\rho_x + \rho u_x = 0,$$

$$(2) \qquad u_t + uu_x + \frac{1}{\rho} p_x = 0.$$

In (2) $p_x$ can be expressed by $\rho_x$ since according to the assumption $p = \varphi(\rho)$

$$(3) \qquad p_x = c^2 \rho_x , \qquad c^2 = \frac{dp}{d\rho} = \varphi'(\rho).$$

As in (13.5) $c$ is again the *sound velocity*; in the case of infinitesimal amplitude $c$ could be considered as a constant, but now $c$ is a function of $\rho$.

[11]Über die Fortpflanzung ebener Luftwellen von endlicher Schwingungsweite. Abh. d. Gött. Ges. d. Wiss. 1860. Also Ges. Werke, p. 156.

[12]From Riemann's abstract of his paper in Göttinger Nachr. 1859, No. 19.

Eq. (2) can be rewritten by the use of (3):

(4) $$u_t + uu_x + \frac{c^2}{\rho}\rho_x = 0.$$

Eqs. (1) and (4) are linear and homogeneous in the differentials of the independent variables $x$ and $t$, while these variables themselves do not occur in the equations. On the other hand, the quantities $\rho$ and $u$ and their differentials occur in non-linear form. We therefore introduce with Riemann the pair $x$, $t$ as dependent and the pair $\rho$, $u$ as independent variables by putting

$$x = x(\rho, u), \qquad t = t(\rho, u).$$

$$dx = x_\rho\, d\rho + x_u\, du, \qquad dt = t_\rho\, d\rho + t_u\, du.$$

If these expressions for $dx$ and $dt$ are introduced in the differentials

$$du = u_x\, dx + u_t\, dt \qquad \text{and} \qquad d\rho = \rho_x\, dx + \rho_t\, dt,$$

the following identities result:

(5) $$du = (u_x x_\rho + u_t t_\rho)\, d\rho + (u_x x_u + u_t t_u)\, du$$

(6) $$d\rho = (\rho_x x_\rho + \rho_t t_\rho)\, d\rho + (\rho_x x_u + \rho_t t_u)\, du.$$

Since $d\rho$ and $du$ are now independent we conclude from (5)

(7) $$\begin{aligned} u_x x_\rho + u_t t_\rho &= 0 \\ u_x x_u + u_t t_u &= 1 \end{aligned} \qquad \text{or} \qquad u_x = \frac{t_\rho}{\Delta}, \qquad u_t = -\frac{x_\rho}{\Delta},$$

where we assume that the functional determinant

(7a) $$\Delta = \begin{vmatrix} x_u & t_u \\ x_\rho & t_\rho \end{vmatrix}$$

does not vanish. In the same way we conclude from (6)

(8) $$\begin{aligned} \rho_x x_\rho + \rho_t t_\rho &= 1 \\ \rho_x x_u + \rho_t t_u &= 0 \end{aligned} \qquad \text{or} \qquad \rho_x = -\frac{t_u}{\Delta}, \qquad \rho_t = \frac{x_u}{\Delta}.$$

Eqs. (7) and (8) represent the formulas for $u_x$, $u_t$, $\rho_x$, $\rho_t$, to be used in transforming (1) and (4). Substituting in (1) and (4), we obtain the system

(9) $$x_u - ut_u + \rho t_\rho = 0, \qquad x_\rho - ut_\rho + \frac{c^2}{\rho}t_u = 0.$$

The second equation can be transformed as follows:

$$\frac{\partial}{\partial \rho}(x - ut) = \frac{-c^2}{\rho}\frac{\partial}{\partial u} t = \frac{\partial^2 V}{\partial \rho\, \partial u},$$

where we have introduced a new dependent variable $V = V(\rho, u)$. The variables $x$ and $t$ are connected with $V$ in the following way ($c$ depends only on $\rho$):

(10) $$x - ut = V_u, \qquad t = -\frac{\rho}{c^2} V_\rho,$$

so that the second equation (9) is identically fulfilled. The new variable $V$ is now introduced [13] in the first equation (9); this gives

(11) $$V_{uu} - \frac{\partial}{\partial \rho}\left(\frac{\rho^2}{c^2} V_\rho\right) = 0,$$

which is a *linear partial differential equation of second order* for the single unknown $V$.

This equation is simplified if we introduce instead of $\rho$ the new independent variable

(12) $$\xi = \int_0^\rho \frac{c}{\rho}\, d\rho; \qquad \frac{d\xi}{d\rho} = \frac{c}{\rho}.$$

Writing at the same time $\eta$ instead of $u$, we have the following relations:

$$\frac{\rho^2}{c^2} V_\rho = \frac{\rho^2}{c^2} V_\xi \frac{d\xi}{d\rho} = \frac{\rho}{c} V_\xi,$$

$$\frac{\partial}{\partial \rho}\left(\frac{\rho^2}{c^2} V_\rho\right) = \frac{c}{\rho}\frac{\partial}{\partial \xi}\left(\frac{\rho}{c} V_\xi\right) = V_{\xi\xi} + \frac{c}{\rho} V_\xi \frac{d}{d\xi}\frac{\rho}{c} = V_{\xi\xi} + V_\xi \frac{d}{d\rho}\frac{\rho}{c}.$$

We now assume the polytropic $p$, $\rho$-relation as in (7.1), such that

$$p = a\rho^n, \qquad \frac{dp}{d\rho} = c^2 = na\rho^{n-1}, \qquad \frac{c}{\rho} = \sqrt{na}\, \rho^{(n-3)/2}.$$

The relation between $\xi$ and $\rho$ and the expression $d(\rho/c)/d\rho$ is then

(12a) $$\xi = \frac{2\sqrt{na}}{n-1} \rho^{(n-1)/2}, \qquad \frac{d}{d\rho}\frac{\rho}{c} = \frac{3-n}{n-1}\frac{1}{\xi},$$

and our differential equation (11) reads in the new variables

(13) $$V_{\xi\xi} - V_{\eta\eta} + \frac{k}{\xi} V_\xi = 0,$$

---

[13]From the first and second equation (10) one has respectively $x_u - ut_u - t = V_{uu}$ and $\rho t_\rho + t = -\partial/\partial\rho(\rho^2 V_\rho/c^2)$.

where for brevity we have put

$$(14) \qquad k = \frac{3-n}{n-1}.$$

In the case $k = 0$, Eq. (13) reduces to the equation of the vibrating string, and we have d'Alembert's solution (13.11)

$$(15) \qquad V = V_0 = F(\xi + \eta) + G(\xi - \eta),$$

where $F$ and $G$ are *arbitrary* functions that have only to meet certain continuity conditions. More generally, we have the following lemma due to Bechert:[14] If $W$ is a solution of the equation

$$(16) \qquad W_{\xi\xi} - W_{\eta\eta} + \frac{l}{\xi} W_\xi = 0$$

then

$$(16a) \qquad V = \frac{1}{\xi} \frac{\partial W}{\partial \xi}$$

is a solution of the equation

$$(16b) \qquad V_{\xi\xi} - V_{\eta\eta} + \frac{l+2}{\xi} V_\xi = 0.$$

This can be shown simply by differentiating (16) partially with respect to $\xi$ and putting $W_\xi = \xi V$ according to (16a). One obtains then

$$\frac{\partial^2}{\partial \xi^2}(\xi V) - \frac{\partial^2}{\partial \eta^2}(\xi V) + l V_\xi = 0$$

which becomes identical with (16b) on carrying out the differentiations and dropping the factor $\xi$.

Since the general solution (15) of Eq. (13) is known in the case $k = 0$, *we have the general solution in the cases* $k = 2, 4, 6 \cdots$ in the following form:

$$(17) \qquad V_2 = \frac{1}{\xi} \frac{\partial V_0}{\partial \xi}, \quad V_4 = \frac{1}{\xi} \frac{\partial}{\partial \xi} \frac{1}{\xi} \frac{\partial V_0}{\partial \xi}, \quad V_6 = \frac{1}{\xi} \frac{\partial}{\partial \xi} \frac{1}{\xi} \frac{\partial}{\partial \xi} \frac{1}{\xi} \frac{\partial V_0}{\partial \xi}, \cdots$$

These elementary integrable cases may be written comprehensively by setting $k$ in (14) equal to an arbitrary even integer $2h$. The polytropic exponent (in the sense of p. 49) associated with $h$ is found from (14) as

$$(18) \qquad n = 1 + \frac{2}{2h+1}.$$

[14]K. Bechert, Zur Theorie ebener Störungen in reibungsfreien Gasen. Ann. d. Phys. 37 (1940) 89; 38 (1940) 1.

It is this rational form of $n$ which makes Bechert's result so interesting since it discloses a surprising thermodynamic side view.

For an adiabatic change of state of a gas we have according to p. 50

$$(18a) \qquad n = 1 + \frac{2}{f},$$

where $f$ is the number of degrees of freedom of the molecule. For diatomic gases (e.g. for air) $f$ equals 5, as already mentioned on p. 50. For monatomic gases $f$ equals 3. In this case there are only the three degrees of freedom of *translation* since those of rotation cannot be excited. In present day physics this follows directly from a quantum-theoretical argument, while Boltzmann had still to make the rather artificial assumption that the monatomic molecules were spheres and the diatomic molecules were ellipsoids of revolution, in order to understand the cases $f = 3$ and $f = 5$. In these two cases there is

$$f = 2h + 1 \begin{cases} h = 1 & \text{for monatomic gases} \\ h = 2 & \text{for diatomic gases} \end{cases}$$

Thus it turns out that just the physically most interesting cases are distinguished in that they can be integrated[15] by the elementary expressions $V_2$ and $V_4$ of Eq. (17).

It is difficult not to be impressed by what appears to be a "pre-established harmony" between mathematics and physics. An idea suggests itself that the performance of mathematics is adapted in some miraculous way to the exigencies of physics. Yet, the scope of the miracle is rather a limited one: For molecules consisting of three and more atoms such as $H_2O$ or $NH_4$ the number $f = 6$: three translatory degrees of freedom plus *three* of rotation. These gases do not belong to the integrable cases in this sense. On the other hand, the case $f = 1$, $k = 0$, which is the simplest integrable case, is not realized in nature; one would have to imagine a strictly "one-dimensional gas" whose atoms can only move in one direction and cannot rotate.

It would be a mistake to think that the integration becomes impossible unless $n$ is one of the rational numbers defined in (18). To show this we[16] write the sequence of equations in (17) once more in general form:

$$(19) \qquad V_{2h} = \left(\frac{\partial}{\partial \zeta}\right)^h V_0 \quad \text{with} \quad \zeta = \frac{\xi^2}{2}, \quad \xi = \sqrt{2\zeta}.$$

[15]Mr. Bechert kindly informed me that this peculiar behavior of the cases (18) had been noticed before by A. E. H. Love and F. B. Pidduck. Cf. Trans. R. Soc. London 222, 167 (1922).

[16]Similarly as in Bechert's second paper quoted on p. 266.

According to a well known extension of Cauchy's integral formula in the theory of complex functions it is possible to express the $h$-fold differential quotient in (19) by a complex integral:

$$(20) \qquad V_{2h} = \frac{1}{2\pi i} \frac{(-1)^h}{h!} \oint \frac{V_0(\sqrt{2w}, \eta)}{(w - \zeta)^{h+1}} \, dw.$$

Here $h$ is supposed to be an integer number, and integration refers to a closed path in the $w$-plane, surrounding the point $w = \zeta$. The argument of $V_0$ as it is written in (20) indicates that the original dependence on $\xi$ is to be replaced by that on $\sqrt{2w}$ while the dependence oh $\eta$ remains unchanged. However, in the representation (20) the fact that $h$ is an integer is unessential. One may replace $2h$ by the arbitrary quantity $k$ the specialization of which led us to $2h$, if one replaces at the same time

$$h + 1 \quad \text{by} \quad \frac{k}{2} + 1, \quad \frac{(-1)^h}{h!} \quad \text{by} \quad \frac{\exp\left(\frac{i\pi}{2} k\right)}{\Gamma\left(\frac{k}{2} + 1\right)}.$$

In this way one has instead of (20)[17]

$$(21) \qquad V_k = \frac{1}{2\pi i} \frac{\exp\left(\frac{i\pi}{2} k\right)}{\Gamma\left(\frac{k}{2} + 1\right)} \oint \frac{V_0(\sqrt{2w}, \eta)}{(w - \zeta)^{(k/2)+1}} \, dw.$$

Evidently (21) satisfies the differential equation in the same way as (20); it represents the general solution since it contains two arbitrary functions $F$ and $G$ in the expression for $V_0$ under the integral sign. Our statement that just the physically interesting cases of the monatomic and diatomic gases are *integrable*, should now be put more precisely: they are distinguished from the other cases by the elementary form of their general integral.

We have limited this discussion to Riemann's case of *one* spatial coordinate. The generalization to two or three variables[18] presents great difficulties. The differential equation which is the analogue of (13) in the general case has so far resisted all attempts at a general solution.

---

[17]The sign $\oint$ does not mean here a *single* loop about the point $w = \zeta$, since this would not be a closed path because of the branch points of $(w - \zeta)^{(k/2)+1}$ at $\zeta$ and at infinity; the path must now be a double loop about $w = \zeta$ and $w = \infty$. Such paths occur in theory of functions at various occasions.

[18]K. Bechert, Ann. Phys. **39** (1941) 169 and 357, **40** (1941) 207. In the first of these papers a class of cylindrical and spherical waves is treated, in the second friction and heat conduction is taken into account.

In the foregoing we have only dealt with the formal side of the problem. Although this presents enough interest, it is now time to turn to the *physical aspect* of the motion as it was described above in Riemann's summary. For this purpose we do not need the integral $V$ in its completeness, but go back to the definition of the variables $\xi$ and $\eta$ [see Eq. (12) and after]. We put with Riemann

(22)
$$r = \frac{1}{2}(\xi + \eta) = \frac{1}{2}\left(\int_0^\rho \frac{c}{\rho}\,d\rho + u\right),$$

$$s = \frac{1}{2}(\xi - \eta) = \frac{1}{2}\left(\int_0^\rho \frac{c}{\rho}\,d\rho - u\right),$$

and note that the assumption of a polytropic $p$, $\rho$-relation is immaterial for what follows. The quantities $r$ and $s$ represent in our particular case the *characteristics* of the differential equation (13) and play a decisive part in the general theory of differential equations of this type. (For further details see Vol. VI).

From (22) we calculate the derivatives

(23)
$$r_t = \frac{1}{2}\left(\frac{c}{\rho}\rho_t + u_t\right), \qquad r_x = \frac{1}{2}\left(\frac{c}{\rho}\rho_x + u_x\right)$$

(differentiations with respect to the original variables $x$ and $t$ are again represented by subscripts.) A glance at the expression

(24)
$$r_t + (u + c)r_x = \frac{c}{2\rho}\left\{\rho_t + (u + c)\rho_x + \frac{\rho}{c}u_t + \frac{\rho}{c}(u + c)u_x\right\}$$

shows that the right member vanishes because of the equations of motion (1) and (4). If we compute in the same way $s_t$ and $s_x$ from (22) and form the expression $s_t + (u - c)s_x$ we find again the right member zero as before. Both results may be indicated in the following way:

(25)
$$\frac{\partial}{\partial t}\begin{Bmatrix}r\\s\end{Bmatrix} + (u \pm c)\frac{\partial}{\partial x}\begin{Bmatrix}r\\s\end{Bmatrix} = 0.$$

If we form the total differentials $dr$ and $ds$ with respect to $x$ and $t$ and express the partial derivatives $r_t$ and $s_t$ by $r_x$ and $s_x$ respectively according to (25), we obtain

$$dr = r_x\,dx - (u + c)r_x\,dt \quad \text{and} \quad ds = s_x\,dx - (u - c)s_x\,dt.$$

These are differential identities between the original independent variables $x$ and $t$ and the variables $r$ and $s$, which are valid if $u$ and $\rho$ (of which $c$ is a known function) fulfill the differential equations (1) and (4). Consider now the point $x$ varying in course of time. The total derivatives of $r$ and $s$ with respect to $t$, if $x = x(t)$, are

$$(26) \qquad \frac{dr}{dt} = \left(\frac{dx}{dt} - u - c\right)r_x , \qquad \frac{ds}{dt} = \left(\frac{dx}{dt} - u + c\right)s_x .$$

These relations can be interpreted as follows: at a point (*not a particle*) that travels in such a way that

$$(27) \qquad\qquad \text{either} \quad \frac{dx}{dt} = u + c \qquad \text{or} \qquad \frac{dx}{dt} = u - c$$

the quantities $r$ or $s$ keep their values. In other words, a certain value of $r$ respectively $s$ travels with the variable velocity $u + c$ respectively $u - c$. As long as $|u| < c$, the first moves in the positive the second in the negative $x$-direction.

As an example take the case of an initial pressure disturbance restricted to a short interval of the $x$-axis. Compression waves issue in both directions from the interval if we assume an initial disturbance in the form of a pressure *increase*. However, the two wave profiles do not preserve their shape as in Fig. 16; the wave front becomes steeper since larger densities travel faster according to (27). Hence it is understood that the compression waves finally become compression *shocks*. At this point our calculation which disregards friction and heat conduction, ceases to be valid. The extensions of the theory that are necessary in the case of shocks have been investigated by Hugoniot and Hadamard.[19]

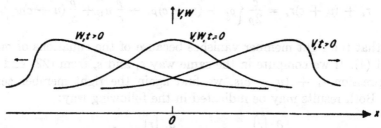

Fig. 64. The waves caused by an initial disturbance progress in both directions and are deformed in this process. The notations $V$ and $W$ of the figure have the same meaning as the parameters of the characteristics, $r$ and $s$, as functions of $x$ and $t$. In comparing Figs. 64 and 16, note that the central curve in the present figure represents the disturbances $V$ and $W$ *individually*, while the corresponding *sum* is represented in Fig. 16.

This may suffice for an exposition of Riemann's results which we quoted on p. 263. Fig. 64 has been taken from Bechert's first paper quoted

---

[19]Cf. the comprehensive representation by G. Zemplén in Enc. d. Math. Wiss. Vol. IV.3, Art. 19, which is to be supplemented by a more recent paper of R. Becker, Z. Phys. 8 (1922) 321.

on p. 266. It refers to the simplest case of a strictly "one-dimensional" gas (see p. 267) which shows, as Bechert points out, the characteristic traits of wave propagation in spite of the extreme idealization.

## 38. On Turbulence

With the present report which is divided in a number of sections we enter the most difficult part of hydrodynamics. Under (A) we have collected a series of *half empirical, half theoretical statements* about the properties of turbulent flow. Inasmuch as they are of a theoretical nature, they are based on the Stokes-Navier equations and the diffusion of momentum which these equations can be considered to express (see p. 76).

Under (B) we approach the fundamental question whether it is at all possible to explain the observed facts on the basis of the Navier-Stokes equations. In particular, one asks whether these equations give account of the instability of the laminar motion that sets in at a certain Reynolds number. Earlier attempts to settle this question by the method of small oscillations failed; it had been impossible to find a critical limit, and the laminar motion appeared to be stable at any value of $R$. The method of small oscillations had been applied to the fully developed laminar flow, that is to the Hagen-Poiseuille flow and to the Couette flow with the aim of investigating the stability of a flow in a pipe and between plates, respectively.

Considering these negative results, such experts in the field of turbulence as Th. v. Kármán and Sir Geoffrey Taylor have occasionally in their papers favored the idea that turbulence like gas theory can only be understood by statistical methods. The underlying physical fact is that a fluid is a system of very many (in principle infinitely many) degrees of freedom, all of which come into play at large Reynolds numbers. The motion becomes then so complicated that one can expect simple answers only to those questions that concern the *average* behavior of the moving particles. This is especially true for the limiting case of the so-called isotropic turbulence; some preliminary remarks according to hitherto unpublished work of C. F. v. Weizsäcker and W. Heisenberg will be made on this topic under (D) where we are concerned with what one may call fully developed *turbulent equilibrium*.

Previous to that we shall briefly discuss under (C) investigations of *the process of generation of laminar flow in its relation to turbulence*: the processes at the inlet in pipe flow experiments and the formation of Prandtl's boundary layer when the plate is set in motion. This was Prandtl's own view of the problem, and was soon taken up successfully by Tollmien, Tietjens, Görtler, and Schlichting, all of whom belong to Prandtl's school.

In these investigations the method of small oscillations is not applied to the velocity profile of the developed laminar flow, but to the intermediate profiles that occur while the laminar flow comes into existence. The significance of the velocity profile for the stability of the flow had been anticipated in much older investigations of Lord Rayleigh which, however, dealt only with inviscid fluids. The more recent investigations of Prandtl's group start from the complete Navier-Stokes equations (i.e. in their abbreviated form for *small* oscillations) and actually result in stability limits, which, to be sure, do not refer to the final laminar motion but to the way of its generation.

Under (E) we shall bring a mathematical example of a system of differential equations that lead to phenomena similar to those of turbulence in fluids. It is a non-linear system that resembles the Navier-Stokes equations in structure, but is much simpler than those. This example, which is due to I. M. Burgers, may strengthen our belief that turbulence does belong to classical hydrodynamics, although to a part of it that presents great difficulties to the mathematical attack.

## A. Some Properties of Turbulent Flow

Empirical data on the turbulent state in pipes and channels are very extensive, having been compiled by hydraulic engineers over a long time. In organizing this material two principles have to be observed: The flow must be referred to equal Reynolds numbers and the wall roughness must be taken into account. As to the second requirement we shall restrict ourselves to *smooth* walls so that we do not have to digress too far into technological details. "Smooth" is, of course, meant in the sense of the limit: wall roughness →0. Under these conditions one obtains the general result: *In the fully developed turbulent state the velocity distribution in pipes with geometrically similar cross-sections is similar*, or, the ratio of the velocity at a certain point to the velocity average over the entire cross-section is independent of the Reynolds number.

Fig. 65 shows this velocity distribution for a circular cross-section in

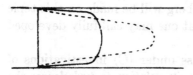

FIG. 65. The velocity profile in a circular pipe for the turbulent and laminar states of flow (solid and dotted lines respectively).

comparison with that of the Hagen-Poiseuille flow. In the latter case we had a parabolic velocity profile; the velocity at the center being equal to twice the average velocity. In the turbulent case the velocity at the axis of the tube is only a little larger than the average velocity, which has been assumed the same for both cases in the diagram. Only in the proximity of the wall the velocity drops quickly to zero; a narrow wall zone develops in this region, where the flow is more or less laminar, and the viscosity the important quantity.

The term velocity as it is used here with reference to the turbulent state is to be understood in the sense of the *time average of the velocity*. The individual particle velocity at a given point is by no means constant in time. The *spectrum* of its temporal and spatial variability covers a wide range; it has been possible to obtain more detailed information about it in a characteristic limiting case [see section (D)]. We denote the time average of the velocity by $U$, a velocity in the direction of the axis of the pipe ($x$-axis); the additional components of the individual velocities are denoted by $u$, $v$, $w$. Likewise, $P$ stands for the time average of the pressure and $p$ is the individual instantaneous pressure difference relative to $P$.

In addition to the law of turbulent velocity distribution we have the **law** of the turbulent *pressure drop*. In the laminar case the pressure gradient was proportional to the velocity, in our case it is proportional to the square or in better approximation to the $7/4^{\text{th}}$ power of the velocity (Blasius Law, see p. 120). When this law is expressed according to (16.11) and (16.13) by means of the similarity invariants $R$ and $S$, it reads

$$\frac{\Pi}{l} = \frac{\rho v^2}{a} S, \qquad S = \lambda R^{-1/4},$$

where $a$ is the radius of the pipe and $\lambda$ a numerical coefficient. Written in our present notation, it takes the form

(1) $$-\frac{dP}{dx} = \frac{\lambda \rho U^2}{a} R^{-1/4} = \frac{\lambda \rho}{a} \left(\frac{\nu}{a}\right)^{1/4} U^{7/4}.$$

This equation shows that the coefficient of viscosity is no longer of decisive importance for the state of the flow; its influence expresses itself by the factor $\nu^{1/4}$ and is rather weak. Another kind of energy loss prevails which is connected with inertia effects and becomes evident in the factor $\rho$, or rather $\rho^{3/4}$ if the dependence of the kinematic viscosity $\nu$ on $\rho$ is also considered.

Prandtl and Kármán were able to deduce the velocity distribution in the wall zone from (1), using certain additional assumptions. Consider a fluid element of thickness $dx$ extending over the pipe cross-section: The difference of the average pressures $P$ acting on the two faces of the element must be balanced by the shear forces acting between fluid and pipe wall. Thus we have

$$-\frac{dP}{dx} dx \, \pi a^2 = \tau_0 \, dx \, 2\pi a,$$

or, according to (1),

$$(2) \qquad \tau_0 = -\frac{dP}{dx}\frac{a}{2} = \frac{\lambda\rho}{2}\left(\frac{\nu}{a}\right)^{1/4} U^{7/4}.$$

Let now $y = a - r$ be the distance between a point inside the wall zone and the wall. According to Prandtl and Kármán there is only one way to construct a dimensionally correct and physically meaningful expression that approximates[20] the shear stress $\tau$ at the point $y$, viz.

$$(2a) \qquad \tau = \text{const } \rho\left(\frac{\nu}{y}\right)^{1/4} U^{7/4}.$$

From (2a) one obtains directly

$$(3) \qquad U = \text{const}\left(\frac{y}{\nu}\right)^{1/7}\left(\frac{\tau}{\rho}\right)^{4/7}.$$

Hence the velocity inside the wall zone increases from zero at the wall with the $1/7^{\text{th}}$ power of the distance from the wall. This agrees well with experimental results. Kármán was also able to develop from the same assumptions a formula for the heat conduction in the wall zone, which is in good agreement with the observations.

We shall now discuss the original approach to the turbulence problem made by Reynolds[21] and, in connection with his work, by H. A. Lorentz.[22] We restrict ourselves to a steady main flow along the $x$-direction with a pressure gradient $-dP/dx$ on which a non-steady quasiperiodic additional motion is superimposed. Let the state be two-dimensional, i.e. depending only on $x$ and $y$. The main flow depends then only on $y$, the turbulent additional motion has the components $u$ and $v$ that depend on $t$, $x$ and $y$. We shall consider the time averages of the total velocity $\mathbf{v}$ and of the total pressure $Q$, that is of the quantities

$$(4) \qquad v_x = U + u, \qquad v_y = v, \qquad Q = P + p$$

for which we stipulate

$$(5) \qquad \bar{v}_x = U, \bar{v}_y = 0, \bar{Q} = P, \text{ hence } \bar{u} = 0, \bar{v} = 0, \bar{p} = 0.$$

---

[20]Note that (2a) does not give (2) for $y = 0$, but becomes infinite. According to Kármán (Z. angew. Mathem. u. Mech. 1, 1921, p. 233) the reason is that (2a) is only asymptotically valid (i.e., for infinitely large Reynolds numbers).

[21]In the paper quoted in footnote 18 on p. 114.

[22]Akad. Amsterdam, Zittingsverlag, 6. p. 28, 1897 and Ges. Abhandlungen, Teubner 1907, p. 43.

This implies

$$\overline{v_x v_x} = U^2 + 2U\overline{u} + \overline{uu} = U^2 + \overline{uu}$$

(5a)
$$\overline{v_x v_y} = U\overline{v} + \overline{uv} = \overline{uv}$$

and

(5b)
$$\frac{\partial}{\partial x}\overline{v_x v_x} = \frac{\partial}{\partial x}\overline{uu}, \qquad \frac{\partial}{\partial y}\overline{v_x v_y} = \frac{\partial}{\partial y}\overline{uv}.$$

Furthermore, in a quasiperiodic motion the averages of the time derivatives such as $\partial u/\partial t$ vanish:

(5c)
$$\frac{\overline{\partial u}}{\partial t} = 0;$$

and because of (4) and (5)

(5d)
$$\frac{\partial U}{\partial t} = \frac{\partial U}{\partial x} = 0, \qquad \overline{\nabla^2 u} = 0, \qquad \overline{\nabla^2 v} = 0, \qquad \frac{\overline{\partial p}}{\partial x} = 0.$$

We now write the equation of motion in the abbreviated vectorial form (16.1) and set $\mathbf{F} = 0$:

(6)
$$\rho\frac{\partial \mathbf{v}}{\partial t} + \rho(\mathbf{v}\,\mathrm{grad})\mathbf{v} = -\,\mathrm{grad}\,Q + \mu\nabla^2 v.$$

As a consequence of the condition div $\mathbf{v} = 0$ we have the identity

$$v_x\frac{\partial v_x}{\partial x} + v_y\frac{\partial v_x}{\partial y} = \frac{\partial}{\partial x}v_x v_x + \frac{\partial}{\partial y}v_x v_y$$

which permits us to write the $x$-component of (6) in the form

(6a)
$$\rho\frac{\partial v_x}{\partial t} + \rho\frac{\partial}{\partial x}v_x v_x + \rho\frac{\partial}{\partial y}v_x v_y = -\frac{\partial Q}{\partial x} + \mu\nabla^2 v_x.$$

Forming the time average of this equation we obtain with the use of (4), (5), and (5a, b, c, d)

(7)
$$\frac{\partial}{\partial x}\rho\overline{uu} + \frac{\partial}{\partial y}\rho\overline{uv} = -\frac{\partial P}{\partial x} + \mu\nabla^2 U.$$

The quantities on the left side are gradients of *friction pressures* which appear here alongside with the gradient of the normal pressure $P$. To be in accordance with our system of notations (10.4) these quantities ought to be denoted as follows:

(7a)
$$\frac{\partial}{\partial x}p_{xx} + \frac{\partial}{\partial y}p_{xy} \qquad \text{where} \qquad p_{xx} = \rho\overline{uu}, \qquad p_{xy} = \rho\overline{uv}.$$

Now it should be remembered that the term $\mu\nabla^2 U$ to the right is also due to friction pressures; this was pointed out in Eqs. (10.6)-(10.8b). But the previous $p_{ik}$ originate in the *molecular* disorder of fluid particles on a microscopic scale, while the present $p_{zz}$ , $p_{zv}$ , in (7a) are caused by the *molar* disorder of fluid domains of macroscopic size engaged in turbulent motion. In either case the physical origin of the friction pressures is the *transfer of momentum*. Take as an example the molar friction pressure $p_{zv}$ : momentum is transferred from the motion in $x$-direction to the side motion $v$, a process that cannot remain without influence on the momentum balance of the main flow $U$. In this sense we should interpret the appearance of the terms (7a) on the left side of Eq. (7) of the main flow. But the average pressures (7a) do not lend themselves to the ordinary methods of differential analysis; this is the reason why in this approach to the turbulence problem recourse is taken to statistical methods.

Of the two friction pressures $p_{zv}$ and $p_{zz}$ the first is the important one, hence we shall indicate the contribution of the second only by $\cdots$ . Let us also adopt in what follows the usual terminology in this branch of hydrodynamics and speak of frictional *shear stress* rather than of frictional pressure; we write accordingly

$$(8) \qquad\qquad \tau = \sigma_{zv} = -p_{zv} = -\rho\overline{uv}$$

Instead of Eq. (7) we have then

$$(9) \qquad\qquad \frac{\partial P}{\partial x} = \mu\nabla^2 U + \frac{\partial\tau}{\partial y} + \cdots ,$$

For $\nabla^2 U$ we may also write $d^2 U/dy^2$ since $U$ is a function of $y$ alone. Eq. (9) is the *determinative equation of the main flow*.

We obtain the equation of the $x$-component of the turbulent additional motion by introducing (4) into (6) and by subsequent reduction with the help of (9) and (5d):

$$(10) \quad \frac{\partial u}{\partial t} + U\frac{\partial u}{\partial x} + v\frac{\partial U}{\partial y} = -\frac{1}{\rho}\left(\frac{\partial p}{\partial x} + \frac{\partial\tau}{\partial y}\right) + \frac{\mu}{\rho}\nabla^2 u + \cdots .$$

Here the sign $\cdots$ covers also the terms of second order in $u$ and $v$. There is a corresponding equation for the $y$-component $v$ of the turbulent motion and a second equation (9) for $\partial P/\partial y$.

Quantitative information about the character of the turbulence can hardly be expected from these very involved equations. We have mentioned them here mainly for the purpose of later reference in section (E) and to point out the significance of the molar friction stress $\tau$. The latter concept leads to the idea of a length $l$ characteristic for the turbulence, viz. the *mixing length*, introduced by Prandtl in 1926; it corresponds to the notion of the mean free path in gases:

Prandtl's argument[23] is this: Suppose a "fluid lump" of average velocity $U(y)$ is shifted from $y$ to $y + l$; in terms of the additional motion, this means a fluctuation of approximately

$$u = l \frac{\partial U}{\partial y}.$$

It is a reasonable assumption that the fluctuation of the side motion $v$ in the mixing process will be of the same order of magnitude. This admitted, one obtains for $\tau$ according to (8) an expression of the following form

$$(11) \qquad \tau = \rho l^2 \left| \frac{\partial U}{\partial y} \right| \frac{\partial U}{\partial y}.$$

If this is introduced in (9), the result is a differential equation that depends only on the quantities of the main flow and can be used for its analysis. However, the results obtainable in this way require continual experimental checking because of the hypothetical character of the relation (11),

## B. Older Attempts at a Mathematical Theory of Reynolds' Turbulence Criterion

Even before Reynolds started his theoretical investigations, Lord Kelvin[24] had treated the problem of flow stability by the method of small oscillations, but did not obtain definite results. Twenty years later the problem was attacked again; the object at this time was to study the stability of the form of the Couette flow[25] represented in Eq. (16.14), which is mathematically so very simple. If a small disturbance $(u, v)$ with stream function $\psi$ is superimposed on the main flow $U$, which has a rectilinear velocity profile, the resulting differential equation of fourth order

$$(12) \qquad \frac{\partial \nabla^2 \psi}{\partial t} + U \frac{\partial \nabla^2 \psi}{\partial x} - \frac{d^2 U}{dy^2} \frac{\partial \psi}{\partial x} = \nu \nabla^2 \nabla^2 \psi$$

that describes the problem in the general case, splits up, because of $d^2 U/dy^2 = 0$, into

$$(12a) \qquad \left( \frac{\partial}{\partial t} + U \frac{\partial}{\partial x} - \nu \nabla^2 \right) \chi = 0, \qquad \nabla^2 \psi = \chi.$$

One resolves now $\psi$ and $\chi$ in Fourier components by putting

$$(12b) \qquad \psi = f(y) e^{i(\alpha x - \beta t)}, \qquad \chi = F(y) e^{i(\alpha x - \beta t)}.$$

---

[23]Strömungslehre, Vieweg 1942, p. 105.

[24]Phil. Mag. 24, 188 and 272 (1887).

[25]W. M. F. Orr, Proc. Irish Acad. 27 (1907). — A. Sommerfeld, Internat. Mathem. Congress, Rome, 1908, Vol. III, p. 116.

The resulting ordinary differential equations for $f$ and $F$ are of second order and can be solved by known functions over the whole range $0 < y < h$ (Bessel functions of order $\frac{1}{3}$ and $\frac{2}{3}$). The boundary conditions at $y = 0$ and $y = h$ give a *transcendental equation* for $\beta$ in the form of a certain second order determinant that must vanish if non trivial values of the integration constants are to be obtained. If for a given real $\alpha$ this equation has a solution with positive-imaginary part, then *instability* prevails in the sense that disturbances of the wave number $\alpha$ increase in time. The *critical Reynolds number* would thus be determined by that $\alpha$-value for which the imaginary part of $\beta$ vanishes for the first time.

However, Mises[26] was able to deduce from the nature of the transcendental equation that such a value of $\beta$ cannot exist, or that the straight Couette flow is seemingly stable toward all disturbances $\alpha$. This conclusion has been confirmed by Hopf[27] in a more direct way by a detailed discussion of the corresponding types of oscillations. Moreover, it seems that according to Noether[28] the stability proof can be extended to all flows between two plane parallel plates (unless special discontinuities occur in the initial velocity profile).

Detailed calculations in the sense of Eqs. (12) and (12b) have been carried out by Sexl[29] not only for the flow between plane plates but also for the Couette flow proper[30] between coaxial cylinders, and likewise for the Hagen-Poiseuille flow. In every case the transcendental equation associated with the problem leads to the conclusion that the flow is stable.

Thus we have striking contradiction between theory and experiment. What conclusions should we draw? Should we suspect the method of small oscillations that has proved reliable in all other domains of mechanics including astronomy? Should we rather assume that for such a proof of instability *finite* instead of (infinitely) small disturbances ought to be resorted to just because the object of investigation is the laminar motion?

Or should we blame the Navier-Stokes equations as inadequate for our problem? This does not seem justifiable either, particularly since in the last analysis they form the foundation for all previous theoretical statements, including Reynolds's laws of similarity.

---

[26]R. v. Mises, Heinrich Weber-Festschrift, 1912, p. 112 (Teubner), Jahresbericht d. D. Mathem. Vereinigung 21 (1912) 241.

[27]L. Hopf, Ann. d. Phys. 43 (1914) 1.

[28]F. Noether, Z. angew. Mathem. u. Mech. 6 (1926) 232.

[29]Th. Sexl, Ann. d. Phys. 83 (1927) 835, 84 (1927) 807, 87 (1928) 570.

[30]In the more refined experiments on Couette flow performed by Sir G. Taylor (cf. p. 121) the deviation from the laminar motion occurring beyond the stability limit turned out to be helical (hence three-dimensional) rather than axially-symmetric.

## C. Reformulation of the Turbulence Problem; the Origin of the Turbulence

It was possible to find a way out of these difficulties when investigators directed their attention, as Prandtl had done it, to the *manner in which turbulence comes into being* in a given flow arrangement (cf. also our discussion of pipe flow experiments on p. 118). Here again the Couette flow, in which the flow field is bounded by a plane plate moving in $x$-direction, proved simpler than the pipe flow. At the beginning of the motion a boundary layer develops along the plate; the other plate bounding the flow field may be supposed at infinite distance, for the time being. The fluid that fills the space between the boundary layer and that plate is considered at rest or in laminar motion according to a preassigned law $U(y)$. The decisive factor for the stability is the velocity profile of the boundary layer, as was first recognized by Tollmien.[31] Tollmien's stability criterion (Gött. Nachr. 1935) states that velocity profiles with inflexion points, such as occur in a boundary layer with positive pressure gradient, are extremely likely to become unstable. To characterize the investigations that aim in this direction (determination of the eigenvalues and approximate calculation of the eigenfunctions in the strip between the plates) we quote a paper[32] of Schlichting: "The main flow is not assumed to have a fully developed linear velocity distribution; one rather investigates one by one the velocity profiles that occur between two parallel plane walls when one wall is suddenly set in motion, changing from a state of rest to one of constant velocity. A boundary layer then forms along the moving wall, the thickness of which increases with time, and if the main flow stays laminar, its velocity profile approaches asymptotically a straight line. In contrast to this, the earliest among these transient profiles have finite critical Reynolds numbers whose magnitudes depend on the shape of the profile. The actual Reynolds number of the Couette flow is then obtained as the lowest stability limit during the initial period".

If the problem is formulated in this way, its solution is of course much more difficult than the problem contained in our Eqs. (12). It can only be solved by the use of step by step approximations, carried out by numerical and graphical methods, which we can only mention here. They lead to a definite, practically meaningful value of the stability limit which depends, however, on how the turbulent state was produced.

---

[31]W. Tollmien, Über die Entstehung der Turbulenz, Göttinger Nachr. 1929.

[32]H. Schlichting, Ann. Phys. 14 (1932) 905. He refers to previous work done by other authors in Prandtl's group: O. Tietjens, Z. angew. Mathem. Mech. 5 (1925) 200, W. Tollmien, l. c., cf. also later papers of Schlichting in Gött. Nachr. 1932, 1933, 1935.

To illustrate once more the changed view point we could perhaps say: Not the instability of the laminar flow as such is proved as it was tried under (B), but it is shown that the road toward it is blocked by unstable states. Thus the instability is not an inherent quality of the final state but of its *history*.

## D. The Limiting Case of Isotropic Turbulence

In view of this complicated state of the problem one can ask whether there exist ideal flow conditions that are approached by the turbulent fluid in the limit of infinitely increasing Reynolds number; this final state could be called turbulent equilibrium. We suppose that no spatial direction is preferred and thus speak of *isotropic turbulence*. In the state of isotropic turbulence the space of the fluid is divided into turbulence elements or turbulence cells of ever changing size that are arranged in some statistical way. The eddies in the cells follow the Navier-Stokes equations, but are too complicated for complete mathematical treatment. Thus the equilibrium between the different turbulence elements can only be described statistically. One asks for the probability that a certain size of the turbulence elements occurs. This brings us back to the search for the spectrum of the turbulence elements which was already mentioned on p. 271.

The kinetic theory of gases may serve as an approximate model of this theory. Although the mathematically exact state of the gas (position and velocity of the gas molecules) is entirely determined by the history of the gas (i.e., the preceding states), one is only interested in the *statistically most probable distribution* (sometimes also in the probability of a given deviation from that distribution) which is given by the Maxwell-Boltzmann law. Similarly in the turbulence problem, where one is interested in the *statistically most probable distribution* of the vortices, which can be described in the form of a statement about the frequency of vortices of a certain size and a certain rotational speed.

Weizsäcker obtains the velocity spectrum of the isotropic turbulence by a similarity consideration in which turbulence cells of different sizes are compared under the assumption of a quasisteady state (i.e., its change in time is small compared with the time needed to establish statistical equilibrium). A further assumption is the random nature of the state. As a result the following spectral law is obtained: The mean velocity of a particle relative to its surrounding neighbors varies as *the third root of its diameter*. As a consequence, the energy density in function of the wave number is proportional to the $-5/3$ power of the wave number. However, these laws are only valid in the middle part of the spectrum, which part is very extended in the case of large Reynolds numbers. For *small* wave numbers (large turbulence elements that are comparable in size with the container) the statistical treatment must be replaced by the consideration of the individual hydrodynamic boundary value problem; for very large wave numbers (small turbulence cells) the Navier-Stokes friction causes the spectral intensity to drop to zero very quickly.

Heisenberg substantiates the same results by the use of Fourier analysis and supplements them by the calculation of the factors of proportionality, by a more exact investigation of the limiting cases of small and large wave numbers, and by the determination of the pressure fluctuations.

For all details and in particular, for the comparison with the experiment we refer to the papers of the two authors.*

## E. A Mathematical Model to Illustrate the Turbulence Problem

To J. M. Burgers who had been trying to explain the facts of turbulence on an statistical basis, we owe a mathematical model[33] from which properties analogous to those observed in actual turbulence can be deduced with mathematical rigor. The simplified system of equations which Burgers puts tentatively in place of preceding Eqs. (9) and (10), is

(13)
$$\frac{dU}{dt} = P - u^2 - \nu U,$$

$$\frac{du}{dt} = Uu - \nu u.$$

These equations are non-linear like the actual equations of turbulence ($u^2$ occurs in the first and $Uu$ in the second equation), each equation contains a dissipative friction term [$\nu U$ in the first and $\nu u$ in the second equation, corresponding to the terms $\mu \nabla^2 U$ in (9) and $(\mu/\rho)\nabla^2 u$ in (10)]. The constant $P$ in (13) replaces the pressure gradient $dP/dx$ in (9) and, like it, represents a driving force. The independent variables are $U$ and $u$ as before; $U$ is the main motion, $u$ the turbulent additional motion. However, Eqs. (9) and (10) refer to a domain of three independent variables $x$, $y$, $t$, while there is only *one independent variable t* in (13).

---

*C. F. v. Weizsäcker, Das Spektrum der Turbulenz bei grossen Reynoldsschen Zahlen; W. Heisenberg, Zur statistischen Theorie der Turbulenz, Z. f. Phys., **124**, (1948) p. 608-657. See also W. Heisenberg, On the theory of statistical and isotropic turbulence, Proc. R. Soc., **195**, (1948) p. 402, and Th. v. Kármán and C. C. Lin, The statistical theory of turbulence, Advances in Applied Mechanics, II, Academic Press, New York, 1950.

[33]Mathematical Examples illustrating relations occurring in the theory of turbulent fluid motion, Akad. Amsterdam **17** (1939)1.

The non-linear terms in (13) have been chosen in such a way that they drop out when the expression for the "energy" is formed; for we require the permanence of the general rule, according to which two equations of the form (13) should be added after multiplication with $U$ and $u$ respectively, and so would yield the energy expression. In our case one obtains

(13a) $$\frac{1}{2} \frac{d}{dt} (U^2 + u^2) = PU - \nu(U^2 + u^2).$$

In the right member, the first term is the work per unit of time performed by the driving force $P$ which, supposedly, acts only upon the main motion; the second term is the energy dissipated by friction.

*There exists a "laminar" solution of (13), viz.*

(14) $$U = U_0 = \frac{P}{\nu}, \qquad u = 0.$$

It is stable as long as

(14a) $$P < \nu^2.$$

This can be seen directly if the method of small oscillations is applied. One puts

$$U = U_0 + \xi, \qquad u = \eta$$

and neglects higher powers of $\xi$ and $\eta$. Then, by (13), one has

$$\frac{d\xi}{dt} = -\nu\xi$$

(14b)

$$\frac{d\eta}{dt} = (U_0 - \nu)\eta.$$

The first of these equations represents an exponentially decreasing disturbance; this is also true for the second equation but only if $U_0 < \nu$; hence by (14) we have condition (14a). For $U_0 > \nu$ and, therefore, $P > \nu^2$, Eq. (14) is still a solution of (13), but it is unstable because of the second Eq. (14b).

*There exist, however, two solutions for $P > \nu^2$ which we characterize as turbulent, since they include a contribution of the additional motion.* They are

(15) $$U = \nu, \qquad u = u_0 = \pm \sqrt{P - \nu^2}.$$

Both solutions are stable for $P > \nu^2$, for, if one puts

$$U = \nu + \xi, \qquad u = u_0 + \eta$$

and again neglects higher powers of $\xi$ and $\eta$ in accordance with the method of small oscillations, (13) becomes

$$\frac{d\xi}{dt} = -\nu\xi - 2u_0\eta,$$

(15a)

$$\frac{d\eta}{dt} = u_0\xi.$$

If one sets up a trial solution in the usual form

(15b)                $\xi = Ae^{\lambda t}, \qquad \eta = Be^{\lambda t},$

one obtains the following conditions which determine $\lambda$ and $B/A$:

$$\lambda A = -\nu A - 2u_0 B,$$

$$\lambda B = u_0 A;$$

they yield on elimination of $B/A$

$$\lambda^2 + \nu\lambda = -2u_0^2$$

(15c)

$$\lambda = -\frac{\nu}{2} \pm \sqrt{\frac{\nu^2}{4} - 2u_0^2} = -\frac{\nu}{2} \pm \sqrt{\frac{9}{4}\nu^2 - 2P}.$$

The roots $\lambda$ are *negative real* for

$$\nu^2 < P < \frac{9}{8}\nu^2$$

and turn complex with *negative real part* when

$$P > \frac{9}{8}\nu^2.$$

The disturbance (15b) thus attenuates exponentially for any $P > \nu^2$, which means that the solution (15) is *stable* for these values of $P$. The stability limit lies at $P = \nu^2$ where one of the roots (15c) vanishes. The same value of $P$ is the stability limit of the laminar solution and thus reflects Reynolds criterion of stability. It is true that, in contrast to the hydrodynamic reality, in our case both the laminar and the turbulent motion are steady, i.e., independent of $t$, but the example should be taken as a model and does not attempt to explain the turbulence. (There is also this difference that in the model *two* forms of turbulent motion are equally possible [see (15)], while in reality the laws of turbulent motion appear to be *uniquely* determined by the Reynolds number.)

Following Burgers, we give a graphic representation of the results by drawing in Figs. 66 and 67 a $u$, $U$-diagram of the two curves

(16)                $\frac{dU}{dt} = 0 \qquad$ and $\qquad \frac{du}{dt} = 0.$

The equations are given by (13), viz.

(16a)    $u^2 + \nu U = P$    and    (16b)    $u = 0, U = \nu$ respectively

Eq. (16a) is a parabola, Eq. (16b) a pair of lines consisting of the $U$-axis and a parallel line to the $u$-axis. Since we have got two steady solutions in (14) and (15), their representations in our diagram are the intersection

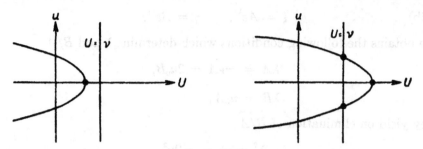

FIG. 66 (left). Diagram of Burger's turbulence model in the "laminar" case $P < \nu^2$.
FIG. 67 (right). Diagram of Burger's turbulence model in the "turbulent" case $P > \nu^2$.

points of (16a) and (16b). In Fig. 66 the parabola lies entirely to the left of the line $U = \nu$, hence there is only one intersection point; it belongs to the solution (14) and is unstable. There are three intersection points in Fig. 67. One of them lies on the $U$-axis. It belongs to solution (14) that is now unstable; the two others belong to the stable solution (15).

Burgers[34] extended his method of "mathematical models" in several steps in the attempt to bring it closer to the actual behavior of a turbulent fluid. He considers a system of three simultaneous differential equations in which the main "flow" $U$ is coupled with two additional motions, $u$ and $v$, and permits these motions to depend not only on time but also on a spatial coordinate. This we shall not discuss here since the simple model of Eq. (13) in our opinion illustrates the mathematical situation sufficiently: *By the non-linear character of the equations the occurrence of (two) entirely different types of motion is made possible.* These types, which we again distinguish by the terms "laminar" and "turbulent" are in some ways analogous to those of the turbulence problem; they branch off from one another at a certain stability limit, which we can compare with Reynolds stability criterion.

[34]Advances in Applied Mechanics, ed. R. v. Mises and Th. v. Kármán, Vol. I., Academic Press, New York, 1948.

## CHAPTER VIII

# SUPPLEMENTS TO THE THEORY OF ELASTICITY

### 39. *Elastic Limit, Proportional Limit, Yield Point, Plasticity, and Strength*

Up to this point we have assumed the solid body to be homogeneous and isotropic, and only permitted small deformations. In the following discussion we shall consider very briefly the principal deviations which appear under more general conditions.

Let us consider a cylindrical steel bar which is clamped at one end and subject to a tensile force $P$ acting at the other end. We shall first refer the tensile stress $\sigma$ to the unloaded cross section of the bar $F_0$, as is customary in engineering practice. Thus

(1)
$$\sigma = \frac{P}{F_0}$$

or the stress is proportional to the load $P$.

If we plot the measured strains $\epsilon$ as abscissas to the tensile stresses $\sigma$ calculated according to (1) as ordinates, we obtain for loads that produce very small deformations $\epsilon$ a curve which is practically an exact straight

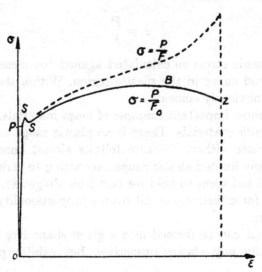

FIG. 68. Stress-strain diagram of a tensile test with a tough material (steel).

line (see solid curve in Fig. 68). This is valid up to a point $P$ that is called the proportional limit.

The elastic limit designates a point on the diagram, below which a decrease or removal of the load produces a decrease or removal of strain. For steel, the elastic limit lies slightly below the proportional limit.

Beyond the proportional limit, the stress-strain line becomes curved and reaches a maximum, which is called the upper yield point (also flow or plasticity limit) $S_u$. The curve drops now to the lower yield point $S_l$; following the point $S_l$ the curve remains in the mean parallel to the $\epsilon$-axis over a short distance, usually presenting a few small waves in this section, the behavior of a given sample depending upon the composition and previous treatment of the steel.

This transition region is followed by the plastic or flow region in which permanent deformations occur. A small change of load causes a rapid change of strain, so that in this range the curve first rises slowly with little curvature only, turning its concave side towards the $\epsilon$-axis. The maximum ordinate $B$ corresponds to the ultimate load or ultimate stress, but fracture occurs only at the point $Z$, where the stress-strain curve ends abruptly. Note in addition that

1. the bar undergoes lateral contraction everywhere and that

2. in the neighborhood of the place where fracture is to occur the reduction of the cross-section is particularly large; the constriction or "neck" formed there becomes quite noticeable once the point $B$ in the diagram is passed. If we refer the load to the (variable) smallest cross-section $F$, that is, if we put

$$(2) \qquad \sigma = \frac{P}{F}$$

and plot the tensile stress so calculated against the measured strain, we obtain the dashed curve in the plastic region. Within the elastic region, the two curves obviously coincide.

Steel is the most important example of *tough* materials, while cast iron is typical for *brittle* materials. There is no plastic range in the latter case and no constriction either. Fracture follows almost immediately at the end of the sharply limited elastic range. According to earlier experiments, Hooke's law did not seem to hold for cast iron altogether, but Grüneisen[1] has shown that for sufficiently small strains proportionality exists between stress and strain.

Materials that can be formed into a given shape like loam, clay, and lead show hardly any elastic properties, but exhibit plastic behavior throughout.

[1] E. Grüneisen, Ber. d. Deutsch. Phys. Ges. 1906.

Similar results as in the tension test described above are obtained in compression, bending, and torsion tests. It should be noted, however, that pure compression causes fracture only in brittle but not in tough materials. It is further assumed that in all these experiments the load is applied slowly, since the deformation depends not only upon the load, but also upon the duration of the application of the load.

In connection with this last point we mention the *elastic aftereffect*. Even in the elastic range the deformation connected with a particular load assumes its final value only after a considerable time. We speak of the *creep* of the material in question.

In analogy to the behavior of ferro-magnetic bodies we find also an *elastic hysteresis*. As in the magnetic case, the area enclosed by the hysteresis loop in the $\epsilon$, $\sigma$-plane represents the input of deformation energy per cycle.

In the previous discussion the load was assumed constant or slowly varying. Of more importance, especially in machine design, is the case of periodic loading, where the stress oscillates above and below a mean value. In such cases fracture occurs at a much lower loading than in the case of stationary load; one speaks of the *fatigue* of the material.

When we plot the number of load cycles until fracture, $z$, against the average stress $\sigma_m$ as ordinate, we obtain a monotonically decreasing curve (so-called Wöhler diagram), which approaches a horizontal asymptote $\sigma_{a_s}$ as $z$ increases. This indicates that the average stress must be kept smaller than the value of $\sigma_{a_s}$ (fatigue limit) in order to prevent a fatigue failure with certainty.

When the load that finally causes the fracture has been applied *slowly*, the cross section at the fracture will exhibit coarse grained structure; in the case of failure under oscillating load, however, a portion of the broken cross-section will be fine grained, possibly showing the spread of an imperfection (see below). The remaining portion of the section has the same appearance as in an ordinary fracture. We may conclude from this that the first part contributes only little to load bearing capacity, so that the remainder simply fails in a fracture by overload. Moreover, we know that the number of cycles $z$ depends greatly upon the quality of the surface of the specimen. Small invisible cracks reduce $z$ considerably. It may be mentioned in this connection that Voigt's measurements of the breaking strength of sodium chloride crystals gave only 1/400 of the value calculated from the lattice theory. In this case, too, small cracks and surface irregularities are the cause of the discrepancy, since tests made under water, where the crystal surface is smoothed out in the process of dissolution, gave 3/4 of the theoretical value of the breaking strength.

Solid bodies used in engineering are mainly metals and metal alloys composed of small crystal grains. Fine grained layers of foreign substances are interspersed between the crystals of the basic material. These intermediate layers can transfer considerable cohesive forces and give the solid body a certain degree of homogeneity. But in the same layers are also found places of lesser strength, imperfections of the structure from which a fracture develops. Organic materials like wood and leather, are composed of chains of crystal-like cells, in which other materials are imbedded. Glass, on the other hand, is an example of an amorphous material.

Concerning the crystalline structure of metals we know today that every crystal is a spatial lattice. The elementary units of the crystal lattice are the atoms of the metal, which are arranged in lattice planes and, within these planes, in lattice lines. The lattice planes can be arranged in three families of parallel planes which are not necessarily orthogonal to each other.

At small loads in the proportional range, the space lattice suffers a continuous deformation without essential changes in shape; the deformation disappears as the load is removed. Within the plastic range, the crystal deforms with a change of shape. The lattice planes of that family which contains the most densely occupied lattice lines, slide in the direction of these lattice lines. In this process the intermediate layers are also distorted.

### 40. Crystal Elasticity

The entire field of physical phenomena in anisotropic bodies has been expertly surveyed in "Lehrbuch der Kristallphysik" by Waldemar Voigt.[2] In this work the most general relations are developed that are compatible with the symmetry properties of the crystal and with the geometrical characteristic of the physical phenomenon under consideration, which may be a scalar field, a vector field or a tensor field of second or higher order (cf. p. 60). In crystal elasticity we are concerned with the relation between stress and strain tensor. Since both are symmetrical tensors of second order and have six components, the most general linear stress-strain relation will depend upon $6 \times 6 = 36$ coefficients.

This number will reduce, however, to

(1)         $$6 + \frac{6 \times 5}{1 \times 2} = 21,$$

when the strain energy $W$ is a *function of state*, as Green first postulated.

---

[2]Teubner 1910, second impression 1928. Beside dielectric and conductive properties such phenomena as pyroelectricity and piezoelectricity are treated; the latter has gained special interest in the meantime. Crystal optics, however, has been considered beyond the scope of this work.

We have pointed out (p. 73) that this assumption is justified both in the cases of an isothermal process (problems of equilibrium) and of an adiabatic change (rapid oscillations). In these cases then, which alone will interest us in the following $dW$ is a total differential. Replacing for the present our general notation $\sigma_{ik}$ and $\epsilon_{ik}$ by a single subscript notation we set[3]

(2)
$$\sigma_{11} = \sigma_1 , \quad \sigma_{22} = \sigma_2 , \quad \sigma_{33} = \sigma_3 , \quad \sigma_{23} = \sigma_4 , \quad \sigma_{31} = \sigma_5 , \quad \sigma_{12} = \sigma$$

$$\epsilon_{11} = \epsilon_1 , \quad \epsilon_{22} = \epsilon_2 , \quad \epsilon_{33} = \epsilon_3 , \quad 2\epsilon_{23} = \epsilon_4 , \quad 2\epsilon_{31} = \epsilon_5 , \quad 2\epsilon_{12} = \epsilon_6 .$$

Then the double sum of Eq. (9.24) becomes

(3)
$$dW = \sum_{k=1}^{6} \sigma_k \, d\epsilon_k$$

and the general stress-strain relation appears in the form

(4)
$$\sigma_k = \sum_{i=1}^{6} s_{ki} \epsilon_i .$$

where the $s_{ik}$ are now the moduli of elasticity.[4] If (3) is to be a total differential, we have

$$\frac{\partial \sigma_k}{\partial \epsilon_i} = \frac{\partial \sigma_i}{\partial \epsilon_k} ,$$

which according to (4) means

(5)
$$s_{ki} = s_{ik} .$$

When we write the coefficients $s$ in a quadratic array of 6 rows, it follows that they are *symmetrical with respect to the diagonal* $s_{11} , \cdots s_{66}$ . This immediately leads us to the count in Eq. (1). The same is of course true for the elasticity constants $c$.

Let us insert here the remark that the older molecular theory of Navier, Cauchy, and Poisson based on the idea of central forces between molecules led to a further reduction of that number to 15. We know, however, that this theory is too narrow in the case of the isotropic body. It would assign the universal value $\frac{1}{4}$ to the Poisson ratio $\nu$ in equation (9.9) and in this way reduce the required number of 2 elastic constants for

---

[3]The strains $\epsilon_4$ , $\epsilon_5$ , $\epsilon_6$ introduced in (2) are defined in the same way as Kirchhoff's $y_z$ , $z_x$ , $x_y$ (cf. the table on p. 000). At this point the omission of the factor $\frac{1}{2}$ which was introduced in our definition of the $\epsilon_{ik}$ with a view to the general structure of tensor algebra proves advantageous.

[4]This notation is due to Voigt who uses it throughout (cf. footnote on p. 64). The coefficients in the strain-stress relations obtained by solving Eqs. (4) for the quantities $\epsilon$ are denoted by $c_{ik}$ according to Voigt (constants of elasticity).

isotropic bodies to 1. Moreover, modern crystal lattice theory, which was already mentioned at the end of the previous article, has shown that the cause of this difficulty lay not so much in the molecular theory as such, as in its too narrow formulation: since the crystal in general consists of several superposed lattices which can move with respect to each other in the elastic deformation, the concept of pure central forces becomes inadequate.

The number 21 mentioned previously refers to a crystal without symmetry, that is to a triclinic crystal. The state of symmetry of a physical process acting on a crystal is determined according to the following principle (Voigt, L.d.K., 53): *The symmetry of the physical process is superimposed on the symmetry of the crystal.* Now, the elastic deformation has a center of symmetry since the strain in any one direction is the same as in the opposite direction. On the other hand, there are two possible cases of symmetry relations in a triclinic crystal, one with, and one without center of symmetry. In the latter case, however, the missing center of symmetry is supplied, as one might say, by the center of symmetry of the strain. Both sub-groups of the triclinic system therefore behave in the same way as far as elasticity is concerned.

There exists a total of 32 groups of symmetry[4a] which make up the 7 *crystal systems*, viz. *the triclinic, monoclinic, orthorhombic, tetragonal, trigonal, hexagonal, and cubic systems.* The principle quoted before implies in our case that the first three systems and the last two are uniform in regard to their elastic constants while the tetragonal as well as the trigonal systems must be divided each into two subclasses with different schemes of elastic constants. We shall give a few examples below of these $5 + 2 \times 2 = 9$ different schemes.

The crystal axes are generally designated by $a$, $b$, $c$. In the triclinic system the axes have different lengths and arbitrary directions. In the monoclinic system, one of the axes is perpendicular to the other two and represents an axis of symmetry for rotations through 180°. If we choose this axis as $z$-axis of an orthogonal coordinate system $x$, $y$, $z$, and call the elastic displacements $\xi$, $\eta$, $\zeta$, a rotation through 180° about the $z$-axis will cause the changes

(a)      $$x, y, z \to -x, -y, z, \qquad \xi, \eta, \zeta \to -\xi, -\eta, \zeta;$$

the associated changes in the strains are according to (2) on the one hand

(b)      $$\epsilon_1, \epsilon_2, \epsilon_3, \epsilon_6 \to \epsilon_1, \epsilon_2, \epsilon_3, \epsilon_6$$

---

[4a] As is well known, they are distinguished from each other according to their symmetry *relative to a point*. The more basic classification according to the *spatial* symmetry character that leads to 230 groups need not interest us here.

on the other hand

(c)                                $\epsilon_4\,,\ \epsilon_5 \rightarrow -\ \epsilon_4\,,\ -\epsilon_5\ ;$

One verifies easily that the expressions

$$\epsilon_4 = \frac{\partial \eta}{\partial z} + \frac{\partial \zeta}{\partial y} \quad \text{and} \quad \epsilon_5 = \frac{\partial \zeta}{\partial x} + \frac{\partial \xi}{\partial z}$$

change their signs in the operation (a), while for example the expression

$$\epsilon_6 = \frac{\partial \xi}{\partial y} + \frac{\partial \eta}{\partial x}$$

keeps its sign. Since the stresses $\sigma$ behave in the same way as the strains $\epsilon$, they must follow the same commutation rules (b) and (c). We now rewrite the stress-strain relations (4), marking the terms whose signs change in the operation (a) by $\pm$ and indicating the remaining terms on the right side by $\cdots$ . In this way we obtain for $k = 1, 2, 3, 6$

(d)                      $\sigma_k = \cdots \pm s_{k4}\epsilon_4 \pm s_{k5}\epsilon_5 \cdots ,$

but for $k = 4, 5$

(e)                  $\pm \sigma_k = \cdots \pm s_{k4}\epsilon_4 \pm s_{k5}\epsilon_5 + \cdots .$

For fixed $k$, (d) represents two equations which, on subtraction, lead to the following relation:

(f)              $s_{k4}\epsilon_4 + s_{k5}\epsilon_5 = 0 \quad \text{for} \quad k = 1, 2, 3, 6.$

Similarly, the pair (e) gives on addition

(g)       $s_{k1}\epsilon_1 + s_{k2}\epsilon_2 + s_{k3}\epsilon_3 + s_{k6}\epsilon_6 = 0 \quad \text{for} \quad k = 4, 5.$

From (f) we conclude

(6)         $0 = s_{14} = s_{24} = s_{34} = s_{64} = s_{15} = s_{25} = s_{35} = s_{65}\,,$

and the same result follows from (g) because of the relation $s_{ik} = s_{ki}$ .

On the basis of (6) the behavior of the moduli of elasticity in the transition from the triclinic to the monoclinic system has now been established:

*Triclinic system*

no symmetry axis

21 constants

| | | | | | |
|---|---|---|---|---|---|
| $s_{11}$ | $s_{12}$ | $s_{13}$ | $s_{14}$ | $s_{15}$ | $s_{16}$ |
| | $s_{22}$ | $s_{23}$ | $s_{24}$ | $s_{25}$ | $s_{26}$ |
| | | $s_{33}$ | $s_{34}$ | $s_{35}$ | $s_{36}$ |
| | | | $s_{44}$ | $s_{45}$ | $s_{46}$ |
| | | | | $s_{55}$ | $s_{56}$ |
| | | | | | $s_{66}$ |

*Monoclinic system*

one digonal axis of symmetry $A_z^2$

13 constants

| $s_{11}$ | $s_{12}$ | $s_{13}$ | $0$ | $0$ | $s_{16}$ |
|---|---|---|---|---|---|
| | $s_{22}$ | $s_{23}$ | $0$ | $0$ | $s_{26}$ |
| | | $s_{33}$ | $0$ | $0$ | $s_{36}$ |
| | | | $s_{44}$ | $s_{45}$ | $0$ |
| | | | | $s_{55}$ | $0$ |
| | | | | | $s_{66}$ |

In the orthorhombic system where $a$, $b$, and $c$ are mutually orthogonal another digonal axis of symmetry $A_x^2$ is added to $A_z^2$ ; the axis $A_y^2$ which is also present can be considered as a consequence of the two other axes and must not be taken into separate account.

By repeating the argument that leads to equation (6) one obtains in the present case

*Orthorhombic system*

two digonal axes of symmetry

$A_z^2$ and $A_x^2$ , 9 constants

| $s_{11}$ | $s_{12}$ | $s_{13}$ | $0$ | $0$ | $0$ |
|---|---|---|---|---|---|
| | $s_{22}$ | $s_{23}$ | $0$ | $0$ | $0$ |
| | | $s_{33}$ | $0$ | $0$ | $0$ |
| | | | $s_{44}$ | $0$ | $0$ |
| | | | | $s_{55}$ | $0$ |
| | | | | | $s_{66}$ |

Similarly, one has for the

*Cubic system*

three interchangeable tetragonal axes of symmetry, 3 constants

| $s_{11}$ | $s_{12}$ | $s_{12}$ | $0$ | $0$ | $0$ |
|---|---|---|---|---|---|
| | $s_{11}$ | $s_{12}$ | $0$ | $0$ | $0$ |
| | | $s_{11}$ | $0$ | $0$ | $0$ |
| | | | $s_{44}$ | $0$ | $0$ |
| | | | | $s_{44}$ | $0$ |
| | | | | | $s_{44}$ |

From the last scheme we conclude that the cubic single crystal has an elastic behavior *different from the isotropic body*. The former possesses only two independent moduli of elasticity, for example, our previous $\mu$ and $\lambda$. In order to adapt the above scheme to that of equation (9.6), we would have to specialize the three constants as follows:

$$s_{11} = 2\mu + \lambda, \qquad s_{12} = \lambda, \qquad s_{44} = \mu,$$

that is, we should add the following condition which is not fulfilled in the cubic single crystal:

$$s_{11} = s_{12} + 2s_{44} .$$

From a physical point of view this is connected with the fact that the so called isotropic solid body has no uniform structure, but is a polycrystal. The elastic behavior which we actually observe is simpler than that of the single crystal, but it is only the result of averaging the elastic reactions of a very large number of single crystals that are in random positions with respect to each other. For an example we refer to the metals discussed in the previous article.

Quite different is the answer to the corresponding question in the field of optics. Cubic single crystals of fluorite or rock salt are frequently preferred as lense materials for optical instruments in certain ranges of the spectrum, where they are superior in transmissivity to ordinary isotropic (amorphous) glass lenses. This is possible, because the cubic single crystal is optically isotropic. In the framework of Voigt's general system the cause for this different behavior must be seen in the following distinction: in elasticity the basic relation is a tensor-tensor relation (stress-strain), in optics it is a vector-vector relation (electric field- dielectric displacement).

We cannot further pursue the interesting details of crystal elasticity. Our principal aim was to characterize the elastic behavior of isotropic bodies as contrasted with that of the uniform anisotropic bodies.

## 41. The Bending of Beams

Loaded and bent beams are among the principal elements of all building constructions. In planning a new structure bending loads are calculated and must be legally approved. The theory employed for this purpose is simple and has been in use for 200 years; its origin dates back to Daniel Bernoulli. We shall discuss below how this theory can be substantially justified through more rigorous methods (St.-Venant), but we shall first follow the historical development.

We limit ourselves to the simplest case of a *straight* beam subject to slight bending. The cross-section of the beam is arbitrary (one may think of a $T$ or $I$ section), but it is assumed as constant over the length of the beam.

Let us first disregard the weight of the beam. At one end the beam supports a vertical load, while the other end is clamped rigidly; thus the upper fibers of the beam are stretched and the lower ones compressed.

Bernoulli assumes that each plane cross-section of the beam remains plane also after bending. Navier concludes from this that the normal stresses transmitted by a cross-section are distributed according to a *linear law*. We consider two neighboring sections originally separated by the distance $dx$ ($x$ is taken from right to left as indicated in Fig. 69a). The

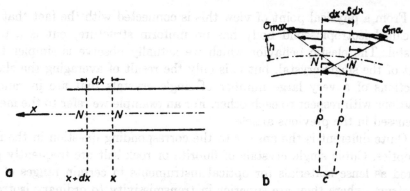

FIG. 69. Illustrating the elementary bending theory. An element of a beam (a) in the original state and (b) in the bent state.

fibers running between the two cross-sections have the length $dx + \delta dx$ after bending. Due to the assumed planeness of both cross sections, the extension of the fibers decreases linearly from the upper face of the beam downward and becomes zero at some particular layer $NN$; $\delta dx$ now stays negative down to the lower face of the beam and represents the contraction of the fibers. The strains are distributed in the same linear way.

$$\epsilon = \frac{\delta \, dx}{dx},$$

and so are the stresses [cf. (9.8)]:

(1)                              $\sigma = E\epsilon.$

When $\sigma_{max}$ is the stress in the (upper or lower) boundary fiber and $h$ its distance from $NN$, (cf. Fig. 69b), the linear law states that

(2)          $\sigma = \dfrac{\sigma_{max}}{h} y \cdots y = $ distance between fiber with stress $\sigma$ and $NN$

In order to determine the maximum stress $\sigma_{max}$, we shall investigate the equilibrium of forces acting on an isolated beam section of length $x$ (cf. Fig. 70); in doing so we may disregard the deflection of the beam. We are mainly concerned with the *equilibrium of moments* taken about an axis normal to the plane of Fig. 70 and containing the point $N$ at the cross-section $x$, but we shall also consider the *equilibrium of force components in the horizontal x-direction*. Equilibrium of forces in the vertical $y$-direction will be considered later.

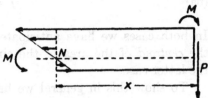

FIG. 70. Equilibrium between the bending moment $M$ and the moment of the normal stresses $\sigma$.

For the particular loading here assumed the moment of the external forces, usually designated as *bending moment M*, is

(3) $$M = Px.$$

According to Eq. (2) the forces $\sigma df$ have the moment

(4) $$\int y\sigma \, df = \frac{\sigma_{max}}{h} I, \qquad I = \int y^2 \, df.$$

The quantity $I$ is called the *moment of inertia of the cross-section*. In Vol. I the moment of inertia of a plane mass distribution has already been introduced and a distinction made between *polar* and *equatorial* moments of inertia. Here, however, we are dealing with a plane area and not with a plane mass distribution, hence the dimension of $I$ is not gcm$^2$ but rather cm$^4$. Since the axis through $N$ for which $I$ is computed lies within the cross-section, $I$ is to be considered as an equatorial moment of inertia. When the opposite direction of the moment arrows shown in Fig. 70 is considered, the moment equilibrium requires according to (3) and (4)

(5) $$\sigma_{max} = \frac{hM}{I}.$$

This relation together with (2) determines the stress $\sigma$ for any other point of the cross-section.

The engineer is particularly interested in $\sigma_{max}$ since it is his task to keep the maximum occurring stress below the *legally admissible limit*, $\sigma_{adm}$. He has to find the cross-section with the largest loading, that is, with the largest bending moment, and proportions the beam in such a way that in this cross-section the stress at the boundary face $\sigma_{max} < \sigma_{adm}$. For beam profiles used in engineering, the values of $I$ and $I/h$ are tabulated in tables of moments of inertia and section moduli. For a circular cross-section of radius $a$ we have[5] (cf. Vol. I, Problem IV. 1)

---

[5]Cf. Synge and Griffith, *op. cit.*, p. 220.

$$(6) \qquad I = \frac{1}{4} Fa^2 = \frac{\pi}{4} a^4;$$

for a rectangular cross-section of width $b$ and depth $2h$

$$(6a) \qquad I = \frac{2}{3} bh^3.$$

In both cases we have anticipated that the axis $y = 0$ passes through the *centroid* of the area of the circle or rectangle, being parallel to $b$ in the latter case.

To show this in general we have to use the *equilibrium condition for force components in $x$-direction*. The external forces do not contribute anything to the equilibrium equation, neither in our particular case where the external force $P$ is vertical, nor in general, since otherwise the problem would not be one of pure bending but of combined bending and tension or compression. The condition is then simply

$$\int \sigma \, df = 0,$$

or, because of (2),

$$\int y \, df = 0.$$

On the other hand, the *centroid* of the cross-sectional area is defined by the fact that the first moment[6] of the area with respect to its centroid is zero, equivalent to $\int y \, df = 0$. The fiber passing through the consecutive centroids of all cross-sections of the beam is called its *neutral fiber* or *neutral line*. The horizontal plane containing the points $y = 0$, is known as the *neutral plane*. This has already been indicated by the notation $NN$.

We now determine the shape of the neutral line in its deflected position. For this purpose we have to set up the *differential equation of the elastic line*, which can be derived on the basis of Fig. 69b: Here the center of curvature $C$ of the elastic line $NN$ has been indicated as the intersection of the prolonged traces of the cross-sections at $x$ and $x + dx$. The sides of the triangle $CNN$ are $\rho$, $\rho$, and $dx$ where $\rho$ is the radius of curvature. If we draw through one of the points $N$ a parallel to the trace of the other cross-section we obtain a triangle similar to $CNN$ the third side of which is the face of the beam. The sides of this triangle are $h$, $h$, $\delta dx_{max} = \epsilon_{max} \, dx$. Hence the following proportion holds:

$$\frac{dx}{\rho} = \frac{\epsilon_{max} \, dx}{h}.$$

---

[6]Sometimes the term static moment is used which is not a very appropriate expression. Likewise, second moment should be preferred to moment of inertia.

By the use of (1) and (5), it may also be written

(7)
$$\frac{1}{\rho} = \frac{M}{EI} .$$

As to the choice of the sign in this equation we refer to Fig. 69b for the sake of brevity and omit the introduction of sign rules for $\rho$. The right hand side is a known function of $x$ or, more generally, *a known point function along line NN*. (The second formulation refers to finite deflections where the abscissa must be replaced by the arc length $s$ of the elastic line.) The left hand side of (7) is the curvature, a differential invariant of the curve; thus equation (7) is the differential equation required.

For sufficiently small deflections we know that approximately

$$\frac{1}{\rho} = \frac{d^2y}{dx^2} ,$$

and consequently

(8)
$$\frac{d^2y}{dx^2} = \frac{M(x)}{EI} .$$

Referring to the particular loading (3), we obtain

(9)
$$\frac{d^2y}{dx^2} = \frac{P}{EI} x$$

and after integration

(9a)
$$y = \frac{P}{EI} \left\{ \frac{x^3}{6} + Ax + B \right\},$$

which is a cubic parabola. The integration constants $A$ and $B$ result from the conditions at the point of support:

(9b)     for     $x = l \left\{ \begin{array}{l} y = 0 \\ \dfrac{dy}{dx} = 0 \end{array} \right\} \cdots A = -\dfrac{l^2}{2}, \qquad B = \dfrac{l^3}{3} .$

The maximum deflection occurs at $x = 0$:

(9c)
$$y_0 = \frac{1}{3} \frac{P}{EI} l^3.$$

We consider now a beam simply supported at each end by a pair of uprights at distance $l$ from each other, and loaded only by its own weight. The (constant) weight per unit length being $p$, the reactions at each support will be $-pl/2$, where the minus sign indicates the upward direction of these forces. The bending moment at any point is composed of the moment of the weight and the moment of one of the end reactions:

$$(10) \qquad M(x) = \frac{p}{2} x(x - l).$$

Integration of (8) then gives

$$(10a) \qquad y = \frac{p}{2EI} \left\{ \frac{x^4}{12} - \frac{x^3 l}{6} + Ax + B \right\},$$

a parabola of the fourth order.

Instead of the conditions (9b) we have now

$$\left. \begin{array}{ll} \text{for} & x = 0 \\ \text{for} & x = l \end{array} \right\} y = 0 \cdots \qquad B = 0, \qquad A = \frac{l^3}{12} .$$

The largest deflection occurs, of course, at the central cross-section and amounts to

$$(10b) \qquad y_{max} = \frac{5}{16 \times 24} \frac{p}{EI} l^4.$$

Of more interest for the following is the case where a *couple* acts at the end of the beam instead of a single force as in Fig. 70. This could be physically realized by two equal forces of opposite direction, acting in close proximity to each other and having such magnitude that their moment has a finite value. Then $M$ can be considered as constant and according to (8)

$$(11) \qquad y = \frac{M}{EI} \left\{ \frac{x^2}{2} + Ax + B \right\}.$$

If the beam is clamped at the other end ($x = l$) we have the same conditions as in (9b), and it follows that $A = -l$, $B = \frac{1}{2}l^2$. The deflection at $x = 0$ is then

$$(11a) \qquad y = \frac{1}{2} \frac{M}{EI} l^2.$$

In this case the shape of the elastic axis can be more accurately described than by the parabola (11): according to (7) it is a circular arc, since its curvature is constant.

This last case is the only one in which we are led to a clear-cut result by our equilibrium conditions. The *third equilibrium condition* which we have to stipulate, viz.

$$(12) \qquad \textbf{Sum of all vertical components} = 0$$

is here satisfied, whereas this condition requires in general that at every cross-section $x$ a vertical shear force opposes the external loads. This shear force results from the combined action of all shear stresses of the section under consideration. If we take the beam of Fig. 70 as an example, shear stresses $\sigma_{zy}$ would have to occur in addition to the normal stresses $\sigma_{zz}$, so that for every cross-section

$$(13) \qquad \int \sigma_{zy} \, df = P.$$

This, however, would imply the occurrence of angular changes $\gamma_{zy}$, causing the originally plane cross-section to warp. In other words, Bernoulli's hypothesis, in asserting the permanent planeness of the cross-sections, contradicts the general fundamentals of statics. *Thus it would seem that Navier's law of linear variation of the normal stress which is based on Bernoulli's hypothesis, is also invalidated.*

Before we show a way to overcome this difficulty, we wish to determine the magnitude of the required shear stresses. The integral condition (13) is not suitable for this purpose, but we can make use of the differential relation (8.11). In the absence of external body forces and with $\sigma_{zz}$ evidently vanishing if for instance a rectangular cross-section is assumed, this relation may be written

$$\frac{\partial \sigma_{zz}}{\partial x} + \frac{\partial \sigma_{zy}}{\partial y} = 0.$$

Notwithstanding the previous critical remarks, the statements of equations (2), (5) and (3) may be considered as approximations. We have then

$$(14) \qquad \frac{\partial \sigma_{zy}}{\partial y} = \frac{-Py}{I}, \qquad \sigma_{zy} = \frac{P}{2I}(A - y^2) = \frac{P}{2I}(h^2 - y^2).$$

The integration constant $A$ in the last equation must equal $h^2$, so that $\sigma_{zy} = 0$ for $y = \pm h$. This must be required since the value of $\sigma_{zy}$ at the upper or lower edge of the cross-section determines the shear stress $\sigma_{yz}$ acting within the face of the beam; the face, however, must be completely free of stress. It is not difficult to determine the warping of the rectangular cross-section that results from the $\sigma_{zy}$ and the associated shear angle $\gamma_{zy}$.

We now wish to point out briefly how the bending theory can be freed from the internal contradiction revealed above. St.-Venant[7] accomplished this in 1855. First of all, it must be emphasized that the relations between stress and external forces used so far are not sufficient to describe the

---

[7]A complete representation is found in de Saint-Venant's translation of Clebsch's lectures: Théorie de l'elasticité des corps solides, trad. par Saint-Venant and Flamant, Paris, 1883.

elastic state completely. One must make recourse to the elastic displacements $\xi$, $\eta$, $\zeta$, as they occur in the fundamental elastic equations (9.18). Then we note that nowhere in the preceding argument we have mentioned the stresses $\sigma_{yy}$, $\sigma_{zz}$, $\sigma_{yz}$. In fact, they do not enter into the consideration of the equilibrium of a bent beam and can therefore be set equal to zero. If we express them in terms of $\xi$, $\eta$, $\zeta$, we obtain simple differential relations between these quantities by the use of which the equations (9.18) can be greatly simplified and finally come out as:[8]

$$(15) \qquad \left\{ \frac{\partial^2}{\partial x^2} = \frac{\partial^2}{\partial y^2} = \frac{\partial^2}{\partial z^2} = \frac{\partial^2}{\partial y \, \partial z} \right\} \frac{\partial \xi}{\partial x} = 0.$$

From this we conclude first: the strain $\partial \xi / \partial x$ of the $x$ fibers and consequently also their stress $\sigma = \sigma_{xx}$ is a linear function of the coordinates in the cross-section $y$ and $z$. The *linear distribution law* for $\sigma_{xx}$ is thus a *necessary consequence of the fundamental laws of elasticity* and Bernoulli's hypothesis is not required for its derivation.

Eq. (15) shows further that the extension of the $x$-fibers is also a linear function of $x$; this implies together with Eq. (5) that the bending moment $M$ should be a linear function of $x$, which is actually the case for one or several single loads or single moments (but not for the continuously distributed load of the weight of the beam itself). The following restriction should be noticed, however: the single load ought to be applied in such a way that it can be balanced by the shear forces $\sigma_{xy}$ that act across the cross-section at which the load acts; in other words, the load should be distributed over the cross-section in the same way as previously computed for $\sigma_{xy}$. This is, of course, never fulfilled in reality. We may, however, take the following view: whether we have a lumped or a distributed load can only make a difference in the neighborhood of the cross-section under consideration, provided, of course, the loads in both cases are statically equivalent. This can be demonstrated by suitably chosen experiments as well as by theoretical examples and is often formulated as a general theorem under the name of *Saint-Venant's principle*.

Thus the simple method of computing beams set forth at the beginning of this article that has given satisfactory results in innumerable cases has been vindicated from a more rigorous point of view.

## 42. Torsion

We first consider the simplest case of the torsion of a rod with *circular* cross-section, which, in a way, is also the most important case. We

---

[8]Cf. for this calculation A. Föppl, Vorl. üb. techn. Mechanik, 4th ed. Leipzig, 1909, Vol. III, 73.

think of the fact that every engine shaft is subject to torsional loading: *a torque about the axis of the shaft* is produced at one end by the driving mechanism, at the other end an *opposite torque* acts which originates in the mechanism that receives the power; in steady motion both torques have the same magnitude. This example shows the general characteristic of torsional loading. For a straight rod of arbitrary cross-section it consists in a torque that acts about the rod axis at one end; whether the other end is kept fixed or subject to the opposite moment makes no difference; an additional tensile loading, e.g., in the direction of the axis of the rod would mean that we deal no longer with pure torsion.

It is easy to imagine the kind of deformation that is established by twisting a rod of circular cross-section whose one end is kept fixed, but we shall have to show that the deformation we think of is theoretically admissible. We assume (cf. Fig. 71) that every cross-section of the rod is

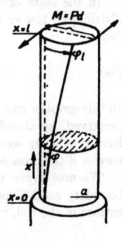

FIG. 71. The torsion of a rod of circular cross-section. The base is rigidly clamped, a torsional moment $M$ acts on the free end.

rotated through a certain angle $\varphi$ which is proportional to the $x$-coordinate. This means

(1)     $\varphi = \alpha x, \quad \alpha = \text{Const.}$

The originally straight generatrices of the rod are transformed into circular helices; the components of the displacement vector **s** are according to (1) in cylindrical polar coordinates $r, \varphi, x$

(2)     $s_r = 0, \quad s_\varphi = \alpha r x, \quad s_z = 0.$

In order to obtain the strains we have simply to introduce the quantities (2) in Eqs. (4.26) and (4.28) for $\delta q_1$, $\delta q_2$, $\delta q_3$, putting, because of $ds^2 = dr^2 + r^2 d\varphi^2 + dx^2$,

$$g_1 = 1, \quad g_2 = r, \quad g_3 = 1.$$

This leads to

$$\epsilon_{rr} = \epsilon_{\varphi\varphi} = \epsilon_{zz} = 0$$

(3)

$$\epsilon_{r\varphi} = 0, \qquad 2\epsilon_{\varphi z} = \alpha r, \qquad \epsilon_{zr} = 0.$$

The associated stresses $\sigma$ are now found according to Eqs. (9.7), which are relations between tensors and therefore valid in arbitrary orthogonal coordinates:

$$\sigma_{rr} = \sigma_{\varphi\varphi} = \sigma_{zz} = \sigma_{zr} = \sigma_{r\varphi} = 0$$

(4)

$$\sigma_{z\varphi} = \tau = \mu\alpha r.$$

In agreement with common engineering usage the notation $\tau$ is here applied to the only shear stress that acts in the cross-section $x = $ const.

In the state of stress described by (4) we have again a *linear* stress distribution law as we had it in bending; in fact, we can write (4) in the form

(4a) $$\tau = \frac{r}{a}\,\tau_{\max} \qquad \text{where} \qquad \tau_{\max} = \mu\alpha a.$$

In this simplest case of torsion the planeness of the cross-sections is strictly preserved, as already anticipated in the assumption $s_z = 0$ in (2). We have previously seen that this is not true in bending; in torsion it is only correct if the cross-section is circular, as we shall see.

We now prove that our system of stresses is in agreement with the boundary conditions of the problem. This is obviously true for the curved surface of the rod, $r = a$, which must be free of stress. There ought to be

$$\sigma_{rr} = \sigma_{r\varphi} = \sigma_{rz} = 0,$$

which, according to (4), is in fact fulfilled. Across the base the condition of vanishing stress must be replaced by the condition of vanishing displacements. This is satisfied by (2), if $x = 0$. For the free end cross-section we should at least stipulate as a *necessary* condition that

(5) $$M = \int r\tau \, df,$$

where $M$ is the moment of the forces $P$ that produce the torsion (cf. Fig. 71). By (4a) this may be rewritten in the following form:

(5a) $$M = \frac{\tau_{\max}}{a} \int r^2 \, df = \frac{\tau_{\max}}{a} J_p = \mu\alpha J_p\,, \qquad J_p = \frac{\pi}{2}\, a^4,$$

where $J_p$ stands for the polar moment of inertia of the cross-section which in the case of a circular area is twice the equatorial moment of inertia given in (41.6).

Strictly speaking, this is not a *sufficient* condition. The external moment $M$ should not only be equal to the total moment of the shear forces about the axis of the rod, but the distribution of $M$ over the cross-section should be point by point identical with the distribution of $\tau df$. The couple $M = Pd$ as it appears in the figure obviously does not satisfy this condition. This difficulty is again taken care of by the Saint-Venant principle (see p. 300), according to which the local disturbance of equilibrium produced by the dissimilarity between the actual and the theoretical load is limited to the proximity of the end face and does not influence the elastic equilibrium anywhere else in the rod.

To be on the safe side we would have to show that the differential equations of elasticity, which the displacements must obey, are satisfied by our special solution (2). These relations written in Cartesian displacement components are given by Eqs. (9.18); in their place we should have to use Eqs. (9.20) which refer to general orthogonal coordinates. A glance at the quantities $\Theta$ and $P_i$ that appear in these equations shows that in our case

$$\Theta = 0, \qquad P_1 = -\alpha r, \qquad P_2 = 0, \qquad P_3 = 2\alpha x.$$

This is easily checked with the aid of Eqs. (9.20a), and it follows immediately that Eqs. (9.20) are satisfied ($\mathbf{F}$ is of course equal to zero).

We have now to study in greater detail Eq. (5a) which determines the only remaining unknown in our problem, viz. the angular torsion per unit of rod length. We may as well determine the total *torsion angle* of the rod which becomes according to (1)

$$\varphi_l = \alpha l, \qquad \text{where } l = \text{length of rod.}$$

From (5a) we obtain

(6)
$$\varphi_l = \frac{Ml}{\mu J_p}.$$

This equation is quite evident as regards the proportionality with the torsion moment $M$ and the length $l$ and also as to the inverse proportionality with the torsion modulus $\mu$. More important is what Eq. (6) says about the dependence on the *dimensions* of the cross-section. This expresses itself in the factor $J_p$ of the denominator. Note, however, that the dependence on the polar moment $J_p$ that occurs here does not hold for arbitrary cross-sections, like the moment $I$ in the bending problem; if the cross-section is not a circle this dependence is by no means expressible through the polar moment $J_p$. This applies e.g. to the elliptic cross-section as will be seen below.

For cross-sections of *unspecified shape* we have no preferred coordinate system, such as the cylindrical polar coordinates in the previous case, and we must take recourse to Cartesian coordinates and displacements $x$, $y$, $z$; $\xi$, $\eta$, $\zeta$, the equilibrium conditions now being given by Eq. (9.18) with $\mathbf{F} = 0$.

In general we shall now have to admit the possibility of warping of the cross-section. Taking the axis of the rod as $x$-axis, there is then no longer $\xi = 0$ [corresponding to the previous $s_z = 0$ in (2)]. As a consequence of the prismatic shape of the rod we may, however, assume that the warping is independent of the location of the cross-section, that is to say, the warping function $\xi$ is independent of the $x$-coordinate:

$$(7) \qquad \xi = \Phi(y, z), \qquad \frac{\partial \xi}{\partial x} = 0.$$

We further assume that the deformation, as in the case of the circular cross-section, is free of dilatation. This would mean $\Theta = 0$; let us, however, introduce the more restricted conditions

$$(8) \qquad \frac{\partial \eta}{\partial y} = \frac{\partial \zeta}{\partial z} = 0,$$

and finally assume that, similarly to $\epsilon_{r\varphi} = 0$ in Eq. (3),

$$(9) \qquad 2\epsilon_{yz} = \frac{\partial \eta}{\partial z} + \frac{\partial \zeta}{\partial y} = 0.$$

These various assumptions will have to be justified in the course of our investigation.

By (7) and (8), the differential equations (9.18) take the following simple form

$$\frac{\partial^2 \xi}{\partial y^2} + \frac{\partial^2 \xi}{\partial z^2} = 0$$

$$(10) \qquad \frac{\partial^2 \eta}{\partial x^2} + \frac{\partial^2 \eta}{\partial z^2} = 0$$

$$\frac{\partial^2 \zeta}{\partial x^2} + \frac{\partial^2 \zeta}{\partial y^2} = 0.$$

The first of these equations states that the warping function $\xi$ introduced in (7) satisfies the *two-dimensional potential equation*.

$$(11) \qquad \nabla^2 \Phi = 0,$$

The two remaining equations (10) can be fulfilled together with (8) if $\eta$ and $\zeta$ are taken as bilinear functions as follows:

(12)
$$\eta = b_0 + b_1 x + b_2 z + b_3 xz,$$
$$\zeta = c_0 + c_1 x + c_2 y + c_3 xy.$$

According to (9), the following relation must hold for the constants $b$ and $c$ identically in $x$:

$$b_2 + b_3 x + c_2 + c_3 x = 0;$$

or

(13)
$$b_2 = -c_2 , \qquad b_3 = -c_3 .$$

As in Fig. 71, the base of the rod is kept fixed, in the sense that lateral displacements are excluded. For $x = 0$ we have therefore identically in $y$ and $z$

$$b_0 + b_2 z = 0, \qquad c_0 + c_2 y = 0,$$

or

(13a)
$$b_0 = c_0 = b_2 = c_2 = 0.$$

If we finally wish to retain the right angle between $x$-axis and end face, we have for $x = 0$

(13b)
$$\frac{\partial \eta}{\partial x} = b_1 = 0, \qquad \frac{\partial \zeta}{\partial x} = c_1 = 0.$$

Because of (13) and (13a, b) the Eqs. (12) take the form

(14)
$$\eta = -\alpha xz, \qquad \zeta = +\alpha xy$$

where $\alpha$ has been written instead of $c_3$ .

The elastic differential equations and the boundary conditions at the end face are certainly satisfied by the assumptions about $\xi, \eta, \zeta$, expressed in (7), (11), and (14), but the question remains whether *the boundary conditions for the curved surface of the rod* can be satisfied in this way. If $n$ denotes the normal at a point $P$ of the curved surface and $s$ the direction tangential to the contour of the cross-section at $P$, the conditions are

(15)
$$\sigma_{ns} = 0 \quad \text{and} \quad \sigma_{nz} = 0.$$

The first of these equations can certainly be satisfied: We may draw the $y$-axis which so far has not been fixed parallel to $n$ so that the $z$-axis becomes parallel to $s$ and $\sigma_{ns}$ coincides with $\sigma_{ys}$ at the point $P$. But since $\epsilon_{ys} = 0$ by (9), it follows that $\sigma_{ys}$ vanishes also at $P$. This argument can obviously be applied to each point $P$ of the curved surface. As to the second condition (15), we base our argument, on the general tensor relation (8.5) which becomes in our case

$$(16) \qquad \sigma_{nx} = \beta_1 \sigma_{yx} + \gamma_1 \sigma_{zx} = 2\mu(\beta_1 \epsilon_{yx} + \gamma_1 \epsilon_{zx})$$

$$= \mu\left(\beta_1\left\{\frac{\partial\xi}{\partial y} + \frac{\partial\eta}{\partial x}\right\} + \gamma_1\left\{\frac{\partial\xi}{\partial z} + \frac{\partial\zeta}{\partial x}\right\}\right).$$

Here $\beta_1$ and $\gamma_1$ are the cosines of angles between the $n$ direction and the $y$ and $z$ directions respectively. If $s$ is oriented in such a way that the sense of rotation $s \rightarrow n$ coincides with $y \rightarrow z$, we have

$$(17) \qquad \begin{aligned} \beta_1 &= \cos(n, y) = -\cos(s, z) = -\frac{dz}{ds}, \\[2mm] \gamma_1 &= \cos(n, z) = +\cos(s, y) = +\frac{dy}{ds}. \end{aligned}$$

Using (16), (17), (7), and (14) the second condition (15) can be written in the following form:

$$(18) \qquad 0 = -\left(\frac{\partial\Phi}{\partial y} - \alpha z\right) dz + \left(\frac{\partial\Phi}{\partial z} + \alpha y\right) dy.$$

This equation must be satisfied at each point of the contour. It can be simplified by introducing the *conjugate* of the potential function $\Phi$ (corresponding to the *stream function* in hydrodynamics), which is connected with $\Phi$ by the Cauchy-Riemann equations (19.5). In our present coordinates ($y, z$ instead of $x, y$) we have therefore

$$(18a) \qquad \frac{\partial\Phi}{\partial y} = \frac{\partial\Psi}{\partial z}, \qquad \frac{\partial\Phi}{\partial z} = -\frac{\partial\Psi}{\partial y}.$$

Eq. (18) thus transforms into

$$\left(\frac{\partial\Psi}{\partial z} - \alpha z\right) dz + \left(\frac{\partial\Psi}{\partial y} - \alpha y\right) dy = 0,$$

or

$$d\Psi = \alpha(y\,dy + z\,dz) = \frac{\alpha}{2} d(y^2 + z^2)$$

or, integrated,

$$(19) \qquad \Psi = \frac{\alpha}{2}(y^2 + z^2) + C \cdots \qquad \text{(boundary condition)}.$$

The conjugate potential $\Psi$ satisfies also the differential Eq. (11):

(19a) $$\nabla^2\Psi = 0.$$

By (19) and (19a) the potential function $\Psi$ is uniquely determined being the (always existing) solution of a simple potential problem. We have therefore proved that our boundary condition (18) can be fulfilled. At the same time the special assumptions made in (8), (9), and (12) have been justified.

Let us now consider once more the circular cross-section. The contour is given by $y^2 + z^2 = a^2$, the origin coinciding with the center of the circle. Eq. (19) becomes therefore

$$\Psi = \text{const.}$$

along the contour and, according to (19a), also in the interior. From (18a) we have then

$$\Phi = \text{Const} \quad \text{and by} \quad (7) \; \xi = \text{Const.}$$

which means: *the cross-sections remain plane.* Conversely, if, for a certain shape of the cross-section, $\xi$ vanishes along the rod, we have $\Phi = 0$ by (7) and therefore $\Psi = \text{const}$ by (18a). From (19) the contour is obtained in the form $y^2 + z^2 = \text{const}$: *The cross-section is circular.*

Take now as a less trivial example the elliptical cross-section, the contour of which may be given as

(20) $$\frac{y^2}{a^2} + \frac{z^2}{b^2} = 1.$$

Since the general solution of (19a) is

$$\Psi = \text{Re } f(y + iz), \quad (\text{Re} = \textit{real part of})$$

we shall assume $f$ as the second power of the argument $y + iz$ so that

(21) $$\Psi = A(y^2 - z^2).$$

Along the contour we have according to (19)

$$A(y^2 - z^2) = \frac{\alpha}{2}(y^2 + z^2) + C.$$

On substituting $z^2$ from (20), an identity in $y$ results from which $A$ follows as

(21a) $$A = \frac{\alpha}{2}\frac{a^2 - b^2}{a^2 + b^2}.$$

From (18a) and (21) we can now calculate

$$\frac{\partial \Phi}{\partial y} = -\alpha \frac{a^2 - b^2}{a^2 + b^2} z, \qquad \frac{\partial \Phi}{\partial z} = -\alpha \frac{a^2 - b^2}{a^2 + b^2} y$$

and obtain the shear stresses according to (7) and (14):

$$
(22) \quad
\begin{aligned}
\sigma_{zy} &= 2\mu\epsilon_{zy} = \mu\left(\frac{\partial \Phi}{\partial y} + \frac{\partial \eta}{\partial x}\right) = -\mu\alpha\left(\frac{a^2 - b^2}{a^2 + b^2} + 1\right)z = -2\mu \frac{\alpha a^2}{a^2 + b^2} z, \\
\sigma_{zz} &= 2\mu\epsilon_{zz} = \mu\left(\frac{\partial \Phi}{\partial z} + \frac{\partial \zeta}{\partial x}\right) = -\mu\alpha\left(\frac{a^2 - b^2}{a^2 + b^2} - 1\right)y = 2\mu \frac{\alpha b^2}{a^2 + b^2} y.
\end{aligned}
$$

The shear stress $\tau$ resulting from $\sigma_{zy}$ and $\sigma_{zz}$ is at any point of the cross-section parallel to the tangent that passes through the intersection point of the contour with the prolonged radius vector. Along the radius vector, $\tau$ follows a linear law. The moment of the shear stresses about the $x$-axis results from (22) in the form of the following formula which is a generalization of our previous formula (5):

$$(23) \qquad M = \iint (y\sigma_{zz} - z\sigma_{zy})\, dy\, dz = \frac{2\mu\alpha}{a^2 + b^2} \iint (b^2 y^2 + a^2 z^2)\, dy\, dz.$$

If the integration is carried out one obtains instead of the factor $J_p$, which appears in (5a) as the coefficient of $\mu a$, the expression

$$(24) \qquad \pi \frac{a^3 b^3}{a^2 + b^2},$$

which is not equal to the polar moment of inertia of the area of an ellipse (cf. problem VIII.4).

The simultaneous consideration of the potential function $\Phi$ and the stream function $\Psi$ suggests a hydrodynamic analogy which has proved useful in the approximate solution of more complicated problems such as the torsion of an $I$-beam.

## 43. Torsion and Bending of a Helical Spring

Imagine a coil consisting of a single layer of thin wire wound on a circular cylinder. On removing the mandrel, we obtain a helical spring which we fasten at one end in some way, loading the other end with a disk, the weight of which is supposed to be small enough so that the resulting pitch of the spring is *small* compared to the radius of the cylinder. If we now put additional weight on the disk so as to have a small load acting along the axis of the spring, this will result in a slightly increased pitch and a corresponding lengthening of the spring. *This kind of loading is accompanied by a torsion of the wire.* If, on the other hand, a couple acts on the disk (e.g., when the disk is turned in its own plane) the pitch remains essentially unchanged and the spring yields to the torque by a slight diminution of the cylinder radius. *In this case the wire is bent.*

Both facts seem paradoxical at first, but can be easily clarified: In *torsion* the individual cross-sections of the wire are loaded by a torque about the *polar* axis while in *bending* the moment acts about an *equatorial* axis, that is an axis in the plane of the cross-section. With this in mind let us examine in Figs. 72a and 72b what happens in the planes that pass in either case through the cylinder axis and the cross-sections marked in the respective figures. The load $P$ acting along the cylinder axis in Fig. 72a has the moment $PR$ about the center of the cross-section; we call it

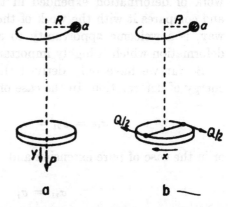

FIG. 72a, b. A helical spring subject (a) to torsion (force $P$, vertical displacement $y$) and (b) to bending (moment $QR$, rotational displacement $x$).

the *torsional moment* $M_T$ since its axial vector is normal to the cross-section. In Fig. 72b the couple $QR$ acts upon the disk whose diameter has been assumed as $2R$ so that the two single forces of the couple can be denoted by $Q/2$. This moment has the character of a *bending moment* $M_B$ since its axial vector coincides with the cylinder axis and can be transferred parallel to itself to the center of the marked cross-section. We denote its position by $NN$ indicating in this way that it coincides with the trace of the neutral layer in the bent cross-section. In the loading according to Fig. 72a there will be *shear stresses* in the plane of the cross-section; they increase proportionally to the distance from the wire axis and will be denoted by $\tau$ as in the previous article. We have therefore

$$(1) \qquad\qquad \int r\tau \, df = M_T \, .$$

In the loading according to Fig. 72b *normal stresses* $\sigma$ act upon the cross-section. They are proportional to the distance $y$ from the equatorial axis $NN$. Again we have

$$(2) \qquad\qquad \int y\sigma \, df = M_B \, .$$

Eqs. (1) and (2) may be evaluated according to (42.4a) and (42.5a), or (41.5):

$$(3) \qquad \frac{\tau_{max}}{a} J_p = M_T \qquad \text{respectively} \qquad \frac{\sigma_{max}}{a} I = M_B .$$

Apart from the maximum stresses $\tau_{max}$ and $\sigma_{max}$ we should like to know the displacement of the spring in vertical direction $y$ and horizontal direction $x$ for the two respective kinds of loading. For this purpose we should have to know the shape of the elastic line of an *originally curved* rod in torsion and bending. There is, however, a simpler way to answer our question when use is made of the *energy principle*: one calculates the work of deformation expended in the torsion or bending of the spring and compares it with the work of the external forces $P$ or $Q$. This, by the way, is a welcome opportunity to come back to the topic of energy of deformation which is highly important in engineering applications.[9]

So far we have only derived the general expression (9.26b) for the energy of deformation. In the case of pure shear, where

$$\sigma_{12} = \sigma_{21} = \tau, \qquad \epsilon_{12} = \epsilon_{21} = \frac{\tau}{2\mu} ,$$

or in the case of pure extension and compression (bending), where

$$\sigma_{11} = \sigma, \qquad \epsilon_{11} = \frac{\sigma}{E} ,$$

(9.26b) reduces to

$$(4) \qquad W = \frac{1}{2}(\sigma_{12}\epsilon_{12} + \sigma_{21}\epsilon_{21}) = \frac{1}{2}\frac{\tau^2}{\mu}$$

or to

$$(5) \qquad W = \frac{1}{2}\sigma_{11}\epsilon_{11} = \frac{1}{2}\frac{\sigma^2}{E} ,$$

respectively. These expressions refer to the unit volume and must be applied in the case of locally variable stresses to each single volume element $df\,ds$, where $df$ is the area element of the cross section and $ds$ the length element of the axis of the rod. Thus we obtain from (4) in the case of torsion

$$(6) \qquad dW = \frac{1}{2}\frac{\tau^2}{\mu} df\,ds, \qquad \tau = \frac{r}{a}\tau_{max} \qquad \text{from (42.4a).}$$

---

[9] In all statically indeterminate systems, in particular in the theory of trusses, Castigliano's theorems about the *minimum of the energy of deformation* play an important role, and so does *Maxwell's theorem of the reciprocity of the displacements*.

Integration over the cross-section yields

(6a)
$$\int dW = \frac{1}{2\mu}\frac{\tau^2_{\text{max}}}{a^2}\int r^2\, df\, ds = \frac{\tau^2_{\text{max}}}{2\mu a^2}\, J_p\, ds,$$

and, by (3),

(6b)
$$\int dW = \frac{M_T^2}{2\mu J_p}\, ds.$$

The entire torsional energy $W_T$ follows by integration over the length $l$ of the rod (in our case the length of the spring):

(7)
$$W_T = \frac{M_T^2 l}{2\mu J_p} = \frac{P^2 R^2 l}{2\mu J_p}.$$

Likewise one obtains in the case of bending from (5)

(8)   $$dW = \frac{1}{2}\frac{\sigma^2}{E}\, df\, ds, \qquad \sigma = \frac{y}{a}\,\sigma_{\text{max}} \qquad \text{according to (41.2),}$$

and by integration over the cross-section

(8a)
$$\int dW = \frac{1}{2E}\frac{\sigma^2_{\text{max}}}{a^2}\int y^2\, df\, ds = \frac{\sigma^2_{\text{max}}}{2Ea^2}\, I\, ds.$$

The entire bending energy contained in the spring is obtained from (8a) and (3) as

(9)
$$W_B = \frac{M_B^2 l}{2EI} = \frac{Q^2 R^2 l}{2EI}.$$

Expressions (7) and (9) represent the energy intake of the spring in an *elastic* deformation in which the spring passes through a sequence of states of equilibrium. It must be compared with the *mechanical* work performed along the paths $y$ (or $x$) when the load $P$ (or the couple $QR$) increases *gradually* from zero to its final value. The work is therefore

(10)   $$W_T = \frac{Py}{2}, \qquad \text{respectively} \qquad W_B = 2\,\frac{(Q/2)x}{2} = \frac{Qx}{2}.$$

Comparison with (7) and (9) gives

(11)
$$y = \frac{R^2 l}{\mu J_p}\, P, \qquad x = \frac{R^2 l}{EI}\, Q.$$

Thus the use of the energy principle leads indeed to a very simple determination of the displacements $x$ and $y$. Note that Eqs. (11) can also be obtained from the first principle of Castigliano in the following way:[10]

$$(11a) \qquad W(P, Q) = W_T + W_B, \qquad y = \frac{\partial W}{\partial P}, \qquad x = \frac{\partial W}{\partial Q}.$$

From these equations we can easily pass to the *free oscillations* of the spring when we replace $P$ and $Q$ by the inertia forces

$$(12) \qquad P = -M \frac{d^2 y}{dt^2}, \qquad Q = -M_{red} \frac{d^2 x}{dt^2}.$$

Here the masses $M$ and $M_{red}$ have to be given the appropriate physical meaning. $M$ is essentially the mass $M_0$ of the disk that oscillates up and down, but since the mass $m$ of the spring partakes also in the oscillation, (its lower portion more strongly than its upper one) a fraction of this mass must be added as a correction to the mass of the disk. According to Lord Rayleigh this fraction is $\frac{1}{3}$, hence $M$ is to be interpreted as $M_0 + m/3$. On the other hand, the inertia force governing the oscillations about the cylinder axis is $-\Theta \, d^2\varphi/dt^2$, where $\varphi = x/R$ is the angular displacement of the disk according to the meaning of $x$ in Fig. 72b. Again $\Theta$ is essentially the moment of inertia $\Theta_0$ of the disk augmented by $\frac{1}{3}$ of the moment of inertia of the coil $mR^2$. Consequently $M_{red}$ is to be defined as

$$M_{red} R^2 = \Theta = \Theta_0 + \frac{1}{3} mR^2.$$

The equations of the free torsional and bending oscillations are according to (12) and (11)

$$(13) \qquad \frac{d^2 y}{dt^2} + \frac{\mu J_p}{MR^2 l} y = 0, \qquad \frac{d^2 x}{dt^2} + \frac{EI}{M_{red} R^2 l} x = 0$$

and the associated circular frequencies

$$(14) \qquad \omega_T = \sqrt{\frac{\mu J_p}{MR^2 l}}, \qquad \omega_B = \sqrt{\frac{EI}{M_{red} R^2 l}}.$$

In general they are different from each other.[11] If they are equal, the two

---

[10]The rule (11a) presupposes that $y$, $x$ and $P$, $Q$ are linearly connected so that $W(P, Q)$ is a quadratic expression in $P$ and $Q$ as in (7) and (9). In the general case one would have to replace $W$ by the "modified" expression for the energy $U(P, Q) = yP + xQ - W(x, y)$. Cf. about this transformation Vol. I. 42.7 and E. L. Ince, Ordinary Differential Equations, Dover Publ., New York, 1944, Chapter 2.5 (Legendre's transformation).

[11]This can be enforced in an actual experiment by loading the spring with an additional weight that acts along its axis and has a small moment of inertia about it (cf. Vol. I, Fig. 37). Thus the mass $M$ that takes part in the torsional oscillation is increased and the frequency $\omega_T$ reduced. $M_{red}$ and $\omega_B$ remain essentially unchanged, however.

oscillations are in *resonance* and the behavior of the spring becomes very sensitive to small changes of the parameters involved. Then the present theory is no longer sufficient: we have to take into account that the two modes of oscillation are coupled through the finite pitch of the spring which so far has been neglected.

For finite pitch $h$ the slope $\alpha$ of the helix is given by

$$\sin \alpha = \frac{h}{2\pi R} \cdot$$

This is also the angle of inclination that the cross-section of the wire forms with the vertical. The two loading moments $PR$ and $QR$ may then be decomposed as follows:

|      | normal to the cross-section | parallel to the cross-section |
|------|-----------------------------|-------------------------------|
| $PR$ | $\cos \alpha \cdot PR$ | $-\sin \alpha \cdot PR$ |
| $QR$ | $\sin \alpha \cdot QR$ | $\cos \alpha \cdot QR$ |

The contributions to the torsional moment are on the left side, those to the bending moment on the right side, therefore

$$M_T = \cos \alpha PR + \sin \alpha QR, \qquad M_B = \cos \alpha QR - \sin \alpha PR.$$

The total bending energy becomes

$$W = W_T + W_B = \frac{M_T^2 l}{2\mu J_p} + \frac{M_B^2 l}{2EI}$$

(15a)

$$= \frac{R^2 l}{2} \left( \frac{(\cos \alpha P + \sin \alpha Q)^2}{\mu J_p} + \frac{(\cos \alpha - \sin \alpha P)^2}{EI} \right),$$

and the displacements can be determined according to the rule (11a):

$$y = \frac{\partial W}{\partial P} = \frac{R^2 l}{\mu J_p} (\cos^2 \alpha P + \sin \alpha \cos \alpha Q)$$

$$+ \frac{R^2 l}{EI} (-\sin \alpha \cos \alpha Q + \sin^2 \alpha P),$$

(15b)

$$x = \frac{\partial W}{\partial Q} = \frac{R^2 l}{\mu J_p} (\sin \alpha \cos \alpha P + \sin^2 \alpha Q)$$

$$+ \frac{R^2 l}{EI} (\cos^2 \alpha Q - \sin \alpha \cos \alpha P).$$

These equations now replace Eqs. (11). Upon replacing the inertia forces (12) by $P$ and $Q$, one obtains the following system of equations which describes *the coupled oscillations of the spring*:

$$(16) \qquad \frac{d^2y}{dt^2} + \omega_T^2(y - kx) = 0, \qquad \frac{d^2x}{dt^2} + \omega_B^2(x - hy) = 0$$

where $\omega_T$, $\omega_B$, $k$, and $h$ are

$$(16a) \qquad \omega_T^2 = \frac{\mu J_p}{MR^2 l}(1 + \nu \sin^2 \alpha), \qquad \omega_B^2 = \frac{\mu J_p}{M_{red}R^2 l}(1 + \nu \cos^2 \alpha),$$

$$(16b) \qquad k = \frac{\nu \sin \alpha \cos \alpha}{1 + \nu \sin^2 \alpha}, \qquad h = \frac{\nu \sin \alpha \cos \alpha}{1 + \nu \cos^2 \alpha}$$

and $\nu$ is Poisson's ratio. The coupling coefficients $k$ and $h$ vanish if $\alpha = 0$ so that the frequencies $\omega_T$ and $\omega_B$ reassume the previous values (14) and Eqs. (16) become identical with Eqs. (13).[12]

The further discussion of Eq. (16) is no longer a problem of theory of elasticity. For the case of resonance, $\omega_T = \omega_B$, the behavior of the spring has already been described in Vol. I 20: the energy passes back and forth between the originally excited mode and the mode that is excited consequential to it, to the extent that the two modes are alternately extinguished. The *beat period* $N$ expressed in time units of single oscillations can be very accurately determined by counting the number of oscillations in a beat. It becomes a maximum in the case of resonance. The large number $N_{max}$ and the small angle $\alpha$ permit a precise determination[13] of Poisson's ratio $\nu$, as shown by the author.

The resonance phenomena discussed here have first been described and explained by Wilberforce,[14] but his determination of Poisson's ratio is founded on the relations (14) rather than on the more elegant resonance method.

### 44. The Elastic Energy Content of a Rectangular Parallelepiped

In 14 we have studied elastic waves in an *infinitely extended* isotropic solid; in the following two articles we shall discuss the elastic oscillations in *finite* bodies, dealing first with a rectangular parallelepiped with sides $a, b, c$. Let us assume that the faces which are situated at $x = 0, a; y = 0, b;$

---

[12]In comparing the values of $\omega_B$ in (16a) and (14), use relation $J_P = 2I$ for circular cross-sections and relation (9.11), $E = 2\mu(1 + \nu)$, which has already been used in the transition from (15) to (16).

[13]Wüllner-Festschrift, Teubner, Leipzig 1905, p. 162.

[14]Phil. Mag. 38 (1894) 386.

$z = 0, c$, are free, in which case we must stipulate that the corresponding normal and shear stresses vanish. The oscillations possible under these conditions are called the *proper oscillations* of the body; they are executed without any *exchange of energy* with the surroundings.

The problem as it has just been outlined presents great mathematical difficulties.[15] A particular case of the general problem, the vibrations of a rectangular plate, has withstood all attempts at a rigorous solution although Chladni's beautiful figures have been known for 150 years and are still a challenge for the best mathematicians. The most successful step so far was taken by Ritz;[16] it does not, however, give a direct solution of the differential equations with their boundary conditions, but establishes a minimum principle by which a stepwise approximation of the solution becomes feasible.

It is, however, possible to prevent the exchange of energy with the surroundings by other sets of boundary conditions. Consider the work performed per unit of $x$-surface in the displacement with components $\xi, \eta, \zeta$:

$$(1) \qquad \delta A = \sigma_{xx}\xi + \sigma_{xy}\eta + \sigma_{xz}\zeta.$$

It is not only zero if

$$(2) \qquad \sigma_{xx} = \sigma_{xy} = \sigma_{xz} = 0 \qquad \text{(surface free of stress)}$$

or if

$$(2a) \qquad \xi = \eta = \zeta = 0 \qquad \text{(surface kept immovable)},$$

but also if for example

$$(3) \qquad \xi = 0, \qquad \sigma_{xy} = \sigma_{xz} = 0$$

or

$$(3a) \qquad \sigma_{xx} = 0, \qquad \eta = \zeta = 0.$$

The assumption of *mixed* boundary conditions as in (3) or (3a) makes an elementary treatment of our problem possible as C. Somigliana[17] showed in 1902 for the case of static loading; they are, on the other hand, equiva-

---

[15]The special case of the *cube* which is usually studied in similar problems of electromagnetic radiation is essentially not simpler than the more general case considered here. One can see that from the final result of Eq. (23), which refers only to the size of the body $V$, but not to its shape.

[16]W. Ritz, Theorie der Transversalschwingungen einer quadratischen Platte mit freien Rändern, Ann. Physik 28 (1909) 737. The circular plate can be directly integrated, as was shown by Kirchhoff.

[17]See the comprehensive article of O. Tedone in Encykl. d. Mathem. Wiss. Vol. IV, 4, Art. 25, No. 2.

lent to the uniform conditions (2) or (2a) as regards the physical applications which we have in mind.

We therefore base our calculation on the boundary conditions (3) and the corresponding conditions for $y$ and $z$-surfaces, obtained from (3) by rotating the letters. Since the first Eq. (3) implies

$$\frac{\partial \xi}{\partial y} = \frac{\partial \xi}{\partial z} = 0, \qquad 2\epsilon_{xy} = \frac{\partial \eta}{\partial x}, \qquad 2\epsilon_{xz} = \frac{\partial \zeta}{\partial x},$$

the boundary conditions can be written:

$$\xi = 0, \qquad \frac{\partial \eta}{\partial x} = \frac{\partial \zeta}{\partial x} = 0 \qquad \text{for} \quad x = \begin{cases} 0 \\ a \end{cases},$$

(4)
$$\eta = 0, \qquad \frac{\partial \zeta}{\partial y} = \frac{\partial \xi}{\partial y} = 0 \qquad \text{for} \quad y = \begin{cases} 0 \\ b \end{cases},$$

$$\zeta = 0, \qquad \frac{\partial \xi}{\partial z} = \frac{\partial \eta}{\partial z} = 0 \qquad \text{for} \quad z = \begin{cases} 0 \\ c \end{cases}.$$

Our differential equations are the Eqs. (14.1b) which because of $\mathbf{F} = 0$ now read

$$\rho \frac{\partial^2 \xi}{\partial t^2} = \mu \nabla^2 \xi + (\lambda + \mu) \frac{\partial \Theta}{\partial x},$$

(5)
$$\rho \frac{\partial^2 \eta}{\partial t^2} = \mu \nabla^2 \eta + (\lambda + \mu) \frac{\partial \Theta}{\partial y},$$

$$\rho \frac{\partial^2 \zeta}{\partial t^2} = \mu \nabla^2 \zeta + (\lambda + \mu) \frac{\partial \Theta}{\partial z}.$$

Eqs. (4) can be satisfied by the following set of expressions for $\xi, \eta, \zeta$:

$$\xi = A \sin \frac{a\pi x}{a} \cos \frac{b\pi y}{b} \cos \frac{c\pi z}{c} e^{-i\omega t},$$

(6)
$$\eta = B \cos \frac{a\pi x}{a} \sin \frac{b\pi y}{b} \cos \frac{c\pi z}{c} e^{-i\omega t},$$

$$\zeta = C \cos \frac{a\pi x}{a} \cos \frac{b\pi y}{b} \sin \frac{c\pi z}{c} e^{-i\omega t}.$$

where $\mathfrak{a}$, $\mathfrak{b}$, $\mathfrak{c}$ are arbitrary integer numbers. $A$, $B$, $C$ are constants whose ratios are to be determined; also the circular frequency $\omega$ must be found.

On introducing the expressions (6) in (5), there appear, in each of the three terms of the equations (5), the same trigonometrical factors as in $\xi$, $\eta$, $\zeta$. One has for instance

$$\rho \frac{\partial^2 \xi}{\partial t^2} = -\rho \omega^2 A \sin \cdots \cos \cdots \cos \cdots e^{-i\omega t},$$

$$\nabla^2 \xi = -A\left(\frac{\mathfrak{a}^2}{a^2} + \frac{\mathfrak{b}^2}{b^2} + \frac{\mathfrak{c}^2}{c^2}\right)\pi^2 \sin \cdots \cos \cdots \cos \cdots e^{-i\omega t},$$

$$\frac{\partial \theta}{\partial x} = -\left(A \frac{\mathfrak{a}^2}{a^2} + B \frac{\mathfrak{a}\mathfrak{b}}{ab} + C \frac{\mathfrak{a}\mathfrak{c}}{ac}\right)\pi^2 \sin \cdots \cos \cdots \cos \cdots e^{-i\omega t}.$$

On omission of these factors, the differential equations (5) give the following linear system of equations for $A$, $B$, $C$

$$A\left(\Omega + \frac{\mathfrak{a}\mathfrak{a}}{a^2}\right) + B \frac{\mathfrak{a}\mathfrak{b}}{ab} + C \frac{\mathfrak{a}\mathfrak{c}}{ac} = 0,$$

(7)
$$A \frac{\mathfrak{b}\mathfrak{a}}{ba} + B\left(\Omega + \frac{\mathfrak{b}\mathfrak{b}}{b^2}\right) + C \frac{\mathfrak{b}\mathfrak{c}}{bc} = 0,$$

$$A \frac{\mathfrak{c}\mathfrak{a}}{ca} + B \frac{\mathfrak{c}\mathfrak{b}}{cb} + C\left(\Omega + \frac{\mathfrak{c}\mathfrak{c}}{c^2}\right) = 0.$$

Here $\Omega$ is defined by

(7a)    $$(\lambda + \mu)\Omega = \mu s - \frac{\rho \omega^2}{\pi^2}, \qquad s = \frac{\mathfrak{a}^2}{a^2} + \frac{\mathfrak{b}^2}{b^2} + \frac{\mathfrak{c}^2}{c^2}.$$

The linear equations (7) have solutions different from zero only, if the determinant

(8)    $$\Delta = \begin{vmatrix} \Omega + \dfrac{\mathfrak{a}\mathfrak{a}}{a^2}, & \dfrac{\mathfrak{a}\mathfrak{b}}{ab}, & \dfrac{\mathfrak{a}\mathfrak{c}}{ac} \\[2mm] \dfrac{\mathfrak{b}\mathfrak{a}}{ba}, & \Omega + \dfrac{\mathfrak{b}\mathfrak{b}}{b^2}, & \dfrac{\mathfrak{b}\mathfrak{c}}{bc} \\[2mm] \dfrac{\mathfrak{c}\mathfrak{a}}{ca}, & \dfrac{\mathfrak{c}\mathfrak{b}}{cb}, & \Omega + \dfrac{\mathfrak{c}\mathfrak{c}}{c^2} \end{vmatrix} = 0$$

On expanding (8) in powers of $\Omega$, one sees easily that the factors of $\Omega^0$

and $\Omega^1$ vanish while those of $\Omega^2$ and $\Omega^3$ are $s$ and 1. Eq. (8) can thus be rewritten in the form

$$(9) \qquad \Omega^3 + s\Omega^2 = \Omega^2(\Omega + s) = 0,$$

an equation that has the double root $\Omega_1 = \Omega_2 = 0$ and the simple root $\Omega_3 = -s$. From (7a) we obtain the corresponding frequency values

$$(10) \qquad \omega_1^2 = \omega_2^2 = \pi^2 \frac{\mu}{\rho} s, \qquad \omega_3^2 = \pi^2 \frac{\lambda + 2\mu}{\rho} s.$$

The ratios $A : B : C$ associated with these values of $\omega$ are found from (7) so that the set (6) has now been determined except for a common amplitude factor. Further arbitrary elements are the numbers $\mathfrak{a}, \mathfrak{b}, \mathfrak{c}$.

What is the physical meaning of this arbitrariness? By the numbers $\mathfrak{a}, \mathfrak{b}, \mathfrak{c}$ the sides $a, b, c$ are subdivided into segments which must be interpreted as wave lengths:

$$(11) \qquad \lambda_a = \frac{a}{\mathfrak{a}}, \qquad \lambda_b = \frac{b}{\mathfrak{b}}, \qquad \lambda_c = \frac{c}{\mathfrak{c}}.$$

*They just fit the boundaries of the parallelepiped.* The associated wave numbers are

$$(11a) \qquad k_a = \frac{2\pi}{\lambda_a} = 2\pi \frac{\mathfrak{a}}{a}, \qquad k_b = \frac{2\pi}{\lambda_b} = 2\pi \frac{\mathfrak{b}}{b}, \qquad k_c = \frac{2\pi}{\lambda_c} = 2\pi \frac{\mathfrak{c}}{c}.$$

With a view to Eqs. (6) and to Eqs. (11) we can write

$$(12) \qquad \xi = A \sin \pi \frac{x}{\lambda_a} e^{-i\omega t}, \qquad \eta = \zeta = 0,$$

if we consider a particular solution depending on $x$ only. Eq. (12) represents a standing wave which has the boundaries $x = 0$ and $x = a = \mathfrak{a}\lambda_a$ as nodal surfaces. Altogether, the Eqs. (6) represent the entirety of standing waves that can be accommodated by the parallelepiped. Unlike the example (12), however, they are in general not plane waves but are composed of waves in the three directions of the edges according to the particular rule set by Eqs. (6).

The structure of Eqs. (10) teaches further that the waves are partly of *transverse*, partly of *longitudinal* nature. According to Eqs. (14.6) and (14.3) we have

$$c_{trans} = \sqrt{\frac{\mu}{\rho}}, \qquad c_{long} = \sqrt{\frac{\lambda + 2\mu}{\rho}},$$

where now $c_{trans}$ and $c_{long}$ stand for the previous notations $b$ and $a$ for the sound velocities. If we further introduce a wave number $k_r$ , resulting from the three wave numbers $k_a$ , $k_b$ , $k_c$ , by

$$k_r = \sqrt{k_a^2 + k_b^2 + k_c^2} \, ,$$

we obtain from (11a) and (7a)

(13)          $k_r = 2\pi \sqrt{\dfrac{a^2}{a^2} + \dfrac{b^2}{b^2} + \dfrac{c^2}{c^2}} = 2\pi \sqrt{s}.$

Eqs. (10) are then simply

(14)          $\omega_1 = \omega_2 = \dfrac{1}{2} c_{\text{trans}} \, k_r \, , \qquad \omega_3 = \dfrac{1}{2} c_{\text{long}} \, k_r \, .$

The fact that the transverse waves are counted twice points to the physical possibility of polarization in two mutually orthogonal directions.

We have thus found a system of infinitely many elastic waves which can be excited in an adiabatically insulated parallelepiped. The system is *complete*, that is, the *only* standing waves that obey our boundary conditions are those of the system (6). The proof of completeness consists in showing that an arbitrarily given initial state can be expanded in terms of the functions (6) for $t = 0$.[18]

Let us now determine the number $z$ of all waves whose frequency is smaller than a given $\nu = \omega/2\pi$, which can be done by geometrical reasoning. We first count the longitudinal waves.

By marking in the rectangular system $k_a$ , $k_b$ , $k_c$ all points that belong to integer values a, b, c, we obtain a three-dimensional orthorhombic lattice, whose elementary cell has according to (11a) the sides

$$\frac{2\pi}{a} \, , \qquad \frac{2\pi}{b} \, , \qquad \frac{2\pi}{c} \, .$$

Consider now the lattice points inside the spherical surface $k_r$ = const, which corresponds to the frequency $\nu$. According to (14) the radius of the sphere $k_r = 4\pi\nu/c_{\text{long}}$ . Because of a, b, c $\geqq$ 0, we are only interested in the lattice points that lie in the positive octant, whose volume is

$$\frac{4\pi}{3} \frac{k_r^3}{8} = \frac{4\pi}{3} \left( \frac{2\pi\nu}{c_{\text{long}}} \right)^3 .$$

---

[18]Cf. for such a proof and for a generalization to anisotropic bodies: R. Ortvay, Ann. Physik. 42 (1913), 745. A treatment of the problem from the point of view of group theory was presented by H. Wierzejewski in his Breslau thesis suggested by E. Fues: Elastische Eigenschwingungen von Kristallen bei gemischten Randbedingungen, Z. f. Kristallographie, Vol. 101, 1939.

On dividing by the volume of the elementary cell

$$(15) \qquad \frac{(2\pi)^3}{abc},$$

we obtain the number of elementary cells or the number of lattice points in the spherical octant which, at the same time, is the number of longitudinal waves with frequencies below $\nu$:

$$(16) \qquad \mathcal{z}_{\text{long}} = \frac{4\pi}{3} \frac{abc\nu^3}{c_{\text{long}}^3}.$$

In exactly the same way one has for either state of polarization of the transversal waves

$$(16a) \qquad \mathcal{z}_{\text{trans}} = \frac{4\pi}{3} \frac{abc\nu^3}{c_{\text{trans}}^3};$$

thus the total number of vibrational states below $\nu$ becomes

$$(16b) \qquad \mathcal{z} = \mathcal{z}_{\text{long}} + 2\mathcal{z}_{\text{trans}} = \frac{4\pi}{3} abc\nu^3 \left( \frac{2}{c_{\text{trans}}^3} + \frac{1}{c_{\text{long}}^3} \right).$$

The number $\mathcal{z}$ just obtained is accurate except for an error which is due to the fact that the spherical surface and the boundaries of the enclosed elementary cells do not coincide. The error is of the order of magnitude $1/abc$ [cf. (15)] and thus negligible for a large parallelepiped. By differentiating (16b) with respect to $\nu$ we can now determine the number of vibrational states between $\nu$ and $\nu + d\nu$:

$$(17) \qquad d\mathcal{z} = 4\pi V \left( \frac{2}{c_{\text{trans}}^3} + \frac{1}{c_{\text{long}}^3} \right) \nu^2 \, d\nu$$

where $V$ is the volume $abc$ of the parallelepiped.

We are now ready to attack the problem which we formulated at the outset. We want to calculate *the energy content of the adiabatically insulated solid body*, which, according to the definition in the older literature, is *identical* with the *vibrational energy of its molecules*. It is preferable, however, to follow Debye[19] who defines the energy content as that of

---

[19]P. Debye, Ann. Physik., **39** (1912) 789. In this fundamental paper Debye calculates the proper vibrations of a sufficiently large *sphere*, the surface of which is considered *free of stresses*, that is, he uses the boundary conditions corresponding to our equations (2). As the circular shape is the only case in which the plate problem can be directly integrated (cf. footnote, p. 315), so the sphere is the only elastic three-dimensional problem whose integration leads in the case of "uniform boundary conditions" to known functions (Bessel functions). But even in this case the solution is very involved and becomes only possible by the use of asymptotic approximations of Bessel functions for large argument and index that are due to Debye. The simplification that can be

the energy of the elastic vibrations of the body. The molecules are strongly coupled, hence it is impossible to excite oscillations of individual molecules; on the other hand, the waves considered in (6) are independent from each other and satisfy individually the differential equations of the elastic body.

Each of these vibrational states may therefore be quantized individually according to the quantum rules first established by Planck in the discovery of his radiation law:

$$(18) \qquad \epsilon_n = nh\nu \begin{cases} n = \text{integer number} \\ h = \text{Planck's constant} \end{cases}$$

The quantum rules are here applied without further explanation; they will be discussed in Vol. V. Also the refinement of the rules due to modern quantum mechanics according to which the integer number $n$ in (18) should be replaced by $n + \frac{1}{2}$ is here disregarded as irrelevant for our purpose.

In addition, we require for the following the Boltzmann factor which was introduced in (7.15c), though without proof. In the present case the Boltzmann factor has this physical meaning: Given $N$ oscillating systems in thermodynamical equilibrium at the temperature $T$, the number $N_n$ of those systems whose energy equals $\epsilon_n$ is then

$$(19) \qquad N_n = A \exp\left(-\frac{\epsilon_n}{kT}\right) = A\alpha^n, \qquad \alpha = \exp\left(-\frac{h\nu}{kT}\right);$$

the constant $A$ can be determined, when the total number $N$ of all oscillating systems is written as the sum of the $N_n$ :

$$(20) \qquad N = \sum_{n=0}^{\infty} N_n = A(1 + \alpha + \alpha^2 + \cdots) = \frac{A}{1 - \alpha}.$$

In our case, $N$ is to be identified with the (large!) number $d_{\frac{1}{3}}$ from (17). We then obtain for $A$

$$(20a) \qquad A = (1 - \alpha)d_{\frac{1}{3}}.$$

---

achieved by "mixed boundary conditions" was first stated by the author in a course on quantum theory in 1912 and then left to Mr. Ortvay for further development and publication (cf. the preceding footnote). It should be noticed that Debye had to take the radius of the sphere sufficiently large to enforce independence of the distribution of the eigenfunctions from the shape of the body. For the same reason the dimensions of the parallelepiped in the present case must be sufficiently large as was already assumed in the text preceding Eq. (17) in order to make the "surface error" negligibly small.

Let us denote by $dU$ that fraction of the total energy $U$ which is contained in these $d\mathfrak{z}$ oscillations. According to (19) and (20a), one has

$$(21) \qquad dU = \sum_{n=0}^{\infty} N_n \epsilon_n = Ah\nu \sum n\alpha^n = (1 - \alpha)h\nu \, d\mathfrak{z} \sum_{n=0}^{\infty} n\alpha^n.$$

The series occurring here can be summed easily (by differentiating the geometrical series in (20) with respect to $\alpha$), and equals $\alpha/(1 - \alpha)^2$. Eq. (21) yields together with the definition of $\alpha$ in (19)

$$(22) \qquad dU = h\nu \frac{\alpha}{1 - \alpha} \, d\mathfrak{z} = \frac{h\nu}{\exp\left(\dfrac{h\nu}{kT}\right) - 1} \, d\mathfrak{z}.$$

The denominator in this formula will reoccur in Planck's radiation law and, in a more general way, in the discussion of the Bose statistics in Vol. V.

We now substitute for $d\mathfrak{z}$ according to (17) and integrate over $\nu$ from $\nu = 0$ up to an upper limit $\nu_g$ :

$$(23) \qquad U = 4\pi V h \left(\frac{2}{C_{\text{trans}}^3} + \frac{1}{C_{\text{long}}^3}\right) \int_0^{\nu_g} \frac{\nu^3 \, d\nu}{\exp\left(\dfrac{h\nu}{kT}\right) - 1}.$$

Our first task is the determination of the limit $\nu_g$ , which we find according to Debye by identifying the total number $\mathfrak{z}_g$ of oscillations with frequencies smaller than $\nu_g$ with the *number of degrees of freedom of the body*. Since a single mass point has three degrees of freedom the number in question equals three times the number of molecules. If in particular the body is one mole of a uniform substance, the number of molecules, according to p. 53, is Loschmidt's number $L$. The number $\mathfrak{z}_g$ thus equals $3L$. From Eq. (16b) we then obtain for the determination of the upper limit $\nu_g$ the relation

$$(24) \qquad 3L = \frac{4\pi}{3} V_{\text{mol}} \nu_g^3 \left(\frac{2}{C_{\text{trans}}^3} + \frac{1}{C_{\text{long}}^3}\right),$$

where the molar volume $V_{\text{mol}}$ has been written instead of $abc$. This equation that defines $\nu_g$ may also be used to simplify (23) which now takes the form

$$(25) \qquad U_{\text{mol}} = \frac{9Lh}{\nu_g^3} \int_0^{\nu_g} \frac{\nu^3 \, d\nu}{\exp\left(\dfrac{h\nu}{kT}\right) - 1}.$$

With the abbreviations

$$(26) \qquad y = \frac{h\nu}{kT}, \qquad x = \frac{h\nu_\theta}{kT} = \frac{\Theta}{T}, \qquad \Theta = \frac{h}{k}\nu_\theta$$

we obtain finally

$$(27) \qquad U_{mol} = 9Lk \frac{T^4}{\Theta^3} \int_0^x \frac{y^3 \, dy}{e^y - 1}.$$

The quantity $\Theta$ here introduced is known as *Debye's characteristic temperature*; for hard materials such as diamond and quartz it is approximately 100° K, for soft metals it amounts only to a few degrees. By and large, Eq. (27) is a good representation of the large number of data collected by Nernst and his co-workers. One can, of course, not expect that such details as depend on the specific structure of the crystal or the molecule should be covered by the general theory here developed.

We examine now Eq. (27) in the case of $T \gg \Theta$. This makes the upper limit $x \ll 1$, and the same is true for the integration variable $y$. Expanding the denominator

$$e^y - 1 = y\left(1 + \frac{y}{2} \cdots \right),$$

we obtain for the integral in (27) with sufficient accuracy

$$\int_0^x y^2 \, dy = \frac{x^3}{3} = \frac{1}{3}\frac{\Theta^3}{T^3}.$$

From this we infer

$$(28) \qquad U_{mol} = 3RT,$$

where the gas constant per mole, $R$, has been written for $kL$ according to Eq. (7.15b).

Relation (28) contains the law of *Dulong-Petit*, since

$$(29) \qquad c_{mol} = \frac{dU_{mol}}{dT} = 3R \cong 6 \text{ cal},$$

where $c_{mol}$ denotes the specific heat per mole of the substance. Eq. (29) should be universally valid for all solids as an asymptotic law for high temperatures, provided one can disregard the relative vibrations of the components of the individual molecule. This restriction is necessary, since in counting the degrees of freedom we have assumed rigid molecules and disregarded the possible internal degrees of freedom. In fact, the rule of Dulong and Petit is particularly well satisfied for monatomic metals where the restriction is irrelevant. (In this case, $c_{mol}$ is the gram atomic heat capacity).

The converse case, $T \ll \theta$, presents greater physical interest. The integral in (27) has now a very large upper limit ($x \gg 1$) and can be approximated by

$$(30) \qquad \int_0^\infty \frac{y^3 e^{-y}}{1 - e^{-y}}\, dy = \int_0^\infty y^3 e^{-y}(1 + e^{-y} + e^{-2y} + \cdots)\, dy.$$

Upon interchanging the order of summation and integration and introducing $z = ny$ in the $n^{\text{th}}$ integral, one obtains instead of (30)

$$(30a) \qquad \sum_{n=1}^\infty \frac{1}{n^4} \int_0^\infty z^3 e^{-z}\, dz = 3! \sum_{n=1}^\infty \frac{1}{n^4} = 3! \frac{\pi^4}{90}.$$

(The summation formula $\sum_{n=1}^\infty n^{-4} = \pi^4/90$ will be derived in Vol. VI in the chapter on Fourier series.)

With the use of the results (30) and (30a), Eq. (27) can be rewritten as

$$(31) \qquad U_{\text{mol}} = \frac{3\pi^4}{5} \frac{RT^4}{\theta^3},$$

and differentiation with respect to $T$ yields

$$(32) \qquad c_{\text{mol}} = \frac{12\pi^4}{5} R\left(\frac{T}{\theta}\right)^3.$$

This is *Debye's third power law for the molar specific heat of a solid body at low temperature.*

Fig. 73 represents the general dependence of the specific heat on the temperature, as it is obtained by differentiation of Debye's equation (27); it also shows the two asymptotic laws for large $T$, i.e. the horizontal

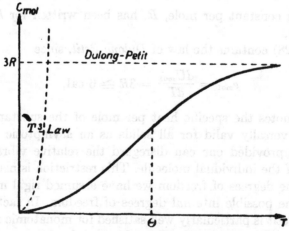

FIG. 73. The specific heat per mole of a solid according to Debye, cf. Eq. (26) for the meaning of $\Theta$

straight line of the Dulong-Petit law, and for small $T$ the $T^3$-parabola of Debye. The latter law is a useful approximation only for extremely small values of $T$. Furthermore, there is a general deviation from this law in the proximity of absolute zero for metallic substances, which is caused by the degrees of freedom of the conduction electrons in this temperature region.

We do not want to leave this topic without mentioning the results that are obtained in the corresponding electromagnetic problem. In that case we are concerned with the energy confined within a cavity containing electromagnetic radiation but entirely devoid of matter. The free oscillations of such a cavity can be described and counted in a fashion similar to the case of the elastic parallelepiped. (Actually, the electromagnetic case is much simpler.) Methods to deal with this problem have been developed by Lord Rayleigh and Jeans, including the procedure of counting lattice points. These investigations have served as a pattern for the elastic theory developed in this article. In the electromagnetic case the result is equivalent to our Eq. (23) with the following simplifications:

1. Longitudinal waves do not occur in the electromagnetic field, so that the term $1/c^3_{long}$ does not occur.

2. Transverse waves travel with the velocity of light $c$; hence the term $2/c^3_{trans}$ is simply $2/c^3$.

3. The number of degrees of freedom of the electromagnetic field is, as far as we know, unlimited (at least, it is not restricted by a molecular structure); $\nu_g$ becomes thus infinitely large.

4. The energy content per unit of volume (energy density) is obtained, if $V = 1$ in (23). In this case, one usually writes $u$ instead of $U$ and obtains from (23)

$$(33) \qquad u = \frac{8\pi h}{c^3} \int_0^\infty \frac{\nu^3\, d\nu}{\exp\left(\dfrac{h\nu}{kT}\right) - 1}.$$

Using again the substitution (26), and the formulas (30) and (30a), one has

$$(34) \qquad u = \frac{8\pi h}{c^3} \frac{\pi^4}{15} \left(\frac{kT}{h}\right)^4 = \frac{8\pi^5}{15 c^3} \frac{k^4}{h^3}\, T^4.$$

This is the law for the black body or cavity radiation in the definitive form due to Planck. *The $T^4$-radiation law of Stefan and Boltzmann has the same physical origin as Debye's $T^4$-law for the energy content of a solid and Debye's $T^3$-law for the specific heat of a solid near absolute zero.*

## 45. The Surface Waves of the Elastic Half-Space.

We consider a homogeneous elastic solid having a plane boundary $y = 0$ and extending to infinity throughout the half-space $y > 0$, and want to study the simplest types of waves that can be propagated in the half-space. Along the boundary plane it is assumed that

$$(1) \qquad \sigma_{yy} = \sigma_{yz} = \sigma_{yx} = 0;$$

then $y = 0$ is a stress-free boundary, as it would be if vacuum (or air) were on the other side.

## A. Reflection of a Plane Transverse Spatial Wave

In the discussion of three-dimensional waves in 14 we distinguished between transverse and longitudinal waves; in the first case, the displacement vector is solenoidal (free of sources and sinks) and in the second case irrotational (cf. 20). Also the terms torsional and compressional waves are in use (cf. 14). The present discussion is restricted to the simplest case, that is, to *plane* waves of either sort.

Let us first consider a plane transverse wave whose direction of propagation forms the angle of incidence $\alpha$ with the positive $y$-axis and is "polarized" in the $x$, $y$-plane. This means that of the three displacements only $\xi$ and $\eta$ are different from zero. In complex notation such a wave is given in the following way:

$$(2) \qquad \left. \begin{matrix} \xi_i \\ \eta_i \end{matrix} \right\} = \left\{ \begin{matrix} \cos \alpha \\ \sin \alpha \end{matrix} \right\} A \exp \left[ ik(x \sin \alpha - y \cos \alpha) \right]$$

where the time-dependent factor $e^{-i\omega t}$ has been omitted for brevity [cf. the remarks to Eq. (13.15) and (13.15a)]; $k$ is the wave number, and the subscript $i$ characterizes the wave as incident. By Eq. (2) the condition of vanishing divergence is automatically fulfilled:

$$\theta = \frac{\partial \xi_i}{\partial x} + \frac{\partial \eta_i}{\partial y} = 0.$$

Also, the planes of constant phase

$$(3) \qquad x \sin \alpha - y \cos \alpha = \text{Const},$$

are normal to the direction of propagation defined by the angle $\alpha$, as they have to be, and the displacement vector lies in this plane. The signs of the trigonometric functions in (2) follow from Fig. 74 and appear in the first section of the following table, which gives the cosines of direction of the propagation and of the displacement vector relative to the positive $x$ and $y$ axes.

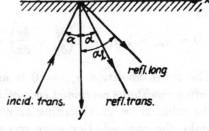

FIG. 74. A longitudinal wave is created in the reflexion of a transverse wave polarized in the plane of incidence.

|  | incid. transv. | | refl. transv. | | refl. longit. | |
|---|---|---|---|---|---|---|
|  | $x$ | $y$ | $x$ | $y$ | $x$ | $y$ |
| dir. of prop. | $\sin \alpha$ | $-\cos \alpha$ | $\sin \alpha$ | $\cos \alpha$ | $\sin \alpha_l$ | $\cos \alpha_l$ |
| dir. of displ. | $\cos \alpha$ | $\sin \alpha$ | $\cos \alpha$ | $-\sin \alpha$ | $\sin \alpha_l$ | $\cos \alpha_l$ |

The differential equations of elasticity, which in the case of a transverse wave (because of $\Theta = 0$) have the simple form

(4)
$$\rho \frac{\partial^2 \xi}{\partial t^2} = \mu \nabla^2 \xi, \qquad \rho \frac{\partial^2 \eta}{\partial t^2} = \mu \nabla^2 \eta,$$

are likewise satisfied by the set (2) provided that $k$ for a given $\omega$ is determined from the well known equation

(5)
$$\frac{\omega^2}{k^2} = \frac{\mu}{\rho} = c^2_{\text{trans}} = b^2$$

($b$ is the notation used for $c_{\text{trans}}$ in 14).

The set (2) as it stands does not, of course, satisfy the boundary conditions (1). One might think that this can be achieved by superimposing a *reflected transversal* wave $\xi_r$, $\eta_r$

(6)
$$\left.\begin{matrix} \xi_r \\ \\ \eta_r \end{matrix}\right\} = \left.\begin{matrix} \cos \alpha \\ \\ -\sin \alpha \end{matrix}\right\} B \exp\left[ik(x \sin \alpha + y \cos \alpha)\right].$$

(the signs of the trigonometric functions now chosen as in the second section of the preceding table). This, however, cannot be done as we shall see now.

The quantities $k$ and $\alpha$ in (6) must be identical with the corresponding quantities in (2). This follows for $k$ from the differential equations (4), and for $\alpha$ as in elementary optics from the boundary conditions. For, the combination of an incident and a reflected wave

(7)
$$\xi = \xi_i + \xi_r, \qquad \eta = \eta_i + \eta_r.$$

is subject to the boundary conditions

$$0 = \sigma_{yy} = 2\mu\epsilon_{yy} = 2\mu\frac{\partial\eta}{\partial y} = -2\mu ik(A + B)\sin\alpha\cos\alpha\, e^{ikx\sin\alpha}$$

(8)

$$0 = \sigma_{xy} = 2\mu\epsilon_{yx} = \mu\left(\frac{\partial\xi}{\partial y} + \frac{\partial\eta}{\partial x}\right) = -\mu ik(A - B)(\cos^2\alpha - \sin^2\alpha)e^{ikx\sin\alpha}$$

The third condition $\sigma_{yz} = 0$ is automatically fulfilled. *But the two preceding conditions contradict each other*, since the one requires $A = -B$ and the other $A = B$. Choosing another angle of reflexion ($\alpha_r \neq \alpha_i$) would make the contradiction even worse, since it would introduce an infinity of contradictory conditions instead of two, viz. a different one for every single $x$-value.

There is only one way to avoid the dilemma: assume that in the process of reflexion of a transverse wave there is generated a *longitudinal* wave in addition to the reflected transverse wave, so as to make available a further amplitude constant $C$. With a view to the third section of the table we put

(9)
$$\left.\begin{array}{c}\xi_l\\ \\ \eta_l\end{array}\right\} = \left.\begin{array}{c}\sin\alpha_l\\ \\ \cos\alpha_l\end{array}\right\}Ce^{ik_l(x\sin\alpha_l + y\cos\alpha_l)}.$$

where $k_l$ is now *different* from the previous $k$. The elastic differential equations (14.1b) require in the present case, because of the term with grad $\Theta$, that

(10)
$$\frac{\omega^2}{k_l^2} = \frac{2\mu + \lambda}{\rho} = c_{\text{long}}^2 = a^2,$$

where $a = c_{\text{long}}$ is the notation already used in 14; the prescribed $\omega$ is of course the same as in (5). From (5) and (10) we have

(11)
$$\frac{k_l}{k} = n,$$

where we have put

(11a)
$$n = \frac{b}{a} = \left(2 + \frac{\lambda}{\mu}\right)^{-1/2}.$$

But the angle of reflexion $\alpha_l$ is now *different* from the previous angle of incidence $\alpha$. In fact, the *reflected longitudinal* wave obeys a law of the same structure as the *refracted transversal* wave in elementary optics, namely

(12)
$$\sin\alpha = n\sin\alpha_l.$$

Let us call this the refraction formula, although it refers in the present case to a reflexion rather than to a refraction, and designate the coefficient $n$ as an index of refraction in the same sense.

In order to prove (12), we first calculate from (9)

$$\theta = \frac{\partial \xi_l}{\partial x} + \frac{\partial \eta_l}{\partial y} = ik_l Ce^{ik_l x \sin \alpha_l}$$

for $y = 0$. In setting up the new boundary conditions the right member of the first equation (8) must be augmented by

$$(12a) \qquad 2\mu \frac{\partial \eta_l}{\partial y} + \lambda \theta = ik_l(2\mu \cos^2 \alpha_l + \lambda)Ce^{ik_l x \sin \alpha_l},$$

and the second Eq. (8) by

$$(12b) \qquad \mu\left(\frac{\partial \xi_l}{\partial y} + \frac{\partial \eta_l}{\partial x}\right) = ik_l 2\mu \sin \alpha_l \cos \alpha_l \, Ce^{ik_l x \sin \alpha_l}.$$

The modified Eqs. (8) should now establish conditions for the $A, B, C$ that are independent of $x$. This is possible only in the case that the same exponential function of $x$ occurs in (8) and (12). *The law of refraction* (12) *must therefore be valid.*

Eqs. (8) modified according to (12) yield now the linear homogeneous system

$$(13) \qquad \begin{aligned} &- 2\mu k \sin \alpha \cos \alpha \, (A + B) + k_l(2\mu \cos^2 \alpha_l + \lambda)C = 0 \\ &- \mu k(\cos^2 \alpha - \sin^2 \alpha)(A - B) + k_l 2\mu \sin \alpha_l \cos \alpha_l \, C = 0 \end{aligned}$$

which determines in a unique way the ratios $B/A$ and $C/A$, that is, the reflected fractions of the incident amplitude $A$; the latter may, of course, be arbitrarily prescribed. One obtains from (13)

$$(14) \qquad \begin{aligned} \frac{C}{A} &= \frac{n \sin 4\alpha}{\cos^2 2\alpha + n^2 \sin 2\alpha \sin 2\alpha_l}, \\ \frac{B}{A} &= \frac{\cos^2 2\alpha - n^2 \sin 2\alpha \sin 2\alpha_l}{\cos^2 2\alpha + n^2 \sin 2\alpha \sin 2\alpha_l}. \end{aligned}$$

This solves the problem of the reflexion of a transverse wave polarized in the plane of incidence. If the incident transverse wave is polarized normal to the plane of incidence, no longitudinal component occurs in addition to the regularly reflected wave. When, on the other hand, an incident longitudinal wave is reflected there appears beside the regularly reflected longitudinal wave a transverse wave which is polarized in the plane of incidence. This will be taken up in problem VIII.5.

In the present discussion we are not so much interested in the *elastic* but in the *optical* interpretation of the mathematical formalism. As is well known, the central idea of 19th century optics is an elastic ether that carries the optical phenomena; this is true for the period of development that leads from Fresnel and Fraunhofer to F. Neumann and ends with H. Hertz. We shall now briefly indicate why the endeavors of these men to build an elastic theory of light resulted in insurmountable contradictions.

Once the polarizability of light had been recognized, the only waves admissible in optics had to be transverse. It is true that Röntgen, under the influence of one of Helmholtz's last papers, was considering the possibility that his newly discovered X-rays might be longitudinal waves. However, experiments on polarization of X-rays, first carried out by Barkla, precluded this possibility.

Now we have seen that even the simplest process of reflexion is elastically impossible without the occurrence of longitudinal waves. Starting with purely *transverse waves*, we have found that *longitudinal waves*—essentially foreign to optics—are generated in the process of reflexion.

The root of this dilemma is in the last analysis an arithmetical one. To see that, let us consider the more general case of reflexion and refraction that occurs in the transition of a wave from a medium 1 into a medium 2. Here we have altogether *six* elastic boundary conditions: three equations for the stresses and three equations that express the continuity of the displacements $\xi$, $\eta$, $\zeta$. For an incident transverse light wave, however, we have only four unknowns at our disposal, viz. two pairs of amplitude ratios for the two modes of polarization in the reflected and the refracted wave. This is a contradiction that can not be resolved.

When we now replace the elastic ether by the quasi-elastic ether of 15, a different set of conditions appears. The stress is no longer a symmetric but an antisymmetric tensor which can be represented by a vector. The quantities $\sigma_{ii}$ are zero and the number of stress conditions to be fulfilled is reduced from 3 to 2; the same is true for the displacement conditions. We then arrive at boundary conditions that are identical with those of the *electromagnetic theory of light*, and the intrinsic difficulties of the elastic theory of light are most beautifully resolved.

## B. Elastic Surface Waves

In dealing with this problem we follow closely the treatment of the analogous hydrodynamic problem in Chapter V. We therefore consider plane waves that progress in the positive $x$-direction with prescribed circular frequency $\omega$; the wave number $k$ is to be determined. When speaking of plane waves, we mean that the excitation is independent of the third coordinate $z$. The wave is composed of a longitudinal and a transversal component.

The longitudinal component is again *irrotational* as in hydrodynamics, and can be derived from a potential $\Phi$. Accordingly we put

$$(15) \qquad \xi_l = -\frac{\partial \Phi}{\partial x}, \qquad \eta_l = -\frac{\partial \Phi}{\partial y}, \qquad \zeta_l = 0,$$

but in the present case $\Phi$ does not satisfy the condition of incompressibility $\nabla^2 \Phi = 0$ as in hydrodynamics; $\Phi$ obeys, however, the relation

$$(16) \qquad \rho \frac{\partial^2 \Phi}{\partial t^2} = (2\mu + \lambda)\nabla^2\Phi,$$

which follows from the elastic differential equations for $\xi$ and $\eta$. We try

$$(17) \qquad \Phi = e^{i(kz - \omega t)}u(y)$$

and obtain from (16) for the unknown function $u$ the equation

$$- \rho\omega^2 u = (2\mu + \lambda)\left(-k^2 u + \frac{d^2u}{dy^2}\right),$$

or

$$(18) \qquad \frac{d^2u}{dy^2} = p^2 u, \qquad \text{where} \qquad \begin{aligned} p^2 &= k^2 - \frac{\omega^2}{a^2}, \\ a^2 &= \frac{2\mu + \lambda}{\rho}. \end{aligned}$$

Let us assume that $p$ is real and take the positive sign of the square root in the determination of $p$. Of the two particular integrals of (18)

$$u = e^{-py} \qquad \text{and} \qquad u = e^{+py}$$

the second must be excluded because of its behavior at $y = \infty$. Thus (17) becomes (if the time-dependent term is omitted as usual)

$$\Phi = Ae^{ikz - py},$$

which yields by (15):

$$(19) \qquad \begin{aligned} \xi_l &= -ikAe^{ikz - py}, \\ \eta_l &= +pAe^{ikz - py}. \end{aligned}$$

The transverse component of the wave to which we now turn is *free of divergence*. We may therefore introduce a function $\Psi$ in analogy to the hydrodynamic stream function and put

$$(20) \qquad \xi_t = -\frac{\partial \Psi}{\partial y}, \qquad \eta_t = +\frac{\partial \Psi}{\partial x}.$$

The function $\Psi$ satisfies the same differential equations as $\xi_t$ and $\eta_t$, viz:

$$(21) \qquad \rho \frac{\partial^2 \Psi}{\partial t^2} = \mu \nabla^2 \Psi.$$

Again we try

$$(22) \qquad \Psi = B e^{i(kz - \omega t)} v(y)$$

and obtain from (21) the following differential equation for $v$:

$$(23) \qquad \frac{d^2 v}{dy^2} = q^2 v \begin{cases} q^2 = k^2 - \dfrac{\omega^2}{b^2}, \\[2ex] b^2 = \dfrac{\mu}{\rho}. \end{cases}$$

With the only admissible solution

$$v = e^{-qy}$$

(where $q$ is supposed to be real and positive) we have from (20) and (22)

$$(24) \qquad \begin{aligned} \xi_t &= qB e^{ikz - qy}, \\[1ex] \eta_t &= ikB e^{ikz - qy}, \end{aligned}$$

where again the time-dependent term has been omitted. By superposition of (19) and (24) we obtain for the total displacement, $\xi$ and $\eta$, expressions in which *two* constants are to be determined: the wave number $k$ and the amplitude ratio $A/B$. They are to be found from the *two* boundary conditions (1) $\sigma_{yy} = \sigma_{ys} = 0$; the third condition (1) is automatically fulfilled because of the planeness of the wave. If one calculates the strains $\epsilon_{yy}$, $\epsilon_{zy}$ and $\Theta$ from the displacements $\xi$, $\eta$, and the stresses $\sigma_{yy}$, $\sigma_{yz}$ from the strains, the first two boundary conditions (1) can be written in the form

$$- 2\mu(p^2 A + ikqB) + \lambda(k^2 - p^2)A = 0$$

$$2ikpA - (q^2 + k^2)B = 0.$$

If one substitutes for $p$ and $q$ by (18) and (23) and expresses the elastic constants $\lambda$ and $\mu$ by the wave velocities $a$ and $b$, one obtains

(25)
$$\left(2k^2 - \frac{\omega^2}{b^2}\right)A + 2ik\sqrt{k^2 - \frac{\omega^2}{b^2}}\,B = 0,$$

$$2ik\sqrt{k^2 - \frac{\omega^2}{a^2}}\,A - \left(2k^2 - \frac{\omega^2}{b^2}\right)B = 0.$$

The determinant of this homogeneous system must vanish, hence

(26)
$$\left(k^2 - \frac{\omega^2}{2b^2}\right)^2 = k^2\sqrt{k^2 - \frac{\omega^2}{a^2}}\sqrt{k^2 - \frac{\omega^2}{b^2}},$$

which is a cubic equation for $k^2$ [the term containing $k^8$ cancels if (26) is squared]. The ratio $A/B$ follows from (25), once (26) has been solved.

We wish to discuss (26) in some detail only in the limiting case $a \to \infty$, that is, in the case of an almost incompressible medium. Eq. (26) is then transformed into

(27)
$$(1-x)^2 = \sqrt{1-2x}, \qquad x = \frac{\omega^2}{2b^2k^2}.$$

Here we are interested in the root close to $x = \frac{1}{2}$. Putting $x = \frac{1}{2} - \delta$, we obtain by expansion of (27)

(28)
$$\left(\frac{1}{2}+\delta\right)^4 = 2\delta, \qquad \frac{1}{16}+\frac{4}{8}\delta + \cdots = 2\delta, \qquad \delta \cong \frac{1}{24}.$$

Hence the root $x$ of (27) is approximately

(29)
$$x = \frac{\omega^2}{2b^2k^2} = \frac{1}{2}\left(1-\frac{1}{12}\right), \qquad \frac{\omega}{k} = b\left(1-\frac{1}{24}\right),$$

where $\omega/k$ is the velocity of propagation of the surface waves. It is slightly smaller than the velocity $b$ of the transversal three-dimensional waves. According to our assumption, the velocity $a$ of the longitudinal three-dimensional waves is much larger.

We finally ask for the depth at which the disturbance due to the surface waves is no longer noticeable. This question is answered by the exponential factors $e^{-qy}$ in (24) and $e^{-py}$ in (19). Characterizing the attenuation in $y$-direction by the depth at which the amplitude decreases to $1/e$ of its surface value, we obtain the depths $y = 1/q$ resp. $y = 1/p$. From (23) and (29) we find for the first of these quantities

(30)
$$\frac{1}{q} = \frac{1}{k}\frac{1}{\sqrt{1-2x}} = \frac{\sqrt{12}}{k} = \frac{\sqrt{12}}{2\pi}l,$$

where $l$ is the wave length; for the second we have from (18) with $a \to \infty$

$$(30a) \qquad \frac{1}{p} = \frac{1}{k} = \frac{l}{2\pi} \, .$$

The depth of penetration is in either case only a fraction of the wave length $l$ with which the wave progresses along the $x$-axis; $l$ is, of course, the same for the two components of the wave since the boundary conditions require the same coordination of the two components everywhere along the surface. The penetrations of the two components are nevertheless different.

The results (30) and (30a) show that we deal here with an elastic process that is restricted to the proximity of the surface and is made possible only by the discontinuity of the elasticity constants at the surface. The particular value of the velocity of the surface waves should be interpreted in the same sense. The elastic reaction at the surface is milder, since yielding is easier than in the interior, whence the smaller velocity of the surface waves in comparison with the spatial waves. (The smaller velocity of the spatial transverse waves as compared with the spatial longitudinal waves can be understood along similar lines; in the second case the material is subject to compressive loading which raises stronger reactive forces than the shear loading that permits yielding in the transverse direction.)

In 1886, when the problem of elastic surface waves was first treated, Lord Rayleigh voiced the belief that they might play an important part in earthquakes. In the subsequent development of seismology, which is due in a large measure to the theoretical and experimental efforts of E. Wiechert, Lord Rayleigh's conjecture was fully confirmed. The relative importance of the two-dimensional waves that spread along the surface of the earth increases with increasing distance from the center of the disturbance in comparison with the spatial waves, the energy of which disperses in three dimensions through the interior of the earth. Hence the surface waves form the *principal part* of the observable seismic effect. Furthermore they are also temporally separate from the spatial waves, since they arrive at the point of observation somewhat later than the transverse spatial waves, partly because of their smaller velocity, partly because of their longer path (arc versus cord). The *longitudinal* spatial waves arrive at the seismograph still earlier than the transverse spatial waves, and form the first part of the *primary wave*.

The seismic signals bring us information not only about the average elastic behavior of the interior of the earth but also about its possible discontinuities. One knows that the analysis of these signals together with other geophysical and astronomical data led Wiechert to the idea of a *"solid"* *core* of the earth *that consists essentially of iron* and is surrounded by lighter layers of sulfides and silicates. This theory is based not only on propagation processes but also on processes of reflexion along successive discontinuity surfaces in the interior of the earth, the simplest form of which has been studied in section A of this article.

# PROBLEMS

## *Chapter I*

I.1. *The definition of curl* **A** *by a linear vector function compared with its definition by a differential operator.* Show that the two definitions of the curl

$$\mathbf{B} = \text{curl } \mathbf{A} \qquad \text{in (2.6)}$$

and

$$2(\mathbf{A} - \mathbf{A_0}) = \mathbf{B} \times \mathbf{r} \qquad \text{in (2.7a)}$$

are equivalent, provided that **r** is considered infinitesimally small as in 1.—It will be noticed that the field vector **A** of Eq. (2.7a) of the text vanishes for **r** = 0 since it corresponds to the rotatory part $\mathbf{v_1}$ of the general motion in 1. To avoid this restriction, we have written $\mathbf{A} - \mathbf{A_0}$ instead of **A**, so that $\mathbf{A_0}$ is the value of the new field vector at **r** = 0. Of course, $\mathbf{A_0}$ corresponds to a velocity $\mathbf{v_0}$ associated by $\mathbf{s_0} = \mathbf{v}\Delta t$ with the translatory part $\mathbf{s_0}$ of the general motion.

I.2. *The vector character of curl* **A**. Investigate the vector character of the curl by applying the transformation formulas (2.2) to the differential definition

$$\mathbf{B} = \text{curl } \mathbf{A}$$

(Note that the components of **A** *and* the "differential operators" $\partial/\partial x_i$ must be transformed.)

I.3. *A table of the most important vector analytic operations in polar coordinates.* Set up the expression for grad, div, curl, $\nabla^2$ and $D$ in cylindrical and spherical coordinates.

I.4. *The symbol* $\nabla^2\mathbf{A}$. Calculate the expression

$$\text{grad div } \mathbf{A} - \text{curl curl } \mathbf{A}$$

in cylindrical coordinates and compare the result with what is obtained when the $\nabla^2$ operator calculated in I.3 is formally applied to the components $A_r$, $A_\varphi$, $A_z$. The two results are identical only in the case of the "Cartesian" component $A_z$ [cf. the remarks following Eq. (3.10).]

I.5. *Concerning the so-called second boundary value problem of potential theory.* Show that a potential problem in which $\partial U/\partial n$ instead of $U$ is prescribed along the boundary of the region has a unique solution except for an additional constant.

I.6. *The geometrical meaning of the tensor invariants.* True geometrical properties of a geometrical object are those that do not change in an orthogonal transformation; conversely, an orthogonal invariant must always allow a geometrical interpretation. The geometrical correlative associated with the tensor concept is the tensor quadric which, for convenience, may be assumed as an ellipsoid; hence it must be possible to interpret the three tensor invariants (4.10) as geometrical quantities connected with that ellipsoid.

Interpret the three tensor invariants as geometrical properties of the tensor ellipsoid.

I.7. *Condition that a given pattern of field lines belongs to a potential field.* If the *field vector* $\mathbf{F}$ is everywhere given, the condition for the existence of a potential $U$ is curl $\mathbf{F} = 0$ according to Eq. (3.7), but if the *field lines* are given, we know only the ratios $F_x : F_y : F_z$, and a scalar multiplier $\lambda$ is left undetermined. Is it possible to choose $\lambda$ in such a way that $\mathbf{F}$ has a potential? Show that this can not be done unless

$$\mathbf{F_0} \perp \text{curl } \mathbf{F_0} ,$$

which is therefore a necessary condition. Here $\mathbf{F_0}$ is any vector that satisfies the given ratio condition, e.g. a unit vector that is everywhere parallel to the field lines.

## Chapter II

II.1. *The altitude of the polytropic atmosphere.* a) Determine the altitude $h$ of the polytropic atmosphere for the following values of the polytropic exponent: $n = 1.4$ (adiabatic case), $n = 1. 2$, and $n = 1$ (isothermal case). b) Find the pressure as a function of the distance from the surface of the earth, the surface being considered spherical rather than plane as in 7. Show that under this condition the pressure in the isothermal atmosphere does not vanish at any finite distance from the earth nor, contrary to what one might expect, at infinity.

II.2. *Separation of a gas mixture in a centrifugal field.* Determine the speed of revolution required to produce a pressure difference equivalent to a water column of 40cm height between the axis and the circumference of a rotating drum that has a diameter of 60cm and contains normal air at 0°C (21% $O_2$ and 79% $N_2$ by volume), and find the composition of the air at the circumference of the drum.

II.3. a) *The two-dimensional analogue to the flow in a capillary tube.* A fluid flows between two parallel plates at a distance $2h$ from each other, the straight streamlines being parallel to the $x$-axis as in Fig. 19. Determine the velocity profile in the $x,y$-plane and the pressure loss $\Pi$ between two cross-sections in the distance $l$. Determine also the force required to keep the plates at rest against the reactive forces of the flow.

b) *River in laminar flow* (cf. Fig. 19a and the remarks on p. 120 about the experimental realization). Show that the velocity profile is identical with the lower half of the velocity profile in part (a) of this problem. (Assume both river depth and angle of inclination $\alpha$ of the bottom constant, and consider the boundary condition at $y = 0$.) Here the external force of gravity replaces the driving pressure of case (a).

II.4. *Straight and cylindrical Couette-flow.* According to (16.14) the so-called Couette-flow in straight streamlines (cf. Fig. 19b) is characterized by the velocity law

$$(1) \qquad u = \frac{y}{h}\, U, \qquad v = w = 0.$$

The equilibrium conditions (10.8a, b) are fulfilled if $p = \mathrm{const}$; no driving pressure is required since the upper plate is kept in motion by the force $\mu U / h$.

Likewise in the case of the Couette flow proper, which takes place between two coaxial cylinders of radii $r_e$ and $r_i$, no driving pressure, but only the moments $M_e$ and $M_i$ are required, the one to maintain the motion of the external cylinder and the other to keep the internal cylinder at rest. Set up and integrate the differential equations of the flow and compare the result with Eq. (1) in the limiting case $r_i \to \infty$. Calculate also the moments $M_e$ and $M_i$ and note that they are not equal.

II.5. *The problem of Boussinesq.* An elastic body bounded by a horizontal plane, but otherwise infinite, is subject to a vertical force $P$ applied at a single point this plane, all other points of which are free of stress. Calculate the state of stress in the body, and in particular the hoop stresses, that is to say, the principal stresses which are tangential to the circles about the direction of $P$.

## Chapter III

III.1. *The convective terms of the hydrodynamic equations in polar coordinates.* Show that there is a difference between the correct calculation of the convective terms according to (11.6) and the result obtained from the "misinterpreted" symbol (**v** grad). Take cylindrical coordinates as an example, where one might be tempted to give the following meaning to the symbol (**v** grad)

$$v_r\, \frac{\partial}{\partial r}, \qquad v_\varphi\, \frac{\partial}{r\, \partial \varphi}, \qquad v_z\, \frac{\partial}{\partial z}.$$

Verify that such a difference occurs in the $r$- and $\varphi$-component but not in the $z$-component. Relate the missing terms to the *centripetal* and the *Coriolis accelerations*.

III.2. *Similarity considerations in pipe flow.* Let the connection between the pressure gradient per unit of length $\Pi/l$ and the average velocity $v$ be given by

$$(1) \qquad\qquad \frac{\Pi}{l} = Cv^n.$$

For the laminar state we have $n = 1$, for the turbulent state $n = 1.722$ according to Reynolds, and $n = 1.75$ according to Hagen (cf. the schematic representation in Fig. 18).

The quantity $C$ can depend on the density, the kinematic viscosity $\nu$, and the radius of the pipe $a$.

a) Following Kármán [Physik. Z.12, (1911) 283], assume $C$ proportional to a product of powers of these quantities and find the exponents by dimensional reasoning.

b) It must also be possible to represent the relation (1) by the dimensionless quantities $R$ and $S$ and the quotient $l/a$ in the general form

$$(2) \qquad\qquad S = f\!\left(\frac{l}{a}, R\right).$$

In particular, if the function $f$ is again assumed as a product of powers and the exponent of $l/a$ is taken as 1, one obtains instead of (2)

$$(3) \qquad\qquad S = \lambda \cdot \frac{l}{a} R^\delta,$$

where $\lambda$ is a dimensionless constant; $\delta$ can be easily expressed by the exponent $n$ occurring in (1).

c) Show finally that the relation $S = \lambda R^{-1/4}$ in (16.13) is in accordance with Hagen's value $n = 1.75$.

III.3. *The free surface of a fluid in the presence of external forces.* Using the principle of virtual work, find the differential equation that determines the shape of the free surface of a liquid subject to capillary forces and external (body) forces that have a potential. The fact that the total volume of the liquid mass does not change in the displacement will have to be used.

III.4. *Rise of a liquid in a capillary tube.* The result of the foregoing problem can be applied to find the elevation (or depression) of a wetting (or non wetting) liquid in a narrow capillary tube. Consider for this purpose the surface of the liquid inside the tube as spherical; outside the tube the surface may be considered everywhere as plane with sufficient accuracy.

# Chapter IV

IV.1. *Calculations in bipolar coordinates.* The bipolar coordinates $\rho$ and $\varphi$ have been defined by Eqs. (19.10a, b). Show that the line element in these coordinates has the form

(1)
$$ds^2 = g^2(d\rho^2 + d\varphi^2), \qquad g = \frac{c}{\cosh \rho - \cos \varphi}.$$

Set up the expression for $\nabla^2$ and the equation of the vibrating membrane (17.8) in these coordinates. Note that the latter is not separable in $\rho$ and $\varphi$, that is, there is no solution of the form

(2)
$$u = f_1(\rho)f_2(\varphi).$$

IV.2. *The field of flow about two vortices of equal strength and opposite orientation.* Eq. (19.10), which is equivalent to Eq. (19.11), yields the velocity field

$$\mathbf{v} = - \operatorname{grad} \Phi = A \operatorname{grad} \varphi, \qquad v_\rho = 0, \qquad v_\varphi = \frac{A}{g}$$

where the value of $g$ is to be taken from Eq. (1) of the foregoing problem. Show by a geometrical argument that this representation of $\mathbf{v}$ is equivalent to Eq. (19.14).

IV.3. *Calculations in elliptic coordinates.* Set up the line element $ds$ and the operator $\nabla^2$ in elliptic coordinates $\xi$, $\eta$. The equation of the vibrating membrane becomes separable in these coordinates, that is, a trial solution of the form (2) in problem IV.1

$$u = f_1(\xi)f_2(\eta)$$

makes the integration possible.

IV.4. *Transition from bipolar to toroidal and "spindle" coordinates.* By rotating the plane of the bipolar coordinates $\rho$ and $\varphi$ about the axis $\rho = 0$ (cf. Fig. 26) one obtains a system of toroidal coordinates; by rotation about the axis $\varphi = 0$ another spatial coordinate system results for which the name "spindle" coordinates seems appropriate. Find the line element in these coordinates, set up the operator $\nabla^2$, and discuss the question of separability.

# Chapters V and VI

V.1. *The kinetic energy of water waves.* As a supplement to the topic of 26, p. 189, compute for gravity waves in deep water the kinetic energy contained in a column of cross-section $\Delta x \cdot \Delta y = \lambda.1$, bounded by the bottom and the free surface; compare this with the amount of potential energy contained in the same column.

Carry out the computations of p. 189 and p. 190 for moderately deep water and show that the value of $S/E$ agrees with the value of $U/N$ in Eq. (26.13).

VI.1. *The drift current.* Winds blowing along the surface of the ocean generate water currents known as drift currents. The surface layer which is exposed to the immediate action of the wind imparts its motion to the adjacent layers by the friction pressures. Consider the ocean infinitely extended and its density and viscosity everywhere constant; introduce a coordinate system $\xi$, $\eta$, $\zeta$ rigidly connected with a surface point of the earth, the $\xi$ and $\eta$ axes being tangential to the meridian and the parallel, respectively, and the $\zeta$ axis coinciding with the outward surface normal, so that $\xi$, $\eta$, $\zeta$ form a right hand system and the $\xi$ axis points south (cf. Fig. 49, Vol. I). The Navier-Stokes equations must now be augmented by the Coriolis terms of Vol. I. 30.2 or 30.5.[1] For an unlimited and homogeneous ocean the components $v_\xi$ and $v_\eta$ as well as $p$ are independent of $\xi$ and $\eta$; the component $v_\zeta$ can be neglected.

Investigate the steady state that results if one assumes the wind velocity vector constant in the $\xi$, $\eta$-system.

VI.2. *Instability of a plane discontinuity surface separating two flow fields with different velocities.* In 30 and 32, pp. 216, 232, it was pointed out that discontinuity surfaces are unstable since they curl into vortices under the influence of friction. However, this instability is also present *without friction*; this is to be shown for the simplest case of a plane discontinuity surface $y = 0$, the fluid being the same on both sides.

Assume the fluid flows in $x$-direction with the constant velocities $U_1$ and $U_2$ for $y > 0$ and for $y < 0$ respectively. The problem is thus restricted to two-dimensions and remains so if a disturbance of the surface is introduced in form of a sinusoidal cylinder with generatrices parallel to the $z$-axis. External forces may be neglected.

Use the velocity potentials

$$\Phi_1 = -U_1 x \Big\}_{y>0} \qquad \Phi_2 = -U_2 x \Big\}_{y<0} \qquad \text{(principal motion)}$$
$$\varphi_1 \qquad\qquad\qquad \varphi_2 \qquad\qquad \text{(disturbance)}$$

and apply Bernoulli's theorem in the form (11.15) valid for non-steady flow. The discussion of the equations of motion so obtained shows that the principal motion is unstable for all wave lengths of the disturbance, and, since every disturbance can be compounded from sinusoidal com-

[1]Cf. Synge and Griffith, *op. cit.*, Sec. 12.3.

ponents according to the Fourier integral theorem, the flow is unstable for any disturbance.

According to Helmholtz (Über atmosphärische Bewegungen, Akad. Ber. Berlin, 1898 and 1899) the formation of high cirrocumulus clouds is connected with the unstability of the boundary between two air currents of equal directions but different velocities. The clouds are generated by the condensation of water vapor in form of droplets or ice crystals in regions of diminished pressure.—The instability of the discontinuity surface to be shown here for incompressible fluids occurs doubtlessly in compressible fluids, too, as long as all velocities are small compared to the sound velocity (cf. appendix to 13).

VI.3. *Stability of a horizontal water surface for varying wind velocities.* Proceeding as in the foregoing problem, investigate the stability of a water surface at rest under the influence of an air current moving parallel to it. The density of the air must not be neglected as was done in Chap. V., although it is small relative to water (0.00129:1). Gravity must be considered in this problem.

## Chapter VII

VII.1. *The shape of the free surface of a rotating liquid derived according to Lagrange's equations.* The free surface of an incompressible fluid rotating with constant angular velocity was determined in 6 p. 44 by the methods of statics and is now to be found by the use of Lagrange's equations. The axis of the drum is vertical as before.

VII.2. *Transition from Euler's to Lagrange's equations in the one-dimensional compressible case.* For the special case of a one-dimensional flow of a compressible fluid in the absence of external forces, derive Lagrange's form of the hydrodynamic equations directly from Euler's equations and the law of conservation of mass.

VII.3. *Simplified determination of the linear differential equations for Riemann's shock wave problem according to Hadamard.* The one-dimensional Lagrange equation derived in the preceding problem permits us, according to J. Hadamard (Leçons sur la propagation des ondes, Paris, 1903), to shorten the transition to the linear differential equation of second order in 37. Assume that the compressible fluid has initially constant density and use the Legendre transformation of Vol. I. Eq. (42.8), cf. also footnote 10 on p. 312.

VII.4. *Stokes' formula for a rotating sphere.* In analogy to the translatory motion of a sphere analyzed in 35, investigate the uniform rotation of a sphere of radius $a$. Assume again laminar motion and an incompressible fluid, and show that the moment $M = 8\pi\mu a^3\omega$ is required to maintain uniform rotation with angular velocity.

## Chapter VIII

VIII.1. *The bending of a beam on two supports under a single load.* A beam of length $l = a + b$ rests with its ends $A$ and $B$ freely on two supports of equal height; a vertical force $P$ acts at a distance $a$ from $A$. Make a diagram of the shear force and the bending moment along the beam and determine the elastic line.

VIII.2. *The vibrations of a rod.* a) A vertically suspended prismatic rod is rigidly clamped at the upper end and carries at the lower end a mass $M$ that is large compared with the mass of the rod $m$. The system is free to oscillate in the direction of the axis of the rod. Determine the circular frequency of the longitudinal oscillations neglecting $m$ as small relative to $M$, and the elasticity of the body $M$ as unimportant relative to that of the rod.

b) The system a) is now allowed to oscillate about the axis of the rod. In calculating the frequency of the torsional oscillations the weight of $M$ that acts in the axis of the rod may be disregarded, since it is balanced by tensile stresses that are constant in time. The appropriate assumption is now that the moment of inertia $\Theta$ corresponding to the mass $M$ is large compared to that of the rod.

c) We finally consider a horizontal rod that is either rigidly clamped at one end and loaded at the other end by a mass $M$, or supported at both ends $A$ and $B$ as in prob. VIII.1 and loaded at a distance $a$ from $A$ (respectively $b$ from $B$) by a mass $M$. The equilibrium of the rod is disturbed by a vertical impulse. Determine the circular frequency of the oscillations normal to the axis of the rod (bending oscillations), if again $M \gg m$.

VIII.3. *Rod subject to an impulsive load.* Assume that the mass $M$ of the rod of prob. VIII. 2a is not a steady load, but was put in contact with the rod at $t = 0$, and that at that instant the velocity of $M$ was $v_0$. Calculate the maximum displacement $\xi_{max}$ of the mass $M$ and the maximum stress $\sigma_{max}$ at the end cross-section of the rod in general and show in particular:

a) for $v_0 = 0$, $\sigma_{max}$ is twice its static value.

b) for large values of $v_0$, $\sigma_{max}$ increases with the kinetic energy of $M$ at the instant of contact, and depends on the individual dimensions of the rod only through its volume.

VIII.4. *A supplement to the torsion problem for an elliptic rod.* Determine the torsional moment $M$ and the angle of torsion $\varphi_l$ for a rod with elliptic cross-section by carrying out the integrations in (42.23), and compare the factor that expresses the influence of the geometry of the rod with the polar moment of inertia of the area of the ellipse.

VIII.5. *The reflexion of a plane transverse and plane longitudinal wave within an elastic solid having a plane boundary.* As a supplement to 45 show that a transverse wave oscillating perpendicularly to the plane of incidence is reflected regularly at a plane surface that is free of forces, and no longitudinal wave is generated simultaneously. Show on the other hand, that a plane longitudinal wave generates, in addition to the regularly reflected longitudinal wave, a transverse wave that oscillates in the plane of incidence.

# ANSWERS AND COMMENTS

I.1. a) Differentiate the $z$-component of

$$2(\mathbf{A} - \mathbf{A}_0) = \mathbf{B} \times \mathbf{r}$$

partially with respect to $y$, the $y$-component partially with respect to $z$ and subtract the results. On disregarding all terms that vanish simultaneously with $r$, the usual expression for the $x$-component of

$$\mathbf{B} = \operatorname{curl} \mathbf{A}$$

is obtained.

b) Note that the omitted terms have the form

$$(\mathbf{r} \operatorname{grad})\mathbf{B}_z - x \operatorname{div} \mathbf{B}.$$

This points to the connection with the general formula

$$\operatorname{curl}(\mathbf{a} \times \mathbf{b}) = (\mathbf{b} \operatorname{grad})\,\mathbf{a} - (\mathbf{a} \operatorname{grad})\,\mathbf{b} + \mathbf{a} \operatorname{div} \mathbf{b} - \mathbf{b} \operatorname{div} \mathbf{a},$$

valid for arbitrary vectors $\mathbf{a}$, $\mathbf{b}$ but restricted to Cartesian components.

I.2. From (2.2) one has

$$x'_i = \sum_k \alpha_{ik} x_k, \qquad x_k = \sum_i \alpha_{ik} x'_i, \qquad A'_l = \sum_i \alpha_{li} A_i$$

and finds by differentiation

$$\frac{\partial}{\partial x'_i} = \sum_k \frac{\partial x_k}{\partial x'_i}\frac{\partial}{\partial x_k} = \sum_k \alpha_{ik}\frac{\partial}{\partial x_k}, \qquad \frac{\partial A'_l}{\partial x'_m} = \cdots, \qquad \frac{\partial A'_m}{\partial x'_l} = \cdots.$$

The difference of the derivatives $\partial A'_m/\partial x'_l - \partial A'_l/\partial x'_m$ gives the 1, 2, 3 components of curl' $A'$, when the pair $(l, m)$ takes the values $(2,3)$, $(3,1)$, $(1,2)$. Each derivative consists of 9 terms. When the difference is formed, 6 of the 18 terms cancel. The remaining 12 terms can be arranged in 6 pairs according to the 6 different $\partial A_i/\partial x_k$ and finally rearranged in 3 terms containing the three differences $\partial A_k/\partial x_i - \partial A_i/\partial x_k$. Hence each component of the transformed curl is a linear function of the components of the original curl; the coefficients are the cofactors of the elements of the transformation determinant as in (2.5). This points again to the axial vector character of the curl.

The treatment of this problem in Eqs. (2.10)-(2.15) of the text deserves preference.

I.3. It is, of course, possible to obtain the required expressions from

345

their definitions in Cartesian coordinates by applying the corresponding orthogonal transformations of coordinates, but this is a highly cumbersome procedure. A more efficient way is to specialize the general formulas (2.24), (2.25), (2.26), (3.9a), (3.9b) for the required type of coordinates. The results of this computation are collected for later use in the following table, which contains also the answers to problems I.4 and III.1 for cylindrical and spherical coordinates. As for the correct meaning of the symbols $\nabla^2 \mathbf{A}$ and $(\mathbf{A}\,\mathrm{grad})\mathbf{A}$, the reader is again referred to Eqs. (3.10a) and (11.6).

It should be noticed that the sign of the curl depends on the orientation of the coordinate system. We have here assumed that the three coordinates in the sequence $r$, $\varphi$, $z$ (and in the sequence $r$, $\vartheta$, $\varphi$) form dexteral systems. Only under this condition the use of formula (2.26) is legitimate. In the case of spherical coordinates the usual definition of $\vartheta$ is such that $\vartheta$ increases from north to south. If this is adopted, $\varphi$ runs contrary to the geographical longitude coordinate. If, however, $\varphi$ is required to coincide with the geographical longitude, then the sign of the curl components would be reversed in accordance with Eq. (2.16), since the curl is an *axial* vector. The same thing happens when the order $r$, $\vartheta$, $\varphi$ is changed into $r$, $\varphi$, $\vartheta$.

Cylindrical coordinates:

$$r, \varphi, z, \qquad ds^2 = dr^2 + r^2\, d\varphi^2 + dz^2$$

$$g_1 = 1, \qquad g_2 = r, \qquad g_3 = 1$$

$$\mathrm{grad}\ U = \frac{\partial U}{\partial r}, \qquad \frac{1}{r}\frac{\partial U}{\partial \varphi}, \qquad \frac{\partial U}{\partial z}$$

$$\mathrm{div}\ \mathbf{A} = \frac{1}{r}\frac{\partial}{\partial r}(rA_r) + \frac{1}{r}\frac{\partial A_\varphi}{\partial \varphi} + \frac{\partial A_z}{\partial z}$$

$$\mathrm{curl}_r\ \mathbf{A} = \frac{1}{r}\frac{\partial A_z}{\partial \varphi} - \frac{\partial A_\varphi}{\partial z}$$

$$\mathrm{curl}_\varphi\ \mathbf{A} = \frac{\partial A_r}{\partial z} - \frac{\partial A_z}{\partial r}$$

$$\mathrm{curl}_z\ \mathbf{A} = \frac{1}{r}\frac{\partial(rA_\varphi)}{\partial r} - \frac{1}{r}\frac{\partial A_r}{\partial \varphi}$$

$$\nabla^2 U = \frac{1}{r}\left\{\frac{\partial}{\partial r}\left(r\frac{\partial U}{\partial r}\right) + \frac{\partial}{\partial \varphi}\left(\frac{1}{r}\frac{\partial U}{\partial \varphi}\right) + \frac{\partial}{\partial z}\left(r\frac{\partial U}{\partial z}\right)\right\}$$

$$= \frac{1}{r}\frac{\partial}{\partial r}\left(r\frac{\partial U}{\partial r}\right) + \frac{1}{r^2}\frac{\partial^2 U}{\partial \varphi^2} + \frac{\partial^2 U}{\partial z^2}$$

$$= \frac{\partial^2 U}{\partial r^2} + \frac{1}{r}\frac{\partial U}{\partial r} + \frac{1}{r^2}\frac{\partial^2 U}{\partial \varphi^2} + \frac{\partial^2 U}{\partial z^2}$$

$$DU = \left(\frac{\partial U}{\partial r}\right)^2 + \frac{1}{r^2}\left(\frac{\partial U}{\partial \varphi}\right)^2 + \left(\frac{\partial U}{\partial z}\right)^2,$$

$$(\mathbf{A}\,\text{grad})\mathbf{A} = \begin{cases} \mathbf{A}\cdot\text{grad}\,A_r - \dfrac{A_\varphi^2}{r} \\[2mm] \mathbf{A}\cdot\text{grad}\,A_\varphi + \dfrac{A_r A_\varphi}{r} \\[2mm] \mathbf{A}\cdot\text{grad}\,A_z \end{cases}$$

$$\nabla^2 \mathbf{A} = \begin{cases} \nabla^2 A_r - \dfrac{A_r}{r^2} - \dfrac{2}{r^2}\dfrac{\partial A_\varphi}{\partial \varphi} \\[2mm] \nabla^2 A_\varphi - \dfrac{A_\varphi}{r^2} + \dfrac{2}{r^2}\dfrac{\partial A_r}{\partial \varphi} \\[2mm] \nabla^2 A_z \end{cases}$$

Spherical coordinates:

$$r,\ \vartheta,\ \varphi,\qquad ds^2 = dr^2 + r^2\,d\vartheta^2 + r^2\sin^2\vartheta\,d\varphi^2$$

$$g_1 = 1,\qquad g_2 = r,\qquad g_3 = r\sin\vartheta$$

$$\text{grad}\,U = \frac{\partial U}{\partial r},\qquad \frac{1}{r}\frac{\partial U}{\partial \vartheta},\qquad \frac{1}{r\sin\vartheta}\frac{\partial U}{\partial \varphi}$$

$$\text{div}\,\mathbf{A} = \frac{1}{r^2}\frac{\partial}{\partial r}(r^2 A_r) + \frac{1}{r\sin\vartheta}\frac{\partial}{\partial \vartheta}(\sin\vartheta A_\vartheta) + \frac{1}{r\sin\vartheta}\frac{\partial A_\varphi}{\partial \varphi}$$

$$\text{curl}_r\,\mathbf{A} = \frac{1}{r\sin\vartheta}\left(\frac{\partial(\sin\vartheta A_\varphi)}{\partial \vartheta} - \frac{\partial A_\vartheta}{\partial \varphi}\right)$$

$$\text{curl}_\vartheta\,\mathbf{A} = \frac{1}{r\sin\vartheta}\frac{\partial A_r}{\partial \varphi} - \frac{1}{r}\frac{\partial(rA_\varphi)}{\partial r}$$

$$\text{curl}_\varphi\,\mathbf{A} = \frac{1}{r}\left(\frac{\partial(rA_\vartheta)}{\partial r} - \frac{\partial A_r}{\partial \vartheta}\right)$$

$$\nabla^2 U = \frac{1}{r^2 \sin \vartheta} \left\{ \frac{\partial}{\partial r} \left( r^2 \sin \vartheta \frac{\partial U}{\partial r} \right) + \frac{\partial}{\partial \vartheta} \left( \sin \vartheta \frac{\partial U}{\partial \vartheta} \right) + \frac{\partial}{\partial \varphi} \left( \frac{1}{\sin \vartheta} \frac{\partial U}{\partial \varphi} \right) \right\}$$

$$= \frac{1}{r^2} \frac{\partial}{\partial r} \left( r^2 \frac{\partial U}{\partial r} \right) + \frac{1}{r^2 \sin \vartheta} \frac{\partial}{\partial \vartheta} \left( \sin \vartheta \frac{\partial U}{\partial \vartheta} \right) + \frac{1}{r^2 \sin^2 \vartheta} \frac{\partial^2 U}{\partial \varphi^2}$$

$$= \frac{\partial^2 U}{\partial r^2} + \frac{2}{r} \frac{\partial U}{\partial r} + \frac{1}{r^2} \frac{\partial^2 U}{\partial \vartheta^2} + \frac{1}{r^2} \operatorname{ctg} \vartheta \cdot \frac{\partial U}{\partial \vartheta} + \frac{1}{r^2 \sin^2 \vartheta} \frac{\partial^2 U}{\partial \varphi^2}$$

$$DU = \left( \frac{\partial U}{\partial r} \right)^2 + \frac{1}{r^2} \left( \frac{\partial U}{\partial \vartheta} \right)^2 + \frac{1}{r^2 \sin^2 \vartheta} \left( \frac{\partial U}{\partial \varphi} \right)^2$$

$$(\mathbf{A} \operatorname{grad})\mathbf{A} = \begin{cases} \nabla^2 A_r - \dfrac{2}{r^2} \left[ A_r + \dfrac{1}{\sin \vartheta} \dfrac{\partial}{\partial \vartheta} (\sin \vartheta\, A_\vartheta) + \dfrac{1}{\sin \vartheta} \dfrac{\partial A_\varphi}{\partial \varphi} \right] \\[3mm] \nabla^2 A_\vartheta + \dfrac{2}{r^2} \left[ \dfrac{\partial A_r}{\partial \vartheta} - \dfrac{A_\vartheta}{2 \sin^2 \vartheta} - \dfrac{\operatorname{ctg} \vartheta}{\sin \vartheta} \dfrac{\partial A_\varphi}{\partial \varphi} \right] \\[3mm] \nabla^2 A_\varphi + \dfrac{2}{r^2 \sin \vartheta} \left[ \dfrac{\partial A_r}{\partial \varphi} + \operatorname{ctg} \vartheta \dfrac{\partial A_\vartheta}{\partial \varphi} - \dfrac{A_\varphi}{2 \sin \vartheta} \right] \end{cases}$$

I.4. The $r$-component of curl curl $\mathbf{A}$ is according to the preceding table

$$\operatorname{curl}_r \operatorname{curl} \mathbf{A} = \frac{1}{r} \frac{\partial}{\partial \varphi} \operatorname{curl}_\varphi \mathbf{A} - \frac{\partial}{\partial z} \operatorname{curl}_\varphi \mathbf{A}$$

$$= \frac{1}{r^2} \frac{\partial}{\partial r} r \frac{\partial A_\varphi}{\partial \varphi} - \frac{1}{r^2} \frac{\partial^2 A_r}{\partial \varphi^2} - \frac{\partial^2 A_r}{\partial z^2} + \frac{\partial^2 A_z}{\partial r \partial z},$$

and, likewise,

$$\operatorname{grad}_r \operatorname{div} \mathbf{A} = \frac{\partial^2 A_r}{\partial r^2} + \frac{\partial}{\partial r} \frac{A_r}{r} + \frac{1}{r} \frac{\partial^2 A_\varphi}{\partial r \partial \varphi} - \frac{1}{r^2} \frac{\partial A_\varphi}{\partial \varphi} + \frac{\partial^2 A_z}{\partial r \partial z}.$$

We obtain therefore for the difference

$$\operatorname{grad}_r \operatorname{div} \mathbf{A} - \operatorname{curl}_r \operatorname{curl} \mathbf{A} = \cdots - \frac{A_r}{r^2} - \frac{2}{r^2} \frac{\partial A_\varphi}{\partial \varphi}.$$

The terms fully written on the right hand side are those that are missed if the expression $\nabla^2 A_r$ is calculated by a schematical application of the $\nabla^2$-expression in cylindrical coordinates. One finds in the same way

$$\operatorname{grad}_\varphi \operatorname{div} \mathbf{A} - \operatorname{curl}_\varphi \operatorname{curl} \mathbf{A} = \cdots - \frac{A_\varphi}{r^2} + \frac{2}{r^2} \frac{\partial A_r}{\partial \varphi}.$$

I.5. For the difference $U = U_2 - U_1$ of two solutions $U_1$ and $U_2$ the Laplacian $\nabla^2 U$ vanishes in the interior, and the normal derivative $\partial U/\partial n$ vanishes on the boundary. From Green's theorem (3.16a) it follows that $U =$ const as in the first boundary value problem so that $U_2 = U_1 +$ const. To conclude that $U_2 = U_1$ is, of course, no longer possible.

I.6. Let the strain ellipsoid with the semi-axes $a$, $b$, $c$ be given by the equation

(1) $$\sum \epsilon_{ik} x_i x_k = 1,$$

referred to a system $x_1$, $x_2$, $x_3$ which is connected with the system of principal axes by some rotation.

a) $\Theta = \Theta'$ : For $x_2 = x_3 = 0$ we obtain $\epsilon_{11} = 1/\bar{x}_1^2$ where $|\bar{x}_1|$ is the radius vector from $O$ to the point at which the $x_1$-axis cuts the ellipsoid. The lengths $|\bar{x}_2|$ and $|\bar{x}_3|$ being correspondingly defined, we obtain for $\Theta$

(2) $$\Theta = \frac{1}{\bar{x}_1^2} + \frac{1}{\bar{x}_2^2} + \frac{1}{\bar{x}_3^2}.$$

The invariance of $\Theta$ permits the following geometric interpretation. *If an ellipsoid is cut by an orthogonal triad whose apex coincides with the center of the ellipsoid, the sum of the reciprocal squares of the intercepts is independent of the position of the triad;* in particular, it is equal to

$$\Theta' = \frac{1}{a^2} + \frac{1}{b^2} + \frac{1}{c^2}.$$

b) $\Delta = \Delta'$ : The plane $x_3 = 0$ intersects the ellipsoid (1) in the ellipse

$$\epsilon_{11} x_1^2 + 2\epsilon_{12} x_1 x_2 + \epsilon_{22} x_2^2 = 1,$$

whose principal semi-axes are $1/\sqrt{\lambda_1}$ and $1/\sqrt{\lambda_2}$, $\lambda_1$ and $\lambda_2$ being the roots of the equation

(3) $$\begin{vmatrix} \epsilon_{11} - \lambda, & \epsilon_{12} \\ \epsilon_{21}, & \epsilon_{22} - \lambda \end{vmatrix} = 0.$$

The area of the ellipse is

(4) $$F_3 = \frac{\pi}{\sqrt{\lambda_1 \lambda_2}}.$$

From (3) and (4) we conclude

$$\lambda_1 \lambda_2 = \epsilon_{11} \epsilon_{22} - \epsilon_{12}^2 = \frac{\pi^2}{F_3^2},$$

and, therefore, ($F_2$, $F_3$ are defined correspondingly)

(5) $$\Delta = \pi^2\left(\frac{1}{F_3^2} + \frac{1}{F_1^2} + \frac{1}{F_2^2}\right).$$

*Hence, if an ellipsoid is cut by an orthogonal trihedron whose apex coincides with the center of the ellipsoid, the sum of the reciprocal squares of the areas cut out by the trihedron is independent of its position; in particular it is equal to*

$$\Delta' = \frac{1}{b^2c^2} + \frac{1}{c^2a^2} + \frac{1}{a^2b^2}.$$

c) $D = D'$: The volume of an ellipsoid with the semi-axes $a$, $b$, $c$ is given by

(6) $$V = \frac{4\pi}{3}\, abc.$$

When the determinant (4.7) is set up in principal coordinates, it reduces to the product of the terms $1/a^2$, $1/b^2$, $1/c^2$ in the main diagonal so that

(7) $$D' = \frac{1}{a^2b^2c^2} = \frac{16\pi^2}{9V^2}.$$

On the other hand, by an orthogonal transformation $D'$ is transformed into $D$ [cf. (4.7)]. We have therefore once more

(8) $$D = \frac{16\pi^2}{9V^2}.$$

*The invariance of $D$ has its geometrical equivalent in the independence of the volume $V$ from the coordinate system.*

I.7. The relation $\mathbf{F} = \lambda\mathbf{F}_0$ implies

$$\text{curl } \mathbf{F} = \lambda \text{ curl } \mathbf{F}_0 + \text{grad } \lambda \times \mathbf{F}_0.$$

By scalar multiplication with $\mathbf{F}_0$ and application of the vector formula

$$\mathbf{A}\cdot(\mathbf{B} \times \mathbf{C}) = \mathbf{B}\cdot(\mathbf{C} \times \mathbf{A}),$$

one obtains an equation from which one concludes that $\mathbf{F}_0 \perp \text{curl } \mathbf{F}_0$, provided curl $\mathbf{F} = 0$.

If $U$ is a potential function that belongs to the pattern of field lines given by $\mathbf{F}_0$ so that $\lambda\mathbf{F}_0 = \text{grad } U$, then any function $V$ of $U$ is also a potential and the associated multiplier $\lambda' = \lambda dV/dU$. Thus the problem under consideration has an infinity of solutions, if it has a solution at all.

II.1. a) With the values of $p_0$ and $\rho_0$ of p. 99 one obtains from (7.9b) the values of $h$ for $n = 1.4$, 1.2 and 1, given on p. 52. The last value is in accordance with the fact that the exponential function in (7.14) vanishes only for $z = \infty$.

In the adiabatic case ($n = 1.4$) the lapse rate of the temperature can be determined from Eq. (7.11). If we assume a ground temperature of $0°C = 273°K$, the lapse rate is just about 1°C per 100m. This same adiabatic lapse rate for dry air is also found in the descending air current of a foehn[2] between two points of *the same stream line*, while in vertical direction, that is, in the transition to a different streamline, only half this value is found (corresponding to the average temperature gradient in the atmosphere).

b) We substitute in the second equation (7.7)

$$V = -\frac{GM}{r},$$

where $G$ = constant of Newton's law of gravitation, $M$ = mass of the earth, $r$ = distance from the center of the earth. The constant in (7.7) is determined by the condition: $p = p_0$ for $r = R$, $R$ being the radius of the earth. In this way one obtains from (6a) and (7)

(1) $$p = p_0 \left[ 1 - \frac{n-1}{n} \frac{p_0}{p_0} GM \left( \frac{1}{R} - \frac{1}{r} \right) \right]^{n/(n-1)}$$

At the surface of the earth we have

(2) $$\frac{GM}{R^2} = g.$$

The limit of the atmosphere is obtained if we let $p = 0$; with $r = R + h$ we obtain from (1)

(3) $$1 = \frac{n-1}{n} \frac{p_0}{p_0} gh \frac{R}{R+h},$$

in accordance with (7.9b), provided $h \ll R$ and $n > 1$.

For the isothermal case, $n = 1$, the two results no longer agree. If we substitute $G = R^2 g / M$ in (1) and pass to the limit $n \to 1$ we obtain

(4) $$p = p_0 \exp \left[ -\frac{p_0}{p_0} \cdot gR \left( 1 - \frac{R}{r} \right) \right], \quad r > R.$$

In this case there is no upper limit of the atmosphere that would be definable by $p = 0$: for $r \to \infty$ we have the paradoxical result of a small but finite pressure value

---

[2]This is a not uncommon meteorological phenomenon in the European Alps, its characteristic feature being the sliding down of air masses along mountain slopes.

(5)
$$p = p_0 \exp\left(-\frac{\rho_0}{p_0} gR\right).$$

II.2. The partial pressures of $N_2$ and $O_2$ are originally ($\omega = 0$)

(1) $\quad p_{01} = p_0 \dfrac{21}{100}, \quad p_{02} = p_0 \dfrac{79}{100}, \quad$ with $\quad p_0 = 1033 \times 981 \dfrac{gr}{cm\ sec^2}$ .

From (7.18) we have the partial pressures at the circumference $r$

(2)
$$p_1 = p_{01} \exp\left(\frac{\mu_1}{2} \frac{r^2\omega^2}{RT}\right) \cdots \mu_1 = 32,$$

$$p_2 = p_{02} \exp\left(\frac{\mu_2}{2} \frac{r^2\omega^2}{RT}\right) \cdots \mu_2 = 28.$$

The pressure difference $(p_1 + p_2) - p_0$ is prescribed as $40 \times 981$ gr cm$^{-1}$ sec$^{-2}$ and equals by (1) and (2)

$$p_1 + p_2 - p_0 = p_{01}\left[\exp\left(\frac{\mu_1}{2} \frac{r^2\omega^2}{RT}\right) - 1\right] + p_{02}\left[\exp\left(\frac{\mu_2}{2} \frac{r^2\omega^2}{RT}\right) - 1\right].$$

The arguments of the exponentials being small, we obtain a good approximation by expanding the exponentials and retaining the linear terms. With $R = 8.31 \times 10^7$ erg/grad and $T = 273°$ K, this leads to

$$40 = 10.33(21 \times 32 + 79 \times 28) \frac{30^2\omega^2}{2 \times 8.31 \times 10^7 \times 273},$$

whence $\omega = 260$ sec$^{-1}$ and $n = 30\,\omega/\pi = 2480$ rpm. The pressure (or density) ratio at the circumference is found as

$$\frac{p_1}{p_2} = \frac{\rho_1}{\rho_2} = \frac{p_{01}}{p_{02}} \cdot \frac{1 + \dfrac{\mu_1}{2} \dfrac{r^2\omega^2}{RT}}{1 + \dfrac{\mu_2}{2} \dfrac{r^2\omega^2}{RT}} = \frac{21}{79} \cdot \frac{1.043}{1.038},$$

or only 0.5% larger than at the center, hence the centrifuge is not an efficient means to separate mixtures of gases or isotopes. Even the ultracentrifuge mentioned on p. 55 separates only aggregates of colloids or, at best, the giant molecules of protein mixtures.

II.3. a) In close analogy with Eqs. (10.11) to (10.17) one obtains

(1) $\quad u = \dfrac{A}{2\mu} (h^2 - y^2) \quad u_m = \dfrac{Q}{2h} = A \dfrac{h^2}{3\mu} \quad \Pi = Al = 3\mu l \dfrac{u_m}{h^2},$

that is, a parabolic velocity profile as in Fig. 14 and a pressure loss proportional to the first power of the average velocity and to the minus second power of the half width $h$, which corresponds to the radius in Eq. (10.17). The reactive force acting on the upper or lower plate in the direction of the flow is $3\mu u_m/h$ per unit of area.

b) The component of gravity in the direction of the $x$-axis of Fig. 19a equals

$$(2) \qquad F_x = \rho g \sin \alpha;$$

the component directed toward the bottom $y = -h$ need not concern us here. At the free surface the pressure equals the atmospheric pressure $(p = 0)$; no pressure gradient in $x$-direction is required since its role is taken over by the body force (2). The differential equation (10.8a) reads under these conditions

$$-\mu \frac{\partial^2 u}{\partial y^2} = \rho g \sin \alpha.$$

The boundary conditions at $y = 0$ require both $p = 0$ and $p_{yx} = -\mu \partial u/\partial y = 0$, but these conditions are satisfied in (a) because of the symmetry of the flow relative to the plane $y = 0$. The velocity profile in the present case is therefore identical with the lower half of the profile in a), the constant $A$ being modified as required.

II.4. We refer to the formulas for cylindrical coordinates developed in the answer to problem I.3 and put

$$v = v_\varphi, \qquad v_r = v_z = 0.$$

The condition of incompressibility

$$\cdots \frac{1}{r} \frac{\partial v}{\partial \varphi} + \cdots = 0$$

requires that $v$ be independent of $\varphi$, so that $v = v(r)$. The same is true for the pressure $p$.

The expression for $\nabla^2 v$, which we find again in the table I.3, yields the differential equation

$$\frac{d^2 v}{dr^2} + \frac{1}{r} \frac{dv}{dr} - \frac{v}{r^2} = 0.$$

The general solution is a linear combination of the particular integrals $v = r$ and $v = 1/r$; the coefficients follow from the boundary conditions

$$v = 0 \quad \text{for} \quad r = r_i, \qquad v = U \quad \text{for} \quad r = r_e.$$

In this way we obtain the solution

$$v = U \frac{r_e}{r} \frac{r^2 - r_i^2}{r_e^2 - r_i^2}$$

which becomes identical with the solution (16.14) for the straight Couette flow if we pass to the limit infinity with $r_i$ and use the notations $r_e = r_i + h$ and $r = r_i + y$.

The reactive moments (per unit of length of the cylinder axis) are given by

$$M_e = 2\pi\mu Ur_e \frac{r_e^2 + r_i^2}{r_e^2 - r_i^2}, \qquad M_i = 2\pi\mu Ur_e \frac{2r_i^2}{r_e^2 - r_i^2}.$$

Their difference per unit of circumferential length of the external cylinder equals $\mu U$; the dimension of this quantity is that of a torque per unit of area. It is in fact equal to the torque per unit of area which acts in the case of the straight Couette flow upon the two boundary plates (stress $\mu U/h \times$ arm $h$).

II.5. This problem shows in a typical way how the fundamental equations of elasticity are applied. The solution requires a somewhat lengthy calculation, the single phases of which we shall clearly indicate. It is left to the reader to supply the connecting steps.

Let the direction of the force $P$ coincide with the $z$-axis and its point of application be the origin of a system of cylindrical coordinates $r$, $\varphi$, $z$. On denoting the components of the displacement vector by

$$\rho = \sqrt{\xi^2 + \eta^2}, 0, \zeta,$$

the dilatation $\Theta$ is obtained according to the table I.3 in the form

(1) $$\Theta = \frac{1}{r}\frac{\partial}{\partial r}(r\rho) + \frac{\partial\zeta}{\partial z}$$

since the problem is symmetrical about the $z$-axis.

a) The differential equation for $\Theta$ is found from (9.18) similarly as in (14.2):

(2) $$\nabla^2\Theta = 0.$$

Solutions that have the required symmetry are to be chosen from the following (cf. p. 248):

(3) $$\Theta = \frac{1}{R}, \frac{\partial}{\partial z}\frac{1}{R}, \frac{\partial^2}{\partial z^2}\frac{1}{R}, \cdots \qquad R^2 = r^2 + z^2.$$

We select the second solution and put therefore

(4) $$\Theta = -A\frac{\partial}{\partial z}\frac{1}{R} = A\frac{z}{R^3}.$$

b) The differential equation for the Cartesian component $\zeta$ is according to (9.18) and the table I.3

(5)
$$\frac{\partial^2 \zeta}{\partial r^2} + \frac{1}{r}\frac{\partial \zeta}{\partial r} + \frac{\partial^2 \zeta}{\partial z^2} = -\frac{\mu+\lambda}{\mu}A\left(\frac{1}{R^3} - \frac{3z^2}{R^5}\right).$$

A particular solution of this equation is

(6)
$$\zeta = B\frac{z^2}{R^3}, \qquad B = -\frac{\mu+\lambda}{\mu}\frac{A}{2},$$

to which we may add solutions of the form (3). Choosing the first solution (3), we assume $\zeta$ in the more general form

(7)
$$\zeta = \frac{C}{R} - \frac{\mu+\lambda}{\mu}\frac{A}{2}\frac{z^2}{R^3}.$$

c) The differential equation for $\rho$ follows easily from Eq. (1), viz.

(8)
$$\frac{1}{r}\frac{\partial}{\partial r}(r\rho) = \Theta - \frac{\partial \zeta}{\partial z} = A\frac{z}{R^3} + C\frac{z}{R^3} + \frac{\mu+\lambda}{\mu}A\frac{z}{R^3}\left(1 - \frac{3}{2}\frac{z^2}{R^2}\right).$$

This may be transformed into

(9)
$$\frac{\partial}{\partial r}(r\rho) = -\left(\frac{2\mu+\lambda}{\mu}A + C\right)z\frac{\partial}{\partial r}\frac{1}{R} + \frac{\mu+\lambda}{\mu}\cdot\frac{A}{2}z^3\frac{\partial}{\partial r}\frac{1}{R^3}$$

and can be integrated to give

(10)
$$r\rho = -\left(\frac{2\mu+\lambda}{\mu}A + C\right)\frac{z}{R} + \frac{\mu+\lambda}{\mu}\frac{A}{2}\frac{z^3}{R^3} + f(z),$$

where $f(z)$ is the integration constant. For $r = 0$ and all positive values of $z$ we must require $r\rho = 0$; since then $R = z$, we have

$$f(z) = C + \frac{3\mu+\lambda}{\mu}\frac{A}{2},$$

and Eq. (10) yields

(11)
$$\rho = \left(C + \frac{3\mu+\lambda}{\mu}\frac{A}{2}\right)\frac{1}{r} - \left(C + \frac{2\mu+\lambda}{\mu}A\right)\frac{z}{rR} + \frac{\mu+\lambda}{\mu}\frac{A}{2}\frac{z^3}{rR^3}.$$

d) The boundary conditions at the surface $z = 0$ are for $r > 0$

(12)
$$\sigma_{zr} = 0, \qquad \sigma_{z\varphi} = 0, \qquad \sigma_{zz} = 0.$$

The first condition requires that

(13)
$$\frac{\partial \zeta}{\partial r} + \frac{\partial \rho}{\partial z} = 0.$$

The second condition is automatically fulfilled for reasons of symmetry, not only at the surface $z = 0$, but everywhere inside the elastic body (see below under $f$). The third condition is equivalent to

$$\frac{\partial \zeta}{\partial z} = 0$$

since $\Theta = 0$ for $z = 0$ and $r > 0$; according to (7) it is also automatically fulfilled. We now substitute in (13) according to (7) and (11) for $z = 0$

$$\frac{\partial \zeta}{\partial r} = -\frac{C}{r^2}, \qquad \frac{\partial \rho}{\partial z} = -\left(C + \frac{2\mu + \lambda}{\mu} A\right)\frac{1}{r^2}$$

and obtain

(14) $$C = -\frac{2\mu + \lambda}{\mu}\frac{A}{2}.$$

e) The state of deformation can thus be described by the following relations

(15)
$$\Theta = A\frac{z}{R^3}, \qquad \zeta = -\frac{A}{2}\left(\frac{2\mu + \lambda}{\mu}\frac{1}{R} + \frac{\mu + \lambda}{\mu}\frac{z^2}{R^3}\right)$$

$$\rho = \frac{A}{2r}\left(1 - \frac{2\mu + \lambda}{\mu}\frac{z}{R} + \frac{\mu + \lambda}{\mu}\frac{z^3}{R^3}\right).$$

It follows that

(16) $$\sigma_{zz} = 2\mu\frac{\partial \zeta}{\partial z} + \lambda\Theta = A(\mu + \lambda)\frac{3z^3}{R^5}.$$

Only the integration constant $A$ remains to be determined; this must be done by relating it to the external force $P$. If the body is cut in two at an arbitrary depth $z_0$ along the plane $z = z_0$, the following condition of equilibrium must hold:

(17) $$P = -2\pi\int_{r=0}^{\infty} \sigma_{zz} r\, dr \qquad \text{when} \qquad z = z_0.$$

On substituting for $\sigma_{zz}$ according to (16) and observing that

$$\int_{r=0}^{\infty} \frac{r\, dr}{R^5} = -\frac{1}{3R^3}\bigg|_{r=0}^{\infty} = \frac{1}{3z^3},$$

one obtains

(18) $$P = -2\pi A(\mu + \lambda).$$

f) Of the three shear stresses $\sigma_{r\varphi}$, $\sigma_{\varphi z}$, $\sigma_{zr}$ the first two obviously vanish everywhere within the elastic body. This implies that the normal stress $\sigma_{\varphi\varphi}$ belonging to any point of the surface $\varphi = $ const is a *principal stress* usually called the *hoop stress*. The associated stress trajectories are circles about the axis given by the forces $P$. In every torus or hoop of infinitesimal cross-section whose axis is a circular stress trajectory, there acts a definite stress $\sigma_{\varphi\varphi}$ that is characteristic of the hoop. The same hoop stresses occur in pipes and boilers subject to internal pressure, and their magnitude is the determining factor for the safety of the construction. In the present case the action of the external force $P$ on the "hoops", of which the semi infinite body can be thought to consist, may be compared with that of an internal pressure. Since $\rho$ denotes the increment of the hoop radius, the change of its length is $2\pi\rho$ and its extension $2\pi\rho/2\pi r = \rho/r$. It follows therefore that

$$(19) \qquad \epsilon_{\varphi\varphi} = \frac{\rho}{r}, \qquad \sigma_{\varphi\varphi} = 2\mu\epsilon_{\varphi\varphi} + \lambda\theta = 2\mu\frac{\rho}{r} + \lambda A\frac{z}{R^3},$$

or, because of (15),

$$\sigma_{\varphi\varphi} = \frac{A}{r^2}\left(\mu - (2\mu + \lambda)\frac{z}{R} + (\mu + \lambda)\frac{z^3}{R^3}\right) + \lambda A\frac{z}{R^3}.$$

This may be rewritten in the following form

$$(20) \qquad \sigma_{\varphi\varphi} = \frac{A\mu}{r^2}\left(1 - 2\frac{z}{R} + \frac{z^3}{R^3}\right).$$

The expression stays finite for $r = 0$ since the factor in parentheses vanishes.

The justification for the use of the particular solutions (4) and (7) must be seen in the fact that they can be made to fit all continuity requirements and boundary conditions of the problem. The vanishing of the displacements and stresses at infinity should also be counted among the boundary conditions. It can be verified directly from Eqs. (15), (16), and (20).

The assumption of a load concentrated at one point is physically not justifiable, but greatly simplifies the mathematics of the problem. The general case of a prescribed load distribution along the surface $z = 0$ can be easily constructed from the foregoing formulas by integration.

Boussinesq published his solution in 1879 and gave a complete representation of it in his book "Applications des potentiels directes, inverses, logarithmiques" Paris 1885.

III.1. The misinterpretation of the operator $(\mathbf{v}\ \mathrm{grad})$ would lead to the following expression for the $r$ component

$$(1) \qquad (\mathbf{v}\ \mathrm{grad})v_r = v_r\frac{\partial v_r}{\partial r} + \frac{v_\varphi}{r}\frac{\partial v_r}{\partial \varphi} + v_z\frac{\partial v_r}{\partial z}.$$

Instead of this one obtains by (11.6)

$$v_r \frac{\partial v_r}{\partial r} + v_\varphi \frac{\partial v_\varphi}{\partial r} + v_z \frac{\partial v_z}{\partial r} - v_\varphi \operatorname{curl}_z v + v_z \operatorname{curl}_\varphi v.$$

With the values for $\operatorname{curl}_\varphi$ $\mathbf{A}$ and $\operatorname{curl}_z$ $\mathbf{A}$ according to table I.3, the last equation yields after cancellation of two pairs of terms

(2) $$v_r \frac{\partial v_r}{\partial r} + \frac{v_\varphi}{r} \frac{\partial v_r}{\partial \varphi} + v_z \frac{\partial v_r}{\partial z} - \frac{v_\varphi^2}{r}.$$

The last term is the *centripetal acceleration* [missing in (1)]. The corresponding computation for the $\varphi$-component of (v grad) v gives the additional term

$$\frac{v_r v_\varphi}{r} = v_r \omega, \qquad \omega = \frac{v_\varphi}{r}.$$

When we here interpret $\omega$ as the amount of the axial vector $\boldsymbol{\omega}$ parallel to the z-axis that represents the angular velocity associated with the linear velocity $v_\varphi$, then the term $v_r \omega = (\boldsymbol{\omega} \times \mathbf{v})_\varphi$ becomes one half of a *Coriolis acceleration* [cf. Vol. I Eq. (29.4a)]. One finds, however, that the two results are identical in the case of the z-component. The formulas have already been tabulated in I.3 for cylindrical and spherical coordinates.

III.2 a) We put in Eq. (1) of the problem

$$C = A \rho^\alpha v^\beta a^\gamma,$$

where the $\alpha$, $\beta$, $\gamma$ are to be determined and $A$ is dimensionless. On substituting the dimensions of $v$, $\rho$, $\Pi/l$ and $v$, the following dimensional identity is obtained

$$\text{gr cm}^{-2} \text{sec}^{-2} = \text{gr}^\alpha \text{ cm}^{-3\alpha + 2\beta + \gamma + n} \text{sec}^{-\beta - n},$$

which yields $\alpha = 1$, $\beta = 2 - n$, and $\gamma = n - 3$, whence

$$C = A \rho v^{2-n} a^{n-3}.$$

b) On substituting in Eq. (3) of the problem the expressions (16.9) and (16.8) for $S$ and $R$, one obtains

$$\frac{\Pi}{\rho v^2} = \lambda \frac{l}{a} \left( \frac{va}{\nu} \right)^\delta.$$

$v$ must occur here in the $n^{\text{th}}$ power since this relation should be equivalent to Eq. (1) of the problem. It follows that $\delta = n - 2$ and therefore

$$\Pi = \lambda \rho \frac{l}{a} v^2 R^{n-2}.$$

c) On using Hagen's value $n = 1.75$ in this equation, one obtains

$$\Pi = \lambda \rho \, \frac{l}{a} \, v^2 R^{-1/4},$$

which agrees with the first form of Eq. (16.13), the exponent $\kappa$ having Blasius's value of $\frac{1}{4}$.

III.3. Let the virtual displacement of the liquid be a motion of the surface elements $dF$ in the direction of their normals by $\delta n$. Since the volume is conserved both positive and negative displacements $\delta n$ (outward and inward displacements) must occur. The virtual work is composed of the work of the capillary forces and that of the external forces; the first contribution follows from (17.11) by multiplication with $\delta n$ and integration over the surface, and the second contribution equals $\int U \delta n \, dF$ where $U$ is the potential energy per unit of volume.

The constancy of the volume $\tau$ as a whole requires that the variation $\delta \tau$ vanish:

$$(1) \qquad \qquad \delta \tau = \int \delta n \, dF = 0.$$

This restriction on $\delta n$ can be taken care of by a Lagrangian multiplier $\lambda$, that is, by adding the term $\lambda \delta \tau$ to the left side of the virtual work and setting the total equal to zero.

The displacements $\delta n$ being now arbitrary, one obtains

$$(2) \qquad \qquad T \! \left( \frac{1}{R_1} + \frac{1}{R_2} \right) + U + \lambda = 0.$$

If the mean curvature is here replaced by the appropriate expressions from differential geometry, one has in (2) a partial differential equation for the determination of the shape of the surface. The hitherto unknown multiplier $\lambda$ is fixed by requiring the surface to pass through a given point with given mean curvature (cf. the next problem).

III.4. Let the $z$-axis coincide with the axis of the vertical capillary tube and count $z$ positive upward from the external liquid level. Then

$$(1) \qquad \qquad U = \rho g z.$$

For a wetting liquid the radius of the concave spherical surface is negative. Eq. (2) of the preceding problem then gives

$$(2) \qquad \qquad -T \frac{2}{R} + \rho g z + \lambda = 0.$$

Since for the external surface $z = 0$ and the radius $R = \infty$, the multiplier

$\lambda$ must be set equal to zero. The radius of the sphere $R$ can be expressed by the radius of the tube $r$ if the angle of contact is known:

(3)                         $$R = r/\cos \vartheta.$$

Defining the elevation $h$ by a suitably chosen mean value of the $z$-ordinates of the meniscus, one obtains from (2) and (3)

(4)                         $$h = \frac{2T}{\rho g r} \cos \vartheta$$

where the value of $\cos \vartheta$ for wetting liquids is nearly 1. For non-wetting liquids $R$ is positive and the right member in Eq. (4) takes therefore a minus sign; $h$ denotes now the capillary depression.

IV.1. From (19.10a, b) one calculates $z = x + iy$ as a function of $\rho + i\varphi$ and obtains by differentiation the differential $dx + idy$ in terms of $d\rho + id\varphi$. The square of the line element $ds^2$ is found as the absolute value of $dx + idy$ which leads directly to relation (1).

The fact that the factors of $d\rho^2$ and $d\varphi^2$ [denoted by $g_1^2$ and $g_2^2$ in (2.22)] are here equal means that the network of the two families of curves $\rho = $ const, and $\varphi = $ const consists of infinitesimal squares (provided the increments $d\rho$ and $d\varphi$ are equal). In other words, the bipolar coordinates are isometric in the sense of p. 139. Note that the same can be achieved with polar coordinates in the plane if one replaces $z = re^{i\varphi}$ by $z = e^{\rho+i\varphi}$, where $\rho = \log r$.

In setting up the operator $\nabla^2$ the three dimensional scheme (3.9b) must be specialized for *two* coordinates $p_1 = \rho$, $p_2 = \varphi$ which amounts to putting

$$g_1 = g_2 = g, \qquad g_3 = 1, \qquad \frac{\partial}{\partial p_3} = 0.$$

In the case of the *potential equation* $\nabla^2 U = 0$ the additional factor $1/g^2$ is immaterial, but its presence destroys the separability in the case of the *wave equation*. With the assumption of a periodic time dependence $U = ue^{-i\omega t}$ the wave equation $\nabla^2 U + k^2 U = 0$ reads

$$\frac{\partial^2 u}{\partial \rho^2} + \frac{\partial^2 u}{\partial \varphi^2} + k^2 g^2 u = 0, \qquad k = \frac{\omega}{c}.$$

On substituting here the trial solution (2) and dividing by $f_1 f_2$, one is led to a contradiction if one differentiates once more with respect to $\rho$ or $\varphi$.

IV.2. We start from Eq. (19.11) which yields

$$\mathbf{v} = A \text{ grad } \varphi.$$

For an arbitrary function of the bipolar coordinates $f(\rho, \varphi)$ we have

$$\operatorname{grad}_\rho f = \frac{1}{g}\frac{df}{d\rho}, \qquad \operatorname{grad}_\varphi f = \frac{1}{g}\frac{df}{d\varphi}.$$

If in particular $f = \varphi$, then $v_\rho = 0$ and $v_\varphi = 1/g$, as already mentioned in the text of the problem. With $A = \mu/\pi$ as in Eq. (19.8a) we have therefore

(1) $$|\mathbf{v}| = v_\varphi = \frac{\mu}{\pi c}\,(\cosh \rho - \cos \varphi).$$

This should be compared with the value of $|\mathbf{v}|$ that follows from (19.14)

(2) $$|\mathbf{v}| = \frac{\mu}{\pi}\left|\frac{\varphi_1}{r_1} - \frac{\varphi_2}{r_2}\right|.$$

The vector $\varphi_1/r_1$ $(\varphi_2/r_2)$ has the amount $1/r_1$ $(1/r_2)$ and is perpendicular to the radius vector from $z = +c$ $(z = -c)$ to the field point. The angle subtended by the two radius vectors $\mathbf{r}_1$, $\mathbf{r}_2$ is therefore $\varphi = \varphi_1 - \varphi_2$. Vectorial addition then yields

$$\sqrt{\frac{1}{r_1^2} + \frac{1}{r_2^2} - \frac{2}{r_1 r_2}\cos \varphi} = \sqrt{\frac{2}{r_1 r_2}}\sqrt{\frac{1}{2}\left(\frac{r_2}{r_1} + \frac{r_1}{r_2}\right) - \cos \varphi}$$

$$= \sqrt{\frac{2}{r_1 r_2}}\sqrt{\cosh \rho - \cos \varphi}.$$

Relations (1) and (2) are equivalent when

$$\frac{1}{c}\sqrt{\cosh \rho - \cos \varphi} = \sqrt{\frac{2}{r_1 r_2}},$$

which is in fact the case, as one sees by applying the law of cosines to the triangle whose corners are $z = \pm c$ and the field point.

IV.3. From the definition of elliptic coordinates in (19.15) we derive similarly as in problem IV.1

$$ds^2 = g^2(d\xi^2 + d\eta^2), \qquad g = \frac{c}{\sqrt{2}}\sqrt{\cosh 2\xi - \cos 2\eta}.$$

The equation of the vibrating membrane has the same form as in IV.1, but the meaning of $g$ is different:

(1) $$\frac{\partial^2 u}{\partial \xi^2} + \frac{\partial^2 u}{\partial \eta^2} + k^2 g^2 u = 0.$$

362 ANSWERS

Substitution of the trial solution and division by $f_1 f_2$ leads to

$$(2) \qquad \frac{f_1''}{f_1} + \frac{c^2 k^2}{2} \cosh 2\xi = -\frac{f_2''}{f_2} + \frac{c^2 k^2}{2} \cos 2\eta = \lambda.$$

Now the quantity $\lambda$ (the "separation parameter") is independent of $\xi$ and $\eta$; to show this differentiate (2) with respect to $\eta$ or $\xi$. (In other words, $\lambda$ is a function of $\xi$ alone according to the first, and of $\eta$ alone according to the second member of Eq. (2) and is therefore constant.) The partial differential equation (2) then yields two ordinary differential equations

$$(3) \qquad f_1'' = \left(\lambda - \frac{c^2 k^2}{2} \cosh 2\xi\right)f_1, \quad f_2'' = \left(-\lambda + \frac{c^2 k^2}{2} \cos 2\eta\right)f_2.$$

The solutions of (3) are called Mathieu's functions.

IV.4a. Toroidal coordinates first occur in a posthumous paper of Riemann "About the potential of a torus", Ges. Werke XXIV, p. 431. We introduce the notations $\rho$, $\varphi$, $\psi$ and restrict the ranges as follows:

$$(1) \qquad 0 \leqq \rho < \infty, \qquad 0 \leqq \varphi < 2\pi, \qquad 0 \leqq \psi < 2\pi.$$

The geometrical meaning of $\rho$ and $\varphi$ follows from Fig. 26, the right half of which need only be considered ($\rho$ positive). The angle $\psi$ is the angle of rotation about the axis $\rho = 0$ (azimuth); the surfaces $\psi = $ const are half planes containing this axis. In the rotation the circles $\rho = $ const yield a system of coaxial tori about the axis $\rho = 0$, and the circular arcs $\varphi = $ const yield spherical segments.

Let us now change the notation of the complex variable $z$ of Eq. (19.10a) into $\bar{z} = \bar{x} + i\bar{y}$; then by (19.10a, b)

$$(2) \qquad \bar{x} = -\frac{c \sinh \rho}{\cosh \rho - \cos \varphi}, \qquad \bar{y} = \frac{c \sin \varphi}{\cosh \rho - \cos \varphi}.$$

If we introduce a spatial Cartesian system $x$, $y$, $z$ such that the $z$-axis coincides with $\rho = 0$, the origin with the center of Fig. 26, and the positive $x$-axis has the azimuth $\psi = 0$, the following relations hold between the coordinates $x$, $y$, $z$ and the coordinates $\bar{x}$, $\bar{y}$, $\psi$ of the same point:

$$(3) \qquad x = \bar{x} \cos \psi, \qquad y = \bar{x} \sin \psi, \qquad z = \bar{y}.$$

The square of the line element follows now by differentiation (cf. prob. IV.1),

$$(4) \qquad ds^2 = dx^2 + dy^2 + dz^2 = g^2(d\rho^2 + d\varphi^2) + g_3^2 \, d\psi^2,$$

where

$$(5) \qquad g = g_1 = g_2 = \frac{c}{\cosh \rho - \cos \varphi}, \qquad g_3 = -\bar{x} = \frac{c \sinh \rho}{\cosh \rho - \cos \varphi}.$$

The Laplacian operator follows by (3.9b):

(6) $$\nabla^2 U = \frac{1}{g^2 g_3}\left\{\frac{\partial}{\partial\rho}\left(g_3\frac{\partial U}{\partial\rho}\right) + \frac{\partial}{\partial\varphi}\left(g_3\frac{\partial U}{\partial\varphi}\right) + \frac{g^2}{g_3}\frac{\partial^2 U}{\partial\psi^2}\right\}.$$

One assumes $U$ in the form

(7) $$U = \sqrt{\cosh\rho - \cos\varphi}\; V(\rho,\varphi)e^{im\psi},$$

and obtains the following form of $\nabla^2 U$ after several steps

(8) $$c^2\nabla^2 U = (\cosh\rho - \cos\varphi)^{5/2}$$
$$\cdot\left\{\frac{\partial^2 V}{\partial\rho^2} + \coth\rho\frac{\partial V}{\partial\rho} + \frac{\partial^2 V}{\partial\varphi^2} + \left(\frac{1}{4} - \frac{m^2}{\sinh^2\rho}\right)V\right\}e^{im\psi}.$$

When we are interested in the potential equation $\nabla^2 U = 0$, the resulting differential equation for $V$ can be separated by setting

(9) $$V = f(\rho)e^{in\varphi}.$$

The ordinary differential equation that results for $f$ is

(10) $$f'' + \coth\rho f' + \left(\frac{1}{4} - n^2 - \frac{m^2}{\sinh^2\rho}\right)f = 0,$$

the solutions of which are generalized spherical harmonics.

In the case of the wave equation, however, the differential equation for $V$, obtained from (7), is

(11) $$\frac{\partial^2 V}{\partial\rho^2} + \coth\rho\frac{\partial V}{\partial\rho} + \frac{\partial^2 V}{\partial\varphi^2}$$
$$+ \left(\frac{1}{4} - \frac{m^2}{\sinh^2\rho} + \frac{k^2 c^2}{(\cosh\rho - \cos\varphi)^{5/2}}\right)V = 0.$$

It is not separable in the coordinates $\rho$ and $\varphi$.

IV.4b. If the diagram in Fig. 26 is made to rotate about the axis $\varphi = 0$ rather than $\rho = 0$, the circles $\rho = $ const generate spheres and the circular arcs $\varphi = $ const generate "spindles" (the latter name is appropriate only for the surfaces $\varphi < \pi/2$). The third coordinate is the angle of rotation $\psi$ about the axis, hence the surfaces $\psi = $ const are half-planes as before. The ranges for the coordinates are now

(1') $$-\infty < \rho < +\infty, \qquad 0 \leqq \varphi < \pi, \qquad 0 \leqq \psi < 2\pi.$$

Introduce again Cartesian coordinates $x, y, z$, so that the $z$-axis coincides with the axis of rotation and the origin with the center of Fig. 26.

With $\bar{x}$ and $\bar{y}$ having the same meaning as before, we have now instead of (3)

(3')                $x = \bar{y} \cos \psi, \qquad y = \bar{y} \sin \psi, \qquad z = \bar{x}.$

By differentiation, a line element of the same form and also with the same value of $g$ as in (4) is obtained, but the value of $g_3$ is now

(5')                $$g_3 = \bar{y} = \frac{c \sin \varphi}{\cosh \rho - \cos \varphi}.$$

The expression (6) for $\nabla^2 U$, which is still correct, yields when the form (7) of $U$ and the present value of $g_3$ are used

(8')
$$c^2 \nabla^2 U = (\cosh \rho - \cos \varphi)^{5/2}$$
$$\cdot \left\{ \frac{\partial^2 V}{\partial \rho^2} + \cot \varphi \, \frac{\partial V}{\partial \varphi} + \frac{\partial^2 V}{\partial \varphi^2} - \left( \frac{1}{4} + \frac{m^2}{\sin^2 \varphi} \right) V \right\} e^{im\psi}.$$

When we are interested in the potential equation $\nabla^2 U = 0$, the resulting differential equation for $V$ becomes separable if one puts

(9')                        $V = f(\varphi) e^{\alpha \rho}.$

The equation for $f$ is

(10')           $f'' + \cot \varphi f' - \left( \frac{1}{4} - \alpha^2 + \frac{m^2}{\sin^2 \varphi} \right) f = 0,$

which can be solved by generalized spherical harmonics like (10).

The wave equation $\nabla^2 U + k^2 U = 0$, however, is also not separable in these coordinates.

V.1. The kinetic energy to be computed is given by

(1)        $E_{\text{kin}} = \frac{\rho}{2} \int_0^\lambda dx \int_\eta^\infty dy \left\{ \left( \frac{\partial \Phi}{\partial x} \right)^2 + \left( \frac{\partial \Phi}{\partial y} \right)^2 + \left( \frac{\partial \Phi}{\partial z} \right)^2 \right\}.$

On substituting for $\Phi$ according to (26.6) one obtains without difficulty

(2)        $E_{\text{kin}} = \rho k \pi A^2 \int_\eta^\infty e^{-2ky} \, dy = \frac{\rho}{2} \pi A^2 e^{-2k\eta}.$

This agrees with the expression (26.7a) for $E_{\text{pot}}$, since the amplitude $A$, which occurs in the expression for $\eta$ [cf. Eq. (26.6)], is considered as a first order infinitesimal, and $e^{-2k\eta}$ may therefore be replaced by 1. Since $E_{\text{kin}}$ is zero if the fluid is at rest, the expression (2) is at the same time the difference in kinetic energy between states of motion and rest for the spatial domain in question.

In the second part of the problem

$$\Phi = C \cos (kx - \omega t) \cosh k(h - y),$$

and one has by (24.3) and (24.8)

$$\eta = - \frac{\omega}{g} \sin (kx - \omega t) \cosh kh;$$

this yields

$$S = \frac{\rho}{4} \pi C^2 (\sinh 2kh + 2kh), \qquad E_{kin} = E_{pot} = \frac{\rho}{4} \pi C^2 \sinh 2kh.$$

VI.1. The Navier-Stokes equations (10.8a, b) augmented by the Coriolis term read

$$(1) \qquad 2\rho \mathbf{v} \times \boldsymbol{\omega} + \mu \nabla^2 \mathbf{v} - \operatorname{grad} p + \mathbf{F} = 0.$$

With the simplification indicated in the problem text the $\xi$ and $\eta$ components of Eq. (1) yield for a point on the northern hemisphere

$$(2) \qquad \begin{aligned} 2\omega \sin \varphi v_\eta + \nu \frac{d^2 v_\xi}{d\zeta^2} &= 0, \\ -2\omega \sin \varphi v_\xi + \nu \frac{d^2 v_\eta}{d\zeta^2} &= 0, \end{aligned}$$

where $\varphi$ is the geographical latitude. (Note that the gravity vector $\mathbf{F}$ is perpendicular to $\xi$ and $\eta$.) Since $v_\zeta = 0$ and the vertical component of the Coriolis acceleration is negligibly small in comparison with $g$, the $\zeta$-component of Eq. (1) gives simply the hydrostatic pressure gradient in $\zeta$ direction

$$-\frac{dp}{d\zeta} = \rho g.$$

With the simplifications introduced, the condition of incompressibility is automatically fulfilled. The two Eqs. (2) can be conveniently written as one complex equation, similarly as in the theory of Foucault's pendulum (cf. Vol. I. 31.5). We put

$$V = v_\xi + i v_\eta$$

and obtain instead of (2)

$$(3) \qquad \frac{d^2 V}{d\zeta^2} - 2i\lambda^2 V = 0, \qquad \lambda^2 = \frac{\omega}{\nu} \sin \varphi.$$

The integration yields

(4)
$$V = Ae^{\lambda(1+i)\zeta}.$$

if due regard is given to the orientation of $\zeta$. The constant $A$ must be determined from the excitation due to the wind pressure, whose components have to balance the friction pressures $p_{\zeta\xi}$ and $p_{\zeta\eta}$ everywhere on the surface $\zeta = 0$. If one introduces a complex friction pressure $P$ by

$$P = p_{\zeta\xi} + i p_{\zeta\eta} = -\mu\left(\frac{dv_\xi}{d\zeta} + i\frac{dv_\eta}{d\zeta}\right) = -\mu\frac{dV}{d\zeta},$$

where $v_\zeta$ has again been neglected, one obtains from (4) for $\zeta = 0$

(5)
$$-P_0 = \mu A\lambda(1 + i).$$

This is the value of the wind pressure on the surface $\zeta = 0$ in complex notation; it must be considered as a known vector. We introduce the notations

$$P_0 = |P_0|\, e^{i\delta}, \qquad v_0 = -\frac{|P_0|}{\mu\lambda\sqrt{2}}, \qquad \zeta = -z,$$

and obtain the velocities $v_\xi$ and $v_\eta$ when we take $A$ from (5), substitute it in (4), and separate real and imaginary parts:

$$v_\xi = v_0 e^{-\lambda z}\cos\left(\lambda z + \frac{\pi}{4} - \delta\right),$$

(6)

$$v_\eta = -v_0 e^{-\lambda z}\sin\left(\lambda z + \frac{\pi}{4} - \delta\right).$$

This result permits the following interpretation: *The drift current on the surface makes an angle of $-45°$ with the wind direction. The velocity of the current $|v|$ attenuates with increasing depth and changes its direction in such a way that at the depth*

(7)
$$z_E = \frac{\pi}{\lambda} = \frac{\pi}{\sqrt{\dfrac{\omega}{\nu}\sin\varphi}}$$

*the current is opposite to the surface current. At that depth $|v|$ is equal to* $v_0 \cdot e^{-\pi}$ (Ekman's friction-depth).

For a justification of our simplifying assumptions and for further discussion see: V. W. Ekman, "Dynamische Gesetze der Meeresströmungen" in the Innsbruck lectures quoted on p. 242. The fact that the actual motion is properly a turbulent one can be accounted for by giving

$\mu$ a larger value than usual; this takes care of the increased internal friction due to turbulence.

VI.2. When higher powers of $\varphi_1$ and $\varphi_2$ are neglected and the constant terms $U_1^2/2$ and $U_2^2/2$ of Bernoulli's equation are absorbed in the constants $C_1$ and $C_2$, the following equations are obtained from (11.15) for the disturbance potentials $\varphi_1$ and $\varphi_2$

$$\frac{p_1}{\rho} = \frac{\partial\varphi_1}{\partial t} + U_1\frac{\partial\varphi_1}{\partial x} - C_1 \cdots y > 0,$$

$$\frac{p_2}{\rho} = \frac{\partial\varphi_2}{\partial t} + U_2\frac{\partial\varphi_2}{\partial x} - C_2 \cdots y < 0.$$

Since the pressure on both sides of the discontinuity surface ($y = 0$) must be the same, one obtains

(1) $$\frac{\partial\varphi_1}{\partial t} + U_1\frac{\partial\varphi_1}{\partial x} = \frac{\partial\varphi_2}{\partial t} + U_2\frac{\partial\varphi_2}{\partial x} + c \cdots y = 0,$$

and the constant $c = C_1 - C_2$ must be zero because of the assumed periodicity of $\varphi_1$ and $\varphi_2$.

We write the sinusoidal disturbance of the surface in the convenient complex form

(2) $$\eta = Ae^{ikx}, \qquad k = \frac{2\pi}{\lambda}, \qquad A = \text{function of } t$$

The $y$-velocity $-\partial\varphi_1/\partial y$ of a particle close to the discontinuity surface is, except for second order terms, given by

$$\frac{d\eta}{dt} = \frac{\partial\eta}{\partial t} + U_1\frac{\partial\eta}{\partial x}.$$

We obtain therefore

(3) $$\frac{\partial\eta}{\partial t} + U_1\frac{\partial\eta}{\partial x} = -\frac{\partial\varphi_1}{\partial y}$$

and, correspondingly,

(4) $$\frac{\partial\eta}{\partial t} + U_2\frac{\partial\eta}{\partial x} = -\frac{\partial\varphi_2}{\partial y}.$$

Since the problem is two-dimensional, we put

(5) $$\varphi_1 = A_1e^{ik(x+iy)}, \qquad \varphi_2 = A_2e^{ik(x-iy)}, \qquad \left.\begin{array}{c} A_1 \\ A_2 \end{array}\right\} = \text{functions of } t$$

Note that (5) fulfills the condition of incompressibility and $\varphi_1$ $(\varphi_2)$ stays finite for $y = +\infty$ $(y = -\infty)$. The problem has a solution only if the wave number $k$ in (5) has the same value as in (2).

Upon substituting (2) and (5) in Eqs. (1), (3), and (4) the amplitudes $A$, $A_1$, $A_2$ are found to obey the following linear system of differential equations

$$(6) \qquad \frac{dA_1}{dt} + ikU_1A_1 = \frac{dA_2}{dt} + ikU_2A_2$$

$$(7) \qquad \frac{dA}{dt} + ikU_1A = kA_1$$

$$(8) \qquad \frac{dA}{dt} + ikU_2A = -kA_2 .$$

The solution can be set up in the form

$$(9) \qquad A = ae^{\alpha t}, \qquad A_1 = a_1e^{\alpha t}, \qquad A_2 = a_2e^{\alpha t}.$$

If one substitutes for $A_1$ and $A_2$ according to (9), one can compute $a_1$ and $a_2$ from (7) and (8), and obtains a quadratic equation for $\alpha$ by introducing these values into (6). The two roots are

$$(10) \qquad \alpha = \frac{k}{2}\left[\pm(U_1 - U_2) - i(U_1 + U_2)\right].$$

Unless $U_1$ equals $U_2$, one of the two roots has a positive real part which implies a continual increase of the amplitudes $A$, $A_1$, $A_2$. At any wave number $k$ there are thus sinusoidal initial disturbances that will grow in the course of time.

VI.3. In the first pair of equations of VI.2, replace $U_1$ by $U$ (wind velocity) and set $U_2 = 0$ (water at rest), add the action of gravity $-g\eta$ ($\eta$ positive upward), and introduce the two densities $\rho_1$ and $\rho_2$ instead of $\rho$. With $s = \rho_1/\rho_2$ and $c = 0$ as before, these equations yield

$$(1') \qquad s\left(\frac{\partial\varphi_1}{\partial t} - g\eta + U\frac{\partial\varphi_1}{\partial x}\right) = \frac{\partial\varphi_2}{\partial t} - g\eta.$$

Eqs. (2) and (5) of VI.2 remain unchanged, and so do (3) and (4) except for the specified values of $U_1$ and $U_2$. One then has the same form of a solution as (9) for the differential equations that correspond to Eqs. (6), (7), (8) in VI.2, and expresses again $a_1$ and $a_2$ by $a$. In this way one obtains for $\alpha$ the quadratic equation

$$(10') \qquad s(\alpha + ikU)^2 + \alpha^2 = -gk(1 - s).$$

One of the two roots has a positive real part it

(11')
$$U^2 > \frac{g}{k}\frac{1 - s^2}{s}.$$

Now, $\sqrt{g/k}$ is, according to (23.16), the phase velocity $V$ of the deep water waves if the inertia of the air is neglected (the value of $V$ is only insignificantly smaller if the inertia of the air is considered, as can be shown by suitably expanding the calculations of 23); hence the condition of instability (11') can be written as

$$U > V\sqrt{\frac{1 - s^2}{s}}.$$

In order to excite disturbances of wave length $\lambda$, the wind velocity $U$ must be considerably larger than the phase velocity $V$ associated with this wave length. The plane water surface is therefore stable with regard to this wave length for all wind velocities

$$U < \frac{V}{\sqrt{0.0013}} = 28V.$$

If in addition to gravity capillarity is considered, the stabilizing tendency of gravity is strengthened and the plane water surface is seen to be absolutely stable for all wave lengths as long as

$$U < 28\,V_{min}; \qquad V_{min} = 23.2 \text{ cm/sec.}$$

The phase velocity $V_{min}$ belongs to the wave length $\lambda_{min} = 1.73$ cm as calculated on p. 184. A *wind velocity smaller than* $28 \times 23.2$ *cm/sec* $\sim$ *6.5m/sec (wind strength 2-3 Beaufort) would not be able to disturb the plane surface of the water*. This result assumes perfect homogeneity of the wind, which in reality is always turbulent.

VII.1. Let $a, b, c$ be the coordinates of a fluid element at the time $t = 0$; in cylindrical coordinates $r, \varphi, z$

$$a = x_0 = r\cos\varphi,$$

(1)
$$b = y_0 = r\sin\varphi,$$

$$c = z_0.$$

Because of the uniform rotation the coordinates of the same element at the time $t$ are obtained if we replace $\varphi$ by $\varphi + \omega t$, where $\omega =$ angular velocity. Hence

$$x = a \cos \omega t - b \sin \omega t,$$

(2) $$\qquad y = b \cos \omega t + a \sin \omega t,$$

$$z = c.$$

It is easy to verify that the condition of incompressibility (34.3) is fulfilled. Lagrange's equations (34.1) give in the present case (gravity being the only external force, and the $z$-axis being counted positive downward as in Fig. 7)

$$\frac{\partial^2 x}{\partial t^2} \cos \omega t + \frac{\partial^2 y}{\partial t^2} \sin \omega t + \frac{1}{\rho} \frac{\partial p}{\partial a} = 0,$$

(3) $$\qquad -\frac{\partial^2 x}{dt^2} \sin \omega t + \frac{\partial^2 y}{dt^2} \cos \omega t + \frac{1}{\rho} \frac{\partial p}{\partial b} = 0,$$

$$\frac{\partial^2 z}{dt^2} \qquad - g \qquad + \frac{1}{\rho} \frac{\partial p}{\partial c} = 0.$$

If the accelerations in (3) are computed according to (2), one obtains for $p$ the system of differential equations

(4) $$\qquad \frac{1}{\rho} \frac{\partial p}{\partial a} = \omega^2 a, \qquad \frac{1}{\rho} \frac{\partial p}{\partial b} = \omega^2 b, \qquad \frac{1}{\rho} \frac{\partial p}{\partial c} = g.$$

The value of $p$ obtained by integration of the system (4) is identical with Eq. (6.10) if (2) is taken into account.

To carry out the same calculation for a compressible gas is quite instructive; the continuity equation gives in this case $\partial \rho / \partial t = 0$ and the equation of motion becomes $\mathcal{P} = gz + r^2 \omega^2 / 2$, which becomes identical with Eq. (7.19) by applying (7.18) (the orientation of the $z$-axis is now different).

VII.2. One starts from the one-dimensional Euler equation (13.1) (no external force)

(1) $$\qquad \frac{du}{dt} = -\frac{1}{\rho} \frac{\partial p}{\partial x}$$

and introduces $t$ and $a$ as independent variables, where $a = x_0$ is the initial coordinate of a volume element whose coordinate at any later time is $x(t)$. The mass contained in this volume element equals

$$\text{for } t = 0 : \rho_0 \, da, \qquad \text{for } t > 0 : \rho \, dx = \rho \frac{\partial x}{\partial a} \, da.$$

Conservation of mass demands

(2) $$\rho \frac{\partial x}{\partial a} = \rho_0 \, .$$

It follows that

(3) $$\frac{\partial}{\partial t} \left( \rho \frac{\partial x}{\partial a} \right) = 0,$$

which is identical with Eq. (34.3b).

According to the remarks preceding Eq. (34.1), we can express the material acceleration in (1) by

(4) $$u = \frac{\partial x}{\partial t} \quad \text{as} \quad \frac{du}{dt} = \frac{\partial^2 x}{\partial t^2} \, ;$$

the pressure gradient may be written

(5) $$\frac{\partial p}{\partial x} = \frac{\partial p}{\partial a} \frac{\partial a}{\partial x} = \frac{\partial p}{\partial a} \Big/ \frac{\partial x}{\partial a} \, .$$

Substitution of these derivatives in (1) yields

(6) $$\frac{\partial^2 x}{\partial t^2} \frac{\partial x}{\partial a} = - \frac{1}{\rho} \frac{\partial p}{\partial a} \, .$$

This agrees with (34.1) if and only if one takes into account the modification of Lagrange's equations for compressible fluids stated in (34.3a).

VII.3. If the compressible fluid has initially the constant density $\rho_0$ one concludes from Eq. (2) of answer VII.2 by differentiation with respect to the independent variable $a$

(1) $$\frac{\partial}{\partial a} \left( \rho \frac{\partial x}{\partial a} \right) = 0,$$

or

(2) $$\frac{1}{\rho} \frac{\partial \rho}{\partial a} = - \frac{1}{\omega} \frac{\delta^2 x}{\partial a^2} \, ,$$

where we have used Hadamard's abbreviation

(3) $$\omega = \frac{\partial x}{\partial a} \, .$$

If now a $p,\rho$-relation $p = \varphi(\rho)$ is assumed one obtains, using (2),

(4) $$\frac{1}{o} \frac{\partial p}{\partial a} = \frac{\varphi'(\rho)}{\rho} \frac{\partial \rho}{\partial a} = - \frac{\varphi'(\rho)}{\omega} \frac{\partial^2 x}{\partial a^2} \, .$$

Eq. (6) of the preceding answer then takes the form

$$(5) \qquad \frac{\partial^2 x}{\partial t^2} = \Omega \frac{\partial^2 x}{\partial a^2}, \qquad \Omega = \frac{\varphi'(\rho)}{\omega^2}.$$

The quantity $\Omega$ is a function of $\omega$ alone since, by Eq. (2) of the preceding answer, $\rho$ is inversely proportional to $\omega$. This suggests the introduction of $\omega$ as an *independent variable*. A second independent variable, suited for our purpose, is

$$(6) \qquad u = \frac{\partial x}{\partial t}.$$

By (3) and (6), one finds

$$(7) \qquad dx = \omega\, da + u\, dt.$$

If one introduces in accordance with the rule of Legendre's transformation the dependent variable

$$(8) \qquad z(u, \omega) = ut + \omega a - x(t, a)$$

and forms its total differential, then one finds

$$(9) \qquad dz = t\, du + a\, d\omega.$$

From (8) and (9) one has

$$(10) \qquad \frac{\partial^2 x}{\partial a^2} = \omega_a, \qquad \frac{\partial^2 x}{\partial t^2} = u_t;$$

$$(11) \qquad \frac{\partial^2 z}{\delta \omega^2} = a_\omega, \qquad \frac{\partial^2 z}{\partial u^2} = t_u,$$

and from (5) and (10) follows

$$(12) \qquad u_t = \Omega \omega_a.$$

Using a well known transformation rule we conclude from (12)

$$(13) \qquad a_\omega = \Omega t_u.$$

Because of (11) this is a *linear differential equation of second order for z* since the coefficient $\Omega$ depends only on *the independent variable* $\omega$. Writing (13) in the variable $z$, we have

$$(14) \qquad \frac{\partial^2 z}{\partial \omega^2} = \Omega \frac{\partial^2 z}{\partial u^2},$$

which is the basis of Hadamard's discussion of Riemann's shock waves. The fact that Eq. (14) is not identical with (37.11) is of course due to the different meanings of the new dependent and independent variables.

VII.4. Introduce spherical coordinates $r$, $\vartheta$, $\varphi$, so that $r = 0$ coincides with the center of the sphere and $\vartheta = 0$ with the axis of rotation; let the constant angular velocity be $\dot{\vartheta} = \omega$. Assume that the fluid moves in circular streamlines about the axis in the sense of increasing $\varphi$, and show that this very simple assumption is permissible.—The first two of the velocity components $v_r$, $v_\vartheta$, $v_\varphi = v$ vanish and, because of the assumed incompressibility,

$$\frac{\partial v}{\partial \varphi} = 0,$$

while $\partial v/\partial \vartheta$ and $\partial v/\partial r$ are different from zero. Eq. (35.6) is again our starting point; developing the expressions for curl curl with the aid of answer I.3 we obtain

(1) $$\mathrm{curl}_r\ \mathrm{curl}\ \mathbf{v} = \mathrm{curl}_\vartheta\ \mathrm{curl}\ \mathbf{v} = 0$$

(2) $$\mathrm{curl}_\varphi\ \mathrm{curl}\ \mathbf{v} = -\frac{1}{r}\frac{\partial^2(rv)}{\partial r^2} - \frac{1}{r}\frac{\partial}{\partial \vartheta}\frac{1}{r \sin \vartheta}\frac{\partial(\sin \vartheta v)}{\partial \vartheta}.$$

From (35.6) and (1) we conclude first

$$\mathrm{grad}_r\ p = \mathrm{grad}_\varphi\ p = 0,$$

and since for reasons of symmetry $\mathrm{grad}_\varphi\ p = 0$, we have altogether

(3) $$p = \mathrm{const.}$$

From (2), the differential equation for $v$ is obtained:

(4) $$\frac{\partial^2 v}{\partial r^2} + \frac{2}{r}\frac{\partial v}{\partial r} + \frac{1}{r^2}\left(\frac{\partial^2 v}{\partial \vartheta^2} + \cot \vartheta \frac{\partial v}{\partial \vartheta} - \frac{v}{\sin^2 \vartheta}\right) = 0.$$

On substituting $v = C \cdot f(r) g(\vartheta)$, one obtains the two equations

(5) $$f'' + \frac{2}{r}f' - \frac{\lambda}{r^2}f = 0, \qquad g'' + \cot \vartheta\ g' + \left(\lambda - \frac{1}{\sin^2 \vartheta}\right)g = 0,$$

$\lambda$ being the parameter of separation. A solution[3] of this equation that vanishes at infinity and remains finite for all values $0 < \vartheta < \pi$ is

(6) $$f = \frac{1}{r^2}, \qquad g = \sin \vartheta, \qquad \lambda = 2.$$

---

[3] A more general solution is

$$f = \frac{1}{r^n}, \qquad g = P_n^1(\cos \vartheta), \qquad \lambda = n(n+1)$$

where $P_n^1$ is Legendre's associated function of the first kind, cf. Vol. VI, Chap. IV.

The velocity law is obtained as

(7) $$v = \omega \frac{a^3}{r^2} \sin \vartheta$$

where the factor $C$ was chosen so as to make $v = \omega a \sin \vartheta$ on the surface of the sphere. Thus, since we can fulfill the boundary conditions, the special choice $\lambda = 2$ in (6) is justified, and it is seen that the more general solutions mentioned in the footnote would be unsuited for our problem.

In order to pass from (7) to the friction pressures, one calculates similarly as on p. 36 from (4.26) and (4.28) with $\dot{\delta q_1} = \dot{\delta q_2} = 0$, $\dot{\delta q_3} = v = \omega \sin \vartheta a^3/r^2$:

$$\dot{\epsilon}_{rr} = \dot{\epsilon}_{\vartheta\vartheta} = \dot{\epsilon}_{\varphi\varphi} = \dot{\epsilon}_{\vartheta\varphi} = \dot{\epsilon}_{r\vartheta} = 0$$

$$\dot{\epsilon}_{r\varphi} = \frac{1}{2g_1}\frac{\partial v}{\partial r} - \frac{1}{2g_3 g_1} v \frac{\partial g_3}{\partial r} = \frac{\omega}{2}\sin\vartheta\left(-2\frac{a^3}{r^3} - \frac{a^3}{r^3}\right) = -\frac{3}{2}\omega\sin\vartheta\,\frac{a^3}{r^3}.$$

Hence all friction pressures vanish except

(8) $$p_{r\varphi} = -2\mu\dot{\epsilon}_{r\varphi} = 3\mu\omega\frac{a^3}{r^3}\sin\vartheta.$$

The moment can now be computed:

(9) $$-M = 2\pi a^3 \int p_{r\varphi}\sin^2\vartheta\,d\vartheta = 6\pi\mu\omega a^3 \int_0^\pi \sin^3\vartheta\,d\vartheta = 8\pi\mu\omega a^3$$

in agreement with (35.21). The moment required for the maintenance of uniform rotation is opposite to the moment of the friction pressures and acts therefore in the positive $\varphi$-direction as expected.

Kirchhoff treats the same problem in a more elegant though less concise manner using Cartesian coordinates, cf. footnote on p. 251.

VIII.1. We denote the reactions at the points $A$ and $B$ by these same letters and have

$$A = P\frac{b}{l}, \qquad B = P\frac{a}{l}.$$

These reactions determine the shear force introduced in the context of Eq. (41.12) in the regions I (between $A$ and $P$) and II (between $P$ and $B$) as

(1) $$S_I = A, \qquad S_{II} = -B, \text{ respectively}$$

Note that the positive $x$-direction is from $A$ to $B$ and that both shears $S_{II}$ and $S_I$ refer to positive $x$-surfaces in the sense of Fig. 6. The shear force is constant throughout either one of the regions I and II, but jumps at the load point by the amount $P$; this discontinuity is, of course, due to the unrealistic assumption of a concentrated load. In a cross-section at a distance $x$ from $A$ (region I) the bending moment equals

(2)
$$M_I = Ax = P\frac{b}{l}x.$$

For the region II, in a cross-section at a distance $\xi$ from $B$, the bending moment is

(3)
$$M_{II} = B\xi = P\frac{a}{l}\xi.$$

$M_{II}$ refers to a positive $\xi$-surface which evidently is equivalent to a negative $x$-surface. A graph of $M$ consists therefore of two straight lines that intersect at $x = a$ (or $\xi = b$). At this point, $M = M_{max} = Pab/l$ according to both (2) and (3).

The deflection of $y_1$ is found for the region I from

(4)
$$\frac{d^2y_I}{dx^2} = \frac{P}{EJ}\frac{b}{l}x,$$

and for the region II from

(5)
$$\frac{d^2y_{II}}{d\xi^2} = \frac{P}{EJ}\frac{a}{l}\xi.$$

The constants occurring in the solutions of (4) and (5) must be determined from the following boundary and continuity conditions:

a) The deflection $y_I$ must vanish at $A$, and $y_{II}$ at $B$.

b) At the point of application of the load both curves must join continuously and with common tangent. These conditions yield

(6)
$$y_I = \frac{P}{EJ}\frac{a^2b^2}{6l}\left(\frac{x^3}{a^2b} - 2\frac{x}{a} - \frac{x}{b}\right),$$

$$y_{II} = \frac{P}{EJ}\frac{a^2b^2}{6l}\left(\frac{\xi^3}{ab^2} - 2\frac{\xi}{b} - \frac{\xi}{a}\right).$$

The deflection at the load point is found from both equations as

(7)
$$y = y_{max} = \frac{P}{EJ}\frac{a^2b^2}{3l}.$$

The shear forces (1) and the bending moments (2) and (3) are always interconnected by the relation

(8)
$$S = \frac{dM}{dx}.$$

If the beam is loaded by several vertical forces at given points, shear force, bending moment, and deflection can be easily obtained by superposition of the corresponding expressions for single loads.

VIII.2. a) Let $F$ be the cross-section of the rod, and count the displacement $\xi$ of $M$ (which is also that of the lower end cross-section) positive downward. $M$ is subject to the downward action of gravity and the elastic force of the rod $\sigma \cdot F = FE \cdot \xi/l$ that acts upward. The equation for the displacement $\xi$ reads

(1) $$M\ddot{\xi} = Mg - k\xi, \qquad k = \frac{FE}{l},$$

and gives, when integrated,

(2) $$\xi = \frac{g}{\omega_0^2} + A \cos \omega_0 t + B \sin \omega_0 t,$$

where

(3) $$\omega_0 = \sqrt{\frac{k}{M}} = \sqrt{\frac{EF}{Ml}};$$

the constants $A$ and $B$ are to be found from the given initial conditions. One sees that $\xi$ oscillates about the constant value $\xi_0 = g/\omega_0^2$ which is nothing else but the static extension obtained from (1) for $\ddot{\xi} = 0$.

b) With the notations $J_p$ = polar moment of inertia of the cross-section of the rod (assumed as circular), $\mu$ = torsion modulus, $\varphi$ = angle of torsion of the mass $M$ about the axis of the rod, we have by (42.6)

(5) $$\Theta\ddot{\varphi} = \frac{-\mu J_p}{l}\varphi,$$

and

(6) $$\omega_0 = \sqrt{\frac{\mu J_p}{\Theta l}}.$$

c) The motion of $M$ can be described in good approximation as a vertical displacement (it is exactly that for a rod on two supports if $a = b = l/2$), hence Eq. (1) may be used again, but $\xi$ means now the transverse displacement of $M$.

The "spring constant" $k$ is

for the rod clamped at one end: $k = 3\,EI/l^3$

for the rod supported at both ends: $k = 3\,EIl/a^2b^2$

vhere $I$ is the equatorial moment of inertia of the cross-section. This leads to the frequencies

(7) $\qquad \omega_0 = \sqrt{\dfrac{3EI}{Ml^2}}$, (8) $\qquad \omega_0 = \sqrt{\dfrac{3EIl}{Ma^2b^2}}$ respectively.

In a more accurate theory of extensional, torsional, and bending oscillations the mass of the rod itself would have to be considered; this would lead to partial differential equations of second order in place of the ordinary equations (1) and (5). In addition to the fundamental oscillation of frequency $\omega$, which was determined in the foregoing, one would obtain an infinite sequence of harmonic oscillations according to the infinitely many degrees of freedom of the elastic rod that are now active.

VIII.3. For $t = 0$, one has $\xi = 0$, $\dot{\xi} = v_0$ ; then, according to Eq. (2) of the preceding answer, one obtains

(1) $$\xi = \frac{g}{\omega_0^2} (1 - \cos \omega_0 t) + \frac{v_0}{\omega_0} \sin \omega_0 t,$$

hence

(2) $$\xi_{max} = \frac{g}{\omega_0^2} + \sqrt{\left(\frac{g}{\omega_0^2}\right)^2 + \left(\frac{v_0}{\omega_0}\right)^2}.$$

The maximum tension is found according to Hooke's law as

(3) $$\sigma_{max} = \frac{E}{l} \xi_{max} .$$

Assume now

a) $v_0 = 0$,

then

$$\xi_{max} = 2\frac{g}{\omega_0^2} = \frac{2gMl}{EF}, \qquad \sigma_{max} = 2\frac{gM}{F} ;$$

b) for large $v_0$

$$\xi_{max} = \frac{v_0}{\omega_0} = \sqrt{\frac{Mv_0^2 l}{EF}}, \qquad \sigma_{max} = \sqrt{\frac{Mv_0^2 E}{Fl}} .$$

Both results verify the statements a) and b) made in the text of the problem.

VIII.4. The equations

$$y = a \cos u, \qquad z = b \sin u, 0 < u < 2\pi;$$

constitute a parametric representation of the elliptic circumference of the cross-section; a family of similar concentric ellipses is given by

$$y = a\lambda \cos u, \qquad z = b\lambda \sin u, \qquad 0 < \lambda < 1.$$

The parameters $\lambda$ and $u$ occurring here are suitable coordinates for the calculation of the integral (42.23). The Jacobian of the transformation $y, z \rightarrow \lambda, u$ is

$$D = \begin{vmatrix} \dfrac{\partial y}{\partial \lambda} & \dfrac{\partial y}{\partial u} \\[2mm] \dfrac{\partial z}{\partial \lambda} & \dfrac{\partial z}{\partial u} \end{vmatrix} = ab\lambda.$$

One obtains therefore

$$M = \mu\alpha\,\frac{\pi a^3 b^3}{a^2 + b^2}, \qquad \varphi_l = \alpha l = \frac{Ml}{\mu}\,\frac{a^2 + b^2}{\pi a^3 b^3}.$$

On the other hand, the polar moment of inertia $J_p$ of the area of the ellipse is conveniently calculated by adding the two equatorial moments of inertia $I_y$ and $I_z$ , which, in turn, can be obtained from the equatorial moment of inertia of a circular area of radius $a$, $I = a^2\pi/4$, by "homogeneous compression". This means that $I_y$ and $I_z$ take up the factors $b^3/a^3$ and $b/a$ respectively. The resulting polar moment of inertia is

$$J_p = \left(\frac{b^3}{a^3} + \frac{b}{a}\right)I = \frac{\pi}{4}\,ab(a^2 + b^2).$$

It is different from the geometrical factor occurring in $M$ and $\varphi_l$ , viz.

$$\frac{\pi a^3 b^3}{a^2 + b^2},$$

the two factors being identical only for a circular cross-section where $a = b$.

VIII.5. Let the plane of incidence be the $x,y$-plane as in 45. Polarization perpendicular to this plane means that of the three displacements $\xi, \eta, \zeta$, only $\zeta$ is excited. According to the table on p. 327 a *transverse wave* incident at the angle $\alpha_1$ and reflected at the angle $\alpha_2$ , is given by

$$\zeta_i = A \exp[ik(x \sin \alpha_1 - y \cos \alpha_1)],$$

$$\zeta_r = B \exp[ik(x \sin \alpha_2 + y \cos \alpha_2)].$$

Since the frequency $\omega$ of both waves is the same this is also true for the wave number $k$, cf. (45.5). The three conditions (45.1) (boundary $y = 0$ free of forces) reduce to one condition

$$\sigma_{yz} = 2\mu\epsilon_{yz} = \mu\,\frac{\partial \zeta}{\partial y} = \mu\left(\frac{\partial \zeta_i}{\partial y} + \frac{\partial \zeta_r}{\partial y}\right) = 0,$$

which can be fulfilled, without assuming an additional longitudinal wave, simply by setting

(1) $$\alpha_1 = \alpha_2 \quad \text{and} \quad B = A.$$

This is a regular reflexion with constant amplitude.

If on the other hand, a *longitudinal* wave strikes at the angle $\alpha_1$ and is reflected at the angle $\alpha_2$, we have to deal with the two components $\xi$, $\eta$ given by

(2) $$\left.\begin{array}{c}\xi_i\\ \eta_i\end{array}\right\} = \left.\begin{array}{c}\sin\alpha_1\\ -\cos\alpha_1\end{array}\right\} A \exp[ik(x\sin\alpha_1 - y\cos\alpha_1)],$$

(3) $$\left.\begin{array}{c}\xi_r\\ \eta_r\end{array}\right\} = \left.\begin{array}{c}\sin\alpha_2\\ \cos\alpha_2\end{array}\right\} B \exp[ik(x\sin\alpha_2 + y\cos\alpha_2)].$$

Since $\omega$ is again the same for both waves [it is given by (45.10)], the wave number is the same too.

Of the three conditions (45.1) only $\sigma_{yz} = 0$ is now identically fulfilled. The two others

(4) $$\sigma_{yz} = 0, \quad \sigma_{yy} = 0$$

call again for regular reflexion, that is, $\alpha_2 = \alpha_1 = \alpha$, but they cannot be satisfied by adjusting the one available constant $B/A$. Again a reflected *transverse* wave polarized parallel to the plane of incidence must be added; this wave can be written according to the central column of the table on p. 327 in the following form

(5) $$\left.\begin{array}{c}\xi_t\\ \eta_t\end{array}\right\} = \left.\begin{array}{c}\cos\alpha_t\\ -\sin\alpha_t\end{array}\right\} C \exp[ik_t(x\sin\alpha_t + y\cos\alpha_t)],$$

The two angles $\alpha_t$ and $\alpha$ are connected by the "law of refraction" (45.12)

$$\sin\alpha = n\sin\alpha_t, \quad n = \frac{k_t}{k} = \left(2 + \frac{\lambda}{\mu}\right)^{1/2}$$

One computes from (2), (3), and (5) the total displacements.

$$\xi = \xi_i + \xi_r + \xi_t, \quad \eta = \eta_i + \eta_r + \eta_t$$

and finds from the two equations (4) two conditions for the two ratios $A : B : C$ [cf. the similar equations (45.13) and (45.14)].

# THE COMPONENTS OF THE THREE-DIMENSIONAL STRAIN TENSOR IN ORTHOGONAL CURVILINEAR COORDINATES

The formulas (4.26) and (4.28) for the strains $\epsilon_{ii}$ and $\epsilon_{ik}$ simplify considerably if instead of the displacements $\delta q$, which have the dimensions of a length, the associated variations of the coordinates $\delta p$ are introduced; the latter do not, in general, have uniform dimensions; e.g. in the case of polar coordinates, the quantities $\delta p$ are (in part) angular changes. On substituting in (4.26) and (4.28) for the $\delta p_k$ according to Eq. (4.19) $\delta q_k = g_k \delta p_k$ and carrying out the differentiations $\partial \delta q_i / \partial p_i$ one obtains directly

$$(1) \qquad \epsilon_{ii} = \frac{\partial \delta p_i}{\partial p_i} + \frac{\delta p_\alpha}{g_i} \frac{\partial g_i}{\partial p_\alpha},$$

$$(2) \qquad 2\epsilon_{ik} = \frac{g_i}{g_k} \frac{\partial \delta p_i}{\partial p_k} + \frac{g_k}{g_i} \frac{\partial \delta p_k}{\partial p_i}.$$

The second term in the right member of Eq. (1) symbolizes the sum of the terms that are obtained by putting $\alpha = 1, 2, \cdots, n$, in accordance with the following summation convention introduced by Einstein: a Greek[1] character that occurs twice in one term is to be considered as a summation index. Since we operate in three-dimensional space, the second term on the right side of Eq. (1) stands for

$$\sum_{\alpha=1}^{3} \frac{\delta p_\alpha}{g_i} \frac{\partial g_i}{\partial p_\alpha}.$$

To simplify the notation further we shall denote the variation of the coordinate, $\delta p_i$, by $\xi^i$. This we do in accordance with the general conventions of tensor calculus, where superscripts indicate contravariant and subscripts covariant components of a vector. With these changes Eq. (1) reads

$$(1a) \qquad \epsilon_{ii} = \frac{\partial \xi^i}{\partial p_i} + \frac{\xi^\alpha}{g_i} \frac{\partial g_i}{\partial p_\alpha} = \xi^i._i.$$

---

[1]By restricting the indices of summation or dummy indices to Greek letters one avoids confusing $\epsilon_{ii}$ or the later occurring $\xi^i._i$ with the first scalar or spur of the corresponding tensors. Without this or a similar restriction the notations of tensor calculus are likely to mislead the student, however concise or convenient they may be.

The *subscript* that occurs here in the symbol $\xi^i_{\cdot i}$ is used to denote the *covariant derivative* of the contravariant vector $\xi^i$ with respect to the co-ordinate $p_i$ (note that the position of the subscript indicating the differentiation is shifted to the right; in this way a definite order of the indices is established). Vector components are called *contravariant*[2] if they transform in the same way as the coordinate differentials $dp_i$ ; covariant vector components transform conversely to the differentials in the sense that the matrix of the transformation $\xi'_i \rightarrow \xi_i$ is the inverse of the matrix of the transformation $dp_i \rightarrow dp'_i$ . The differential quotients of a scalar point function with respect to the coordinates form examples of covariant vector components (gradient of a scalar). The same is *not* true for the differential quotients of a vector like $\partial \xi'/\partial p_i$ . The latter quantity must be augmented by the sum over $\alpha$ that appears in (1a) in order to become a covariant derivative.

We shall first verify that the factor that goes with $\xi^\alpha$ in (1a) is a special case of what is generally called a Christoffel symbol[3] (of the second kind).

The general definition of this symbol (for arbitrary coordinates $x_1 x_2 x_3$) is

$$(3) \qquad \{\mu\nu, \sigma\} = \frac{1}{2} g^{\sigma\lambda}\left(\frac{\partial g_{\mu\lambda}}{\partial x_\nu} + \frac{\partial g_{\nu\lambda}}{\partial x_\mu} - \frac{\partial g_{\mu\nu}}{\partial x_\lambda}\right),$$

where $g_{\sigma\lambda}$ are the covariant components of the metric tensor. In the case of orthogonal coordinates to which we restrict ourselves in the following, we have

$$g_{ik} = \delta_{ik} g_i g_k , \qquad g^{ik} = \frac{\delta_{ik}}{g_i g_k}$$

(for the definition of $\delta_{ik}$ cf. p. 10). It is now easy to verify that in this case

$$(3a) \qquad \{\alpha i, i\} = \frac{1}{g_i}\frac{\partial g_i}{\partial p_\alpha} ,$$

---

[2]The terms contravariant and covariant were introduced by Einstein to replace the terms *cogredient* and *contragredient* (respectively) which had been used before. It will be noticed that for reasons of consistency the coordinates and their differentials should also be written with superscripts, that is, $p_i$ and $dp_i$ should be replaced by $p^i$ and $dp^i$ respectively; this notation has actually been adopted by some authors.—The differential operation $\xi^i_{\cdot i}$ is covariant in the sense defined before, but we shall not prove that here, nor shall we prove formula (4) or the rule concerning the raising or lowering of indices. For these matters the reader is referred to the very careful although condensed representation by A. S. Eddington in his Mathematical Theory of Relativity, Cambridge, 1923 and to The absolute differential Calculus by T. Levi-Civita, (engl. transl. London, 1927). Our Eqs. (3b) and (4) are derived in 29, Eq. (4) of Eddington's book.

[3]Instead of $\{\mu\nu, \sigma\}$ one often finds $\left\{\begin{matrix}\mu\nu\\\sigma\end{matrix}\right\}$; note also Einstein's symbol $\Gamma^\sigma_{\mu\nu}$.

which permits us to write instead of (1a)

(3b) $$\xi^i_{\cdot i} = \frac{\partial \xi^i}{\partial p_i} + \{\alpha i, i\}\xi^\alpha.$$

We supplement this by the general expression for the covariant derivative

(4) $$\xi^i_{\cdot k} = \frac{\partial \xi^i}{\partial p_k} + \{\alpha k, i\}\xi^\alpha.$$

In the present case $\{\alpha k, i\}$ is only different from zero for $\alpha = i$ and $\alpha = k$, therefore

(5) $$\xi^i_{\cdot k} = \frac{\partial \xi^i}{\partial p_k} + \frac{1}{g_i}\frac{\partial g_i}{\partial p_k}\xi^i - \frac{g_k}{g_i^2}\frac{\partial g_k}{\partial p_i}\xi^k.$$

One might think that the last expression represents the strain $\epsilon_{ik}$ [Eq. (2)], but this cannot be correct, since it is not symmetrical in the indices $i$ and $k$. To satisfy the symmetry condition we have to consider the covariant vector $\xi_i$ and its contravariant derivative $\xi^{\cdot k}_i$. The latter is obtained by lowering the superscript $i$ and raising the subscript $k$ in $\xi^i_{\cdot k}$. In the case of an orthogonal line element these operations require only a multiplication by $g_i^2$ and division by $g_k^2$. In this way we obtain from (5)

$$\xi^{\cdot k}_i = \frac{g_i^2}{g_k^2}\frac{\partial \xi^i}{\partial p_k} + \frac{g_i}{g_k^2}\frac{\partial g_i}{\partial p_k}\xi^i - \frac{1}{g_k}\frac{\partial g_k}{\partial p_i}\xi^k.$$

We finally interchange $i$ and $k$, which is entirely a matter of notation, and obtain,

(6) $$\xi^{\cdot i}_k = \frac{g_k^2}{g_i^2}\frac{\delta \xi^k}{\partial p_i} + \frac{g_k}{g_i^2}\frac{\partial g_k}{\partial p_i}\xi^k - \frac{1}{g_i}\frac{\delta g_i}{\partial p_k}\xi^i.$$

The last two terms in (6) are opposite in sign to the corresponding terms in (5), hence we obtain by adding (5) and (6)

$$\xi^i_{\cdot k} + \xi^{\cdot i}_k = \frac{\partial \xi^i}{\partial p_k} + \frac{g_k^2}{g_i^2}\frac{\partial \xi^k}{\partial p_i}.$$

or, after multiplication with $g_i/g_k$ ,

(7) $$\frac{g_i}{g_k}(\xi^i_{\cdot k} + \xi^{\cdot i}_k) = \frac{g_i}{g_k}\frac{\partial \xi^i}{\partial p_k} + \frac{g_k}{g_i}\frac{\partial \xi^k}{\partial p_i}.$$

The right member of this equation agrees with the right member of (2) if one returns to the original notation $\delta p_i = \xi^i$, $\delta p_k = \xi^k$. In the notation of tensor calculus the strains are thus given by the representation

(8) $$2\epsilon_{ik} = \frac{g_i}{g_k}(\xi^i_{\cdot k} + \xi^{\cdot i}_k),$$

which includes the representation (1a) when $i = k$:

(8a) $$\epsilon_{ii} = \xi^i_{.i}$$

Note that the last two equations are in complete analogy with the original representation (1.11) in rectangular coordinates. This is in accordance with the general aim of tensor calculus in which one tries to introduce symbols and operational rules of such a nature that the equations written in curvilinear coordinates correspond as closely as possible to those written in Cartesian coordinates.

## APPENDIX II

# THE CONVECTIVE TERMS OF THE ACCELERATION VECTOR IN TENSOR NOTATION

From the contravariant displacement vector $\delta p_i = \xi^i$ (Appendix I) we pass now to the contravariant velocity vector $\delta\dot{p}_i = v^i$; the general form of the contravariant vector of the material acceleration is

(1) $$\frac{dv^i}{dt} = \frac{\partial v^i}{\partial t} + v^\alpha v^i_{.\alpha} ,$$

where the Greek index again indicates the summation with respect to $\alpha$. This formula corresponds to the elementary vector equation

(2) $$\frac{d\mathbf{v}}{dt} = \frac{\partial \mathbf{v}}{\partial t} + (\mathbf{v}\ \mathrm{grad})\mathbf{v};$$

but (2) is only correct in Cartesian coordinates, while (1) is correct for any form of the line element and yields automatically the additional inertia terms that were computed in Problem III.1 in an indirect way for the case of cylindrical coordinates.

Let us take up this example once more to examine the structure of the convective terms $v^\alpha v^i_{.\alpha}$ for $i = 1, 2, 3$. With $p_1 = r$, $p_2 = \varphi$, $p_3 = z$ and $g_1 = 1$, $g_2 = r$, $g_3 = 1$ we have

(3) $$v^1 = \frac{dr}{dt} = v_r , \qquad v^2 = \frac{d\varphi}{dt} = \frac{v_\varphi}{r} , \qquad v^3 = \frac{dz}{dt} = v_z .$$

According to Eqs. (3a, b) of Appendix I we calculate the first line and according to Eq. (5) the second and the third lines of the following table:

$$v^1_{.1} = \frac{\partial v_r}{\partial r}, \qquad v^2_{.2} = \frac{\partial}{\partial\varphi}\frac{v_\varphi}{r} + \frac{v_r}{r}, \qquad v^3_{.3} = \frac{\partial v_z}{\partial z},$$

(4) $\qquad v^1_{.2} = \frac{\partial v_r}{\partial\varphi} - v_\varphi, \qquad v^2_{.1} = \frac{\partial}{\partial r}\frac{v_\varphi}{r} + \frac{v_\varphi}{r^2}, \qquad v^3_{.1} = \frac{\partial v_z}{\partial r},$

$$v^1_{.3} = \frac{\partial v_r}{\partial z}, \qquad v^2_{.3} = \frac{\partial}{\partial z}\frac{v_\varphi}{r}, \qquad v^3_{.2} = \frac{\partial v_z}{\partial\varphi}.$$

Multiplication of the terms of the $1^{st}(2^d, 3^d)$ column with those terms of (3) that have the same index $\alpha$, and summation over $\alpha$ yield the $1^{st}(2^d, 3^d)$ component of the convective acceleration:

(5) $\qquad v^\alpha v^i_{.\alpha} = \begin{cases} v_r \dfrac{\partial v_r}{\partial r} + v_\varphi \dfrac{\partial v_r}{r\partial\varphi} + v_z \dfrac{\partial v_r}{\partial z} - \dfrac{v_\varphi^2}{r} \\[3mm] \dfrac{1}{r}\left( v_r \dfrac{\partial v_\varphi}{\partial r} + v_\varphi \dfrac{\partial v_\varphi}{r\partial\varphi} + v_z \dfrac{\partial v_\varphi}{\partial z} + \dfrac{v_r v_\varphi}{r}\right) \\[3mm] v_r \dfrac{\partial v_z}{\partial r} + v_\varphi \dfrac{\partial v_\varphi}{r\partial\varphi} + v_z \dfrac{\partial v_z}{\partial z} + *. \end{cases}$

In the second line the common denominator of all terms, $1/r$, has been factored out in preparation for the final equation of motion in Appendix IV.

Note again the *"inertia terms"* that have been placed in the fourth column:

$$-\frac{v_\varphi^2}{r}, \qquad \frac{v_r v_\varphi}{r}, \qquad *;$$

*they are identical with those of Prob.* III.1.

While only the first three terms are correctly obtained by applying the pseudo-vectorial operation $(\mathbf{v}\,\text{grad})\mathbf{v}$ (cf. also the table on p. 347 where these terms have been indicated in the form $\mathbf{A}.\,\text{grad}\,A_r$ , etc.), the present computation yields all four terms at once without discrimination. The correspondence between the "additional" terms on the one hand, the centripetal and half of the Coriolis acceleration on the other hand, was already pointed out.

If the same calculation is made in spherical coordinates $r, \vartheta, \varphi$ where $g_1 = 1$, $g_2 = r$, $g_3 = r \sin\vartheta$, one obtains instead of (3), (4), (5)

(3a) $\qquad v^1 \doteq \dfrac{dr}{dt} = v_r, \qquad v^2 = \dfrac{d\vartheta}{dt} = \dfrac{v_\vartheta}{r}, \qquad v^3 = \dfrac{d\varphi}{dt} = \dfrac{v_\varphi}{r \sin\vartheta},$

$$(4a) \begin{cases} v^1_{.1} = \dfrac{\partial v_r}{\partial r}\,, \qquad v^2_{.2} = \dfrac{\partial v_\vartheta}{r\partial\vartheta} + \dfrac{v_r}{r}\,, \qquad v^3_{.3} = \dfrac{\partial v_\varphi}{r\sin\vartheta\,\partial\varphi} + \dfrac{v_r}{r} + \dfrac{\cos\vartheta\,v_\vartheta}{r\sin\vartheta} \\[2ex] v^1_{.2} = \dfrac{\partial v_r}{\partial\vartheta} - v_\vartheta\,, \qquad v^2_{.1} = \dfrac{\partial v_\vartheta}{r\partial r}\,, \qquad\qquad v^3_{.1} = \dfrac{\partial v_\varphi}{r\sin\vartheta\,\partial r}\,, \\[2ex] v^1_{.3} = \dfrac{\partial v_r}{\partial\varphi} - \sin\vartheta\,v_\varphi\,, \quad v^2_{.3} = \dfrac{\partial v_\varphi}{r\partial\varphi} - \dfrac{\cos\vartheta}{r}v_\omega\,, \quad v^3_{.2} = \dfrac{\partial v_\varphi}{r\sin\vartheta\,\partial\vartheta}\,. \end{cases}$$

$$(5a) \qquad v^\alpha v^i_{.\alpha} = \begin{cases} \mathbf{v}\cdot\mathrm{grad}\, v_r - \dfrac{v^2_\vartheta}{r} - \dfrac{v^2_\varphi}{r} \\[2ex] \dfrac{1}{r}\left\{\mathbf{v}\cdot\mathrm{grad}\, v_\vartheta + \dfrac{v_r v_\vartheta}{r} - \dfrac{\cos\vartheta\,v^2_\varphi}{r\sin\vartheta}\right\} \\[2ex] \dfrac{1}{r\sin\vartheta}\left\{\mathbf{v}\cdot\mathrm{grad}\, v_\varphi + \dfrac{v_r v_\varphi}{r} + \dfrac{\cos\vartheta v_\vartheta v_\varphi}{r\sin\vartheta}\right\}. \end{cases}$$

With $1/r$ and $1/r\sin\vartheta$ factored out (in the second and third line), the result shows the presence of the terms corresponding to the elementary vector formula $\mathbf{v}\,(\mathrm{grad}\,\mathbf{v})$. Let us again consider the additional "inertia terms": in the first line we find the ordinary centripetal accelerations $-v^2_\vartheta/r$ and $-v^2_\varphi/r$. The full value of the latter term that corresponds to the motion along the parallel of radius $r\sin\vartheta$ is $-v^2_\varphi/r\sin\vartheta$, but the component in $r$-direction is obtained by multiplying with $\sin\vartheta$ so that this factor cancels. In the second line we find the other component, which has the factor $\cos\vartheta$. We further recognize in the middle term of the second line and also in the last two terms of the third line one half of a Coriolis acceleration. We need not attempt to visualize these terms in a direct kinematic way, nor do we need the geometrical arguments that single out the components, since we trust the validity of the general formulations of tensor calculus.—It is, of course, possible to obtain the same result by vector analytic methods (cf. the table on p. 348)

## APPENDIX III

# THE TENSOR OF FRICTION PRESSURES AND ITS DIVERGENCE

The connection between the tensor of the friction pressure and the rate of strain is given by Eq. (10.2), $p_{ik} = 2\mu\dot{\epsilon}_{ik}$, when we restrict our argument to incompressible fluids. To obtain $\dot{\epsilon}_{ik}$ , we differentiate Eq. (8) of Appendix I with respect to $t$ and obtain

$$(1) \qquad \dot{}_{ik} = -\mu \frac{g_i}{g_k} (v^i_{\cdot k} + v^{\cdot i}_k).$$

This equation expresses the friction pressure in terms of general tensor analysis (Eqs. (4) and (4a) of Appendix II can again be used for the particular cases of cyl. and sph. coordinates). In the equations of motion, however, the friction pressures do not occur individually, but as arguments of the "vector divergence" [cf. Eq. (10.5)]. In Cartesian coordinates, the latter could be transformed into the pseudo-vector $-\mu \nabla^2 \mathbf{v}$ of Eq. (10.8) (see problem I.4 for the correct form of $\nabla^2 \mathbf{v}$ in cylindrical coordinates). In order to carry out the corresponding computation in terms of general tensor calculus, we have to replace $\nabla^2 \mathbf{v}$ by the differentiation and subsequent contraction of the tensor $v^i_{\cdot k} + v^{\cdot i}_k$ of Eq. (1), in other words, we have to carry out the operation

$$(2) \qquad v^{i\cdot\alpha}_{\cdot\alpha} + v^{\cdot i\alpha}_{\alpha}.$$

The dummy index $\alpha$ indicates here the summation as well as contravariant differentiation. The second term in (2) vanishes, however, when the summation is carried out as a consequence of the assumed incompressibility. The $i^{\text{th}}$ component of the vector so obtained (we denote it the following simply by $\text{Div}^i\, p$) is therefore

$$(3) \qquad \text{Div}^i\, p = -\mu v^{i\cdot\alpha}_{\cdot\alpha}.$$

This calculation requires the contravariant differentiation of a tensor[4] rather than a vector, which is dealt with in Eddington's book, loc. cit. §30. In the case of cylindrical coordinates one obtains on the basis of Eq. (4) Appendix II

$$(4) \qquad v^{1\cdot\alpha}_{\cdot\alpha} = \frac{1}{r}\frac{\partial}{\partial r}\left(r\frac{\partial v_r}{\partial r}\right) + \frac{1}{r^2}\frac{\partial^2 v_r}{\partial\varphi^2} + \frac{\partial^2 v_r}{\partial z^2} - \frac{v_r}{r^2} - 2\frac{\partial v_\varphi}{r^2\,\partial\varphi}.$$

The first three terms of the right member represent the ordinary Laplacian of Prob. I.3, p. 346f applied to $v_r$ and may be denoted by $\nabla^2 v_r$ ;

---

[4]The formula in question is

$$v^i_{\cdot kl} = \frac{Dv^i_{\cdot k}}{Dx_l} = \frac{\partial v^i_{\cdot k}}{\partial x_l} - \{kl,\,\beta\}v^i_{\cdot\beta} + \{\beta l,\, i\}v^\beta_{\cdot k}\, ,$$

and, of course, $v^{i\cdot l}_{\cdot k} = g^{l\lambda} v^i_{\cdot k\lambda}$ .

the last two terms are due to the fact that $v_r$ is not a scalar but a vector component. *These terms and the corresponding terms in $v_{.\alpha}^{2:\alpha}$ agree with those of Prob. I.4 as expected.* In the "Cartesian" component $v_{.\alpha}^{3:\alpha}$ *no such terms occur.* Applying the same $\nabla^2$-operator to the other components we can write the result as

$$v_{.\alpha}^{1:\alpha} = \nabla^2 v_r - \frac{v_r}{r^2} - 2\frac{\partial v_\varphi}{r^2\,\partial\varphi}, \qquad v_{.\alpha}^{2:\alpha} = \frac{1}{r}\left(\nabla^2 v_\varphi - \frac{v_\varphi}{r^2} + 2\frac{\partial v_r}{r^2\,\partial\varphi}\right),$$

(5)

$$v_{.\alpha}^{3:\alpha} = \nabla^2 v_z .$$

Note that in the second equation $1/r$ has been factored out as in Appendix II.

In the case of spherical coordinates one obtains in the same way instead of (4)

$$v_{.\alpha}^{1:\alpha} = \frac{1}{r^2}\frac{\partial}{\partial r}\left(r^2\frac{\partial v_r}{\partial r}\right) + \frac{1}{r^2\sin\vartheta}\frac{\partial}{\partial\vartheta}\left(\sin\vartheta\,\frac{\partial v_r}{\partial\vartheta}\right) + \frac{1}{r^2\sin^2\vartheta}\frac{\partial^2 v_r}{\partial\varphi^2}$$

(4a)

$$-\frac{2}{r^2}\left(v_r + \frac{1}{\sin\vartheta}\frac{\partial}{\partial\vartheta}\left(\sin\vartheta\,v_\vartheta\right) + \frac{1}{\sin\vartheta}\frac{\partial v_\varphi}{\partial\varphi}\right).$$

The first line is again the Laplacian $\nabla^2 v_r$ in the form valid for the present coordinates. The additional terms in the second line can be reduced if the condition of incompressibility is used. Giving $\nabla^2 v_\vartheta$ amd $\nabla^2 v_\varphi$ the appropriate meaning and factoring out $1/r$ and $1/r\sin\vartheta$ as in Appendix II, Eq. (5a), we obtain

$$v_{.\alpha}^{1:\alpha} = \nabla^2 v_r + \frac{2}{r^2}\frac{\partial}{\partial r}(rv_r),$$

(5a) $$v_{.\alpha}^{2:\alpha} = \frac{1}{r}\left[\nabla^2 v_\vartheta + \frac{1}{r^2}\left(2\frac{\partial v_r}{\partial\vartheta} - \frac{v_\vartheta}{\sin^2\vartheta} - \frac{2\cos\vartheta}{\sin^2\vartheta}\frac{\partial v_\varphi}{\partial\varphi}\right)\right],$$

$$v_{.\alpha}^{3:\alpha} = \frac{1}{r\sin\vartheta}\left[\nabla^2 v_\varphi + \frac{1}{r^2\sin\vartheta}\left(2\frac{\partial v_r}{\partial\varphi} - \frac{v_\varphi}{\sin\vartheta} + 2\frac{\cos\vartheta}{\sin^2\vartheta}\frac{\partial v_\vartheta}{\partial\varphi}\right)\right].$$

## APPENDIX IV

# THE EQUATIONS OF MOTION OF A VISCOUS INCOMPRESSIBLE FLUID IN CYLINDRICAL AND SPHERICAL COORDINATES

The general form of the Navier-Stokes equations for arbitrary coordinates is:

$$(1) \qquad \rho\left(\frac{\partial v^i}{\partial t} + v^\alpha v^i_{:\alpha}\right) - \mu v^{i\ \alpha}_{:\alpha} + (p + V)^i = 0.$$

The individual terms correspond exactly to the vector equation (16.1), provided we write the external force **F** as the negative gradient of the potential energy $V$ and collect the latter together with the pressure in one term. The notation $(p + V)^i$ in (1) means contravariant differentiation with respect to the $i^{\text{th}}$ coordinate.

Eq. (1) presents an opportunity to point out a homogeneity law of tensor calculus that refers to the degree in which the individual indices occur. One sees that all terms of Eq. (1) are of degree $+1$ in the index $i$ and of degree 0 in the index $\alpha$, provided we count the occurrence of a certain index positive, if it is a superscript and negative if it is a subscript. It is this circumstance that requires the last term in (1) to be a contravariant differentiation, and is responsible for the factor $1/g_i$ that will occur later, when we pass over to the ordinary gradient operation. (Remember here that the derivative of a *scalar* with respect to $x_i$ is in itself a covariant vector, as pointed out in Appendix I.) It should be kept in mind that $v^i$ is the rate of change of the $i^{\text{th}}$ *coordinate*, but not of the corresponding component of the path; likewise $\partial v^i/\partial t$ is the local acceleration of the coordinate which differs from the corresponding acceleration along the path by the factor $1/g_i$. The condition of incompressibility, div **v** $= 0$, appears now in the form $v^\alpha_{:\alpha} = 0$; it has already been used in setting up the friction term in (1), cf. the simplification of Eq. (2) in Appendix III.

Eq. (1) is perfectly general and does not presume an orthogonal system of coordinates; the simplicity is lost, however, when Eq. (1) is spelled out for any particular system of non-Cartesian coordinates. Even if we assume an orthogonal line element, we should have to introduce the quantities $g_i$ on which the Christoffel symbols depend that are implicitly present in (1). Thus it is hardly worthwhile to develop Eq. (1) in terms of general orthogonal coordinates, and we shall give only the results for our two special cases:

*Cylindrical coordinates.*

We substitute expressions (5) of Appendix II for the convective terms and expressions (4) of Appendix III for the friction terms, and multiply the second component of Eq. (1) with the factor $r$ that appears in the denominators of the second line of III.4 and II.5. The first and third terms in the second component of Eq. (1), viz.

$$\frac{\partial v^2}{\partial t} = \frac{\partial}{\partial t} \frac{v_\varphi}{r} \quad \text{and} \quad (p + V)^2 = \frac{1}{r^2} \frac{\partial(p + V)}{\partial \varphi}$$

are then transformed into the ordinary acceleration and the ordinary $\varphi$ component of the gradient

$$\frac{\partial v_\varphi}{\partial t} \quad \text{and} \quad \frac{\partial(p + V)}{r \, \partial \varphi} \, .$$

Using again the appropriate (scalar) $\nabla^2$ operating on the components of $\mathbf{v}$ and the notation $\mathbf{v} \cdot \text{grad } v_i$ we obtain

$$\rho\left(\frac{\partial v_r}{\partial t} + \mathbf{v} \cdot \text{grad } v_r - \frac{v_\varphi^2}{r}\right)$$

$$- \mu\left(\nabla^2 v_r - \frac{v_r}{r^2} - 2 \frac{\partial v_\varphi}{r \, \partial \varphi}\right) + \frac{\partial(p + V)}{\partial r} = 0,$$

(2)
$$\rho\left(\frac{\partial v_\varphi}{\partial t} + \mathbf{v} \cdot \text{grad } v_\varphi + \frac{v_r v_\varphi}{r}\right)$$

$$- \mu\left(\nabla^2 v_\varphi - \frac{v_\varphi}{r^2} + 2 \frac{\partial v_r}{r \, \partial \varphi}\right) + \frac{\partial(p + V)}{r \, \partial \varphi} = 0,$$

$$\rho\left(\frac{\partial v_z}{\partial t} + \mathbf{v} \cdot \text{grad } v_z\right) - \mu\nabla^2 v_z + \frac{\partial(p + V)}{\partial z} = 0.$$

This must be supplemented by the condition of incompressibility $v^\alpha_{.\alpha} = 0$, viz.

(3)
$$\frac{1}{r} \frac{\partial}{\partial r}(r v_r) + \frac{\partial v_\varphi}{r \, \partial \varphi} + \frac{\partial v_z}{\partial z} = 0.$$

The corresponding calculation in *spherical coordinates* uses (II.5a) and (III.4a). We again have to multiply with the denominators factored out in the second and third lines of these equations and obtain with the symbols $\nabla^2$ and $\mathbf{v} \cdot \mathrm{grad}\ v_i$ , the meaning of which is now different, the following relation:

$$\rho\left(\frac{\partial v_r}{\partial t} + \mathbf{v} \cdot \mathrm{grad}\ v_r - \frac{v_\vartheta^2}{r} - \frac{v_\varphi^2}{r}\right)$$

$$- \mu\left(\nabla^2 v_r + \frac{2}{r^2}\frac{\partial}{\partial r}(rv_r)\right) + \frac{\partial(p + V)}{\partial r} = 0,$$

$$\rho\left(\frac{\partial v_\vartheta}{\partial t} + \mathbf{v} \cdot \mathrm{grad}\ v_\vartheta + \frac{v_r v_\vartheta}{r} - \frac{\cos \vartheta\ v_\varphi^2}{r \sin \vartheta}\right)$$

(2a)

$$- \mu\left(\nabla^2 v_\vartheta + \frac{2}{r^2}\left[\frac{\partial v_r}{\partial \vartheta} - \frac{v_\vartheta}{2 \sin^2 \vartheta} - \frac{\cos \vartheta}{\sin^2 \vartheta}\frac{\partial v_\varphi}{\partial \varphi}\right]\right) + \frac{\partial(p + V)}{r\ \partial \vartheta} = 0,$$

$$\rho\left(\frac{\partial v_\varphi}{\partial t} + \mathbf{v} \cdot \mathrm{grad}\ v_\varphi + \frac{v_r v_\varphi}{r} + \frac{\cos \vartheta\ v_\vartheta v_\varphi}{r \sin \vartheta}\right)$$

$$- \mu\left(\nabla^2 v_\varphi + \frac{2}{r^2 \sin \vartheta}\left[\frac{\partial v_r}{\partial \varphi} + \frac{\cos \vartheta}{\sin \vartheta}\frac{\partial v_\vartheta}{\partial \varphi} - \frac{v_\varphi}{2 \sin \vartheta}\right]\right) + \frac{\partial(p + V)}{r \sin \vartheta\ \partial \varphi} = 0.$$

We add the condition of incompressibility $v^\alpha_{.\,\alpha} = 0$, viz.

(3a) $$\frac{1}{r^2}\frac{\partial}{\partial r}(r^2 v_r) + \frac{1}{r \sin \vartheta}\frac{\partial}{\partial \vartheta}(\sin \vartheta\ v_\vartheta) + \frac{1}{r \sin \vartheta}\frac{\partial v_\varphi}{\partial \varphi} = 0.$$

The appearance of these equations is rather unsymmetric, although we have tried to write the inertia terms that accompany the $\nabla^2$ expressions as uniformly as possible. The only perfectly symmetric form that can compete with the form of the Stokes-Navier equations in Cartesian coordinates is Eq. (1) itself.—It cannot be denied that the calculation of the foregoing results by tensor methods is rather involved; the elementary vector methods of Prob. I.4 and III.1 yield the same results more quickly. In more general cases, however, such as the oblique four-dimensional coordinates indispensable in the general theory of relativity, there is no other method available but that of tensor calculus.

# INDEX

## A

Acceleration, local, 83
  material, 83
Acoustic waves, 95 f.
Adiabatic state, 50
d'Alembert's paradox, 210
d'Alembert's solution of the wave equation, 97, 137, 188, 263, 266
  for spherical waves, 100 f.
Anomalous dispersion, 182
Antisymmetric tensor (def.), 5
Archimedes principle, 46
Atmosphere (polytropic, adiabatic, isothermal), 52, 337, 350 f.
Atmospheric pressure, 44
Atmospheric vortices, 160
Avogadro's number, 49, 53

## B

Barkla, C. G., 330
Barometric formula, 52 f.
Beam theory (Bernoulli), 293 ff.
Beats, 186
Bechert, K., 266 f., 268, 270 f.
Becker, R., 270
Beltrami, E., 31
Bending moment (def.), 295
Bernoulli, D., 86, 293, 294 f.
Bernoulli's equation (incompressible, steady), 86 f.
  (compressible), 102
  examples to, 87 ff.
  (incompressible, non-steady), 89
  modified, 89
  in wave motion, 169 f., 180
Bernoulli's hypothesis (bending), 294, 299 f.
Bessel functions, 191 ff.
Betz, A., 243
Biot-Savart law, 153
Bjerknes, V., 161

Blasius formula (pipe flow), 119 f., 273
Blumenthal, O., 147
Boat equilibrium, 45
  oscillations, 46
  rolling period of, 47
Boltzmann, L., 53 f., 112, 267
Boltzmann factor, 54, 321
Boundary conditions in elasticity, 70
Boundary layer, 241 ff.
Boundary value problem of pot. th., 25
Boussinesq's problem, 338, 354 ff.
Boys, C. V., 127
Branch point (two-dim. flow), 209
Broglie, L. de, 186
Buchwald, E., 128
Burgers, I. M., 272, 281 ff.
Burkhardt, H., 199

## C

Capillarity, 122 ff.
Capillary waves, 180 ff.
Castigliano's first principle, 311 f.
Cauchy, A., 199, 289
Cauchy's integral formula, generalized, 268
Cauchy-Riemann equations, 138 f., 146 f.
Centrifugal separation of gases, 55, 337, 352
Characteristics (diff. eq.), 269
Chladni's figures, 315
Christoffel, E., 224
Christoffel symbol (def.), 381
Circular vortex rings, 161 ff.
  interaction with wall, 162 f.
  interaction with each other, 164 f.
  translatory motion of, 165 f.
Circular waves, 191 ff.
Circulation (def.), 17
Circulation theorem (Kelvin), 134
Clebsch, A., 299
Compatibility conditions, 61
Compressible fluid, statics of, 49 ff.

391

图书在版编目（CIP）数据

变形介质力学 = Lectures on Theoretical Physics: Mechanics of Deformable Bodies : 英文 / (德) 阿诺德·索末菲 (Arnold Sommerfeld) 著 . — 北京 : 世界图书出版有限公司北京分公司 , 2023.1
索末菲理论物理教程
ISBN 978-7-5192-9679-7

Ⅰ . ①变… Ⅱ . ①阿… Ⅲ . ①连续介质力学—教材—英文 Ⅳ . ① O33

中国版本图书馆 CIP 数据核字（2022）第 131081 号

| | |
|---|---|
| 中文书名 | 索末菲理论物理教程：变形介质力学 |
| 英文书名 | Lectures on Theoretical Physics: Mechanics of Deformable Bodies |
| 著　　者 | ［德］阿诺德·索末菲（Arnold Sommerfeld） |
| 策划编辑 | 陈　亮 |
| 责任编辑 | 陈　亮 |
| 出版发行 | 世界图书出版有限公司北京分公司 |
| 地　　址 | 北京市东城区朝内大街 137 号 |
| 邮　　编 | 100010 |
| 电　　话 | 010-64038355（发行）　 64033507（总编室） |
| 网　　址 | http://www.wpcbj.com.cn |
| 邮　　箱 | wpcbjst@vip.163.com |
| 销　　售 | 新华书店 |
| 印　　刷 | 北京建宏印刷有限公司 |
| 开　　本 | 711mm×1245mm　 1/24 |
| 印　　张 | 17.75 |
| 字　　数 | 396 千字 |
| 版　　次 | 2023 年 1 月第 1 版 |
| 印　　次 | 2023 年 1 月第 1 次印刷 |
| 版权登记 | 01-2022-0409 |
| 国际书号 | ISBN 978-7-5192-9679-7 |
| 定　　价 | 129.00 元 |

扫码加群
世图新书发布
每周福利赠书
绝版好书淘宝
（加群备注：世图科技）